group 4	group 5	group 6	group 7	group 8
				2 Helium He 4.00
6 Carbon C 12.01	7 Nitrogen N 14.01	8 Oxygen O 16.00	9 Fluorine F 19.00	10 Neon Ne 20.18
14 Silicon Si 28.09	15 Phosphorus P 30.97	16 Sulphur S 32.06	17 Chlorine Cl 35.45	18 Argon Ar 39.45
32 Germanium Ge 72.59	33 Arsenic As 74.92	34 Selenium Se 78.96	35 Bromine Br 79.90	36 Krypton Kr 83.80
50 Tin Sn 118.69	51 Antimony Sb 121.75	52 Tellurium Te 127.60	53 Iodine I 126.90	54 Xenon Xe 131.30
82 Lead Pb 207.2	83 Bismuth Bi 208.98	84 Polonium Po (209)	85 Astatine At (210)	86 Radon Rn (222)

PERIODIC TABLE OF ELEMENTS

26 Iron Fe 55.85	27 Cobalt Co 58.99	28 Nickel Ni 58.70	29 Copper Cu 63.55	30 Zinc Zn 65.38
44 Ruthenium Ru 101.07	45 Rhodium Rh 102.91	46 Palladium Pd 106.4	47 Silver Ag 107.87	48 Cadmium Cd 112.41
76 Osmium Os 190.2	77 Iridium Ir 192.22	78 Platinum Pt 195.09	79 Gold	80 Mercury

ARTHUR GODMAN

LONGMAN ILLUSTRATED DICTIONARY OF CHEMISTRY

the fundamentals of chemistry explained and illustrated

LONGMAN YORK PRESS

YORK PRESS
Immeuble Esseily, Place Riad Solh, Beirut.

LONGMAN GROUP LIMITED
Longman House, Burnt Mill, Harlow,
Essex CM20 2JE, England
and Associated Companies throughout the world.

First published 1982
Second impression 1983
ISBN 0 582 55550 7

Diagrams by Rosemary Vane-Wright
Colour origination by Chris Willcock Reproductions
Photocomposed in Britain by Composing Operations Ltd.
Printed in Spain by Heraclio Fournier SA

Contents

How to use the dictionary *page* 5

Properties of substance 8
Change of state, physical properties, liquids & precipitates

Chemical change 19
Chemical properties, chemicals, properties of gases

Chemistry apparatus 23

General techniques 30
Liquids & solutions, liquids & solids, physical techniques, chromatography, chemical techniques, experiment

Inorganic chemical names 44
Acid radicals

Mixtures 54

Air & water 56
Combustion

Chemical reactions 62
Catalysis, reactivity

Atomic theory 76
Dalton's theory, formulae, the mole, analysis, molecular structure, descriptive terms

Solutions 86
Solubility, concentration

Crystals 90
Crystal structure, crystal systems

Colloids 98
Properties, types of colloids

Gas laws 102
Pressure & temperature, conditions, kinetic theory

Atomic structure 110
Atomic particles, electron orbitals, isotopes

Periodic system 116
Metals & non-metals, allotropy, periodic table, classification

Electrolysis 122
Electrodes, ionization, electrolytes, electrolytic processes, electrolytic cells, complex ions

Chemical bonds 133
Valency electrons, ionic bonds, covalent bonds, other bonds

Radioactivity 138
Radiation, particle emission, disintegration, radioactive series, observations, reactions, mass spectrography

Chemical energetics 146
Heat of reaction, thermochemistry, rate of reaction, chemical equilibrium, energy levels, energy measurement

Raw materials 154
Mines & ores, minerals, natural products

Commercial processes 157
Industrial processes, chemical industry

Organic compounds 172
Hydrocarbons, alcohols, ketones & acids, esters, amines & cyanides, aromatic compounds, organic radicals

Structure of organic compounds 181
Formulae, isomerism

Organic reactivity 185
Reactive groups

Organic reactions 188
Tests

Organic techniques 201
Distillation

Polymer chemistry 204
Oils, carbohydrates, polymerization, plastics

Important words in chemistry 211

Appendixes:
One: International System of Units 235
Two: Symbols used in chemistry 237
Three: Important constants 238
Four: Common alloys 239
Five: Common abbreviations in chemistry 240
Six: Understanding scientific words 241

Index 247

How to use the dictionary

This dictionary contains some 1500 words used in chemistry. These are arranged in groups under the main headings listed on pp.3–4. The entries are grouped according to the meaning of the words to help the reader to obtain a broad understanding of the subject.

At the top of each page the subject is shown in bold type and the part of the subject in lighter type. For example, on pp.12 and 13:

12 · PROPERTIES OF SUBSTANCE/CHANGE OF STATE

PROPERTIES OF SUBSTANCE/PHYSICAL PROPERTIES · **13**

In the definitions the words used have been limited so far as possible to about 2000 words in common use. These words are those listed in the 'defining vocabulary' in the *New Method English Dictionary* (fifth edition) by M. West and J.G. Endicott (Longman 1976). Words closely related to these words are also used: for example, *characteristic*, defined under *character* in West's *Dictionary*. For some definitions other words have been needed. Some of these are everyday words that will be familiar to most readers; others are scientific words that are not central to chemistry. These are contained on pp.211–33 and will be found listed in the alphabetical index.

1. To find the meaning of a word

Look for the word in the alphabetical index at the end of the book, then turn to the page number listed.

The description of the word may contain some words with arrows in brackets (parentheses) after them. This shows that the words with arrows are defined near by.

(↑) means that the related word appears above or on the facing page;

(↓) means that the related word appears below or on the facing page.

A word with a page number in brackets (parentheses) after it is defined elsewhere in the dictionary on the page indicated. Looking up the words referred to in either of these two ways may help in understanding the meaning of the word that is being defined.

The explanation of each word usually depends on knowing the meaning of a word or words above it. For example, on pp.91–2 the meaning of *crystalloid, crystallization* and *polymorphism*, and the words that follow depends on the meaning of the word *crystal*, which appears above them. Once the earlier words are understood those that follow become easier to understand.

2. To find related words

Look in the index for the word you are starting from and turn to the page number shown. Because this dictionary is arranged by ideas, related words will be found in a set on that page or one near by. The illustrations will also help here.

For example, words relating to crystal systems are on pp.96–7. On p.96 *crystal systems* is followed by words used to describe various kinds of systems and types of structures; pp.98–100 give words for the related subject of colloids; p.101 lists words for the properties of different types of colloidal dispersions.

3. As an aid to studying or revising

There are two methods of using the dictionary in studying or revising a topic. You may wish to see if you know the words used in that topic or you may wish to revise your knowledge of a topic.

(*a*) To find the words used in connection with crystals look up *crystal* in the alphabetical index. Turning to the page indicated, p.91, you would find *crystal, crystalloid, crystallization, recrystallization*, and so on. Turning over to p.93 you would find *pattern, symmetry*, etc.

(*b*) Suppose that you wished to revise your knowledge of a topic; e.g. *isotopes*. If, say, the only term you could remember was *relative molecular mass* you could look it up in the alphabetical index. The page reference is to p.114. There you would find the words *isotope, isotopic ratio* and *relative isotopic mass*, etc. If you next looked at p.113, you would find words relating to the structure of the atom which would help you to understand the descriptions given for the words connected with isotopes.

4. To find a word to fit a required meaning

It is almost impossible to find a word to fit a meaning in most dictionaries, but it is easy with this book. For example, if you had forgotten the word for the substance from which metals are obtained, but you knew such substances came from a mine, all you would have to do would be to look up the word *mine* and turn to the page indicated, p.154. There you would find the word you wanted, which is *ore*, and also related words such as *deposit, seam* and *lode*.

THE DICTIONARY

material (*n*) a material has some basic properties (↓) by which it can be recognized. Other properties can vary between different kinds of the same material. Examples of materials are wood, leather, rubber and brass. Different kinds of wood have slightly different properties: their colours vary, their densities (p.12) vary, their hardness varies. The variation of the properties, from one kind of the same material to another, is small. The chemical composition (p.82) of a material can also vary, but the variation is small. **material** (*adj*).

substance (*n*) a substance has properties (↓) by which it can be recognized. These properties do not vary from one piece of the substance to another. The chemical composition (p.82) of a substance does not vary. Examples of substances are: iron, cane-sugar, salt. Many substances are compounds (↓); some substances are elements (↓).

compound (*n*) a substance (↑) that can be decomposed by chemical action (p.19) into simpler substances. The chemical constitution (p.82) of the substance is known, and a chemical formula (p.78) can be given for it. For example, lime is a compound of calcium and oxygen; one atom (p.110) of calcium combines with one atom of oxygen to form one molecule (p.77) of lime (calcium oxide); the chemical formula is CaO. To contrast *material* (↑), *substance* (↑), and *compound*: a *material* (e.g. wood) has a chemical composition (p.82) and properties (↓) which may vary between limits; a *substance* (e.g. a protein) has a definite chemical composition, but its constitution may be too complex to be described; a *compound* (e.g. sulphuric acid) has a definite chemical composition, a known chemical constitution, and can be given an exact chemical formula.

element[1] (*n*) a substance (↑) that cannot be decomposed by normal chemical action (p.19) into simpler substances. All substances and materials (↑) are composed of elements which are chemically combined. *See page 116 for a modern definition of element.* **elementary** (*adj*).

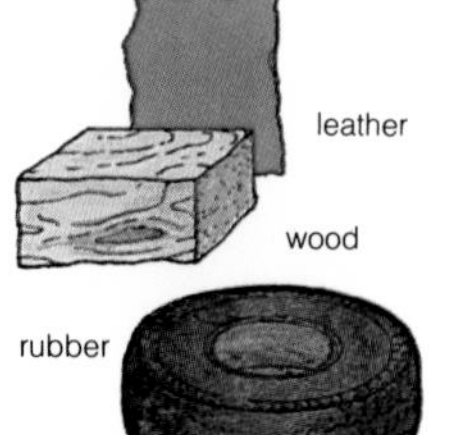

different materials

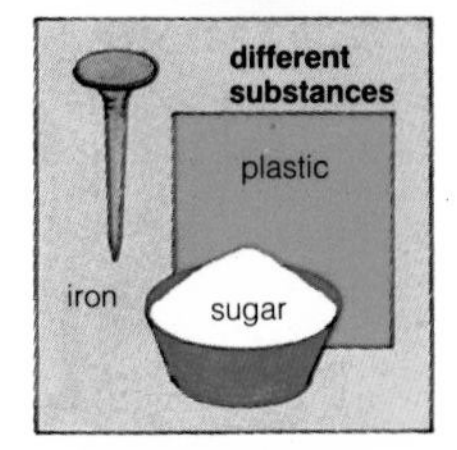

different substances

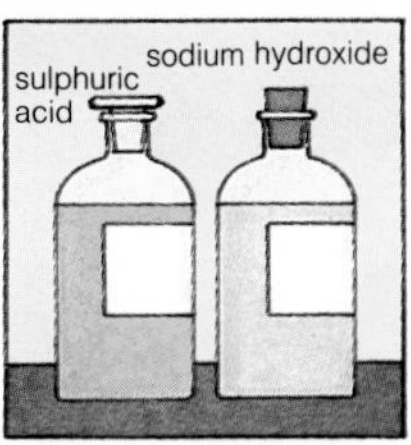

different compounds

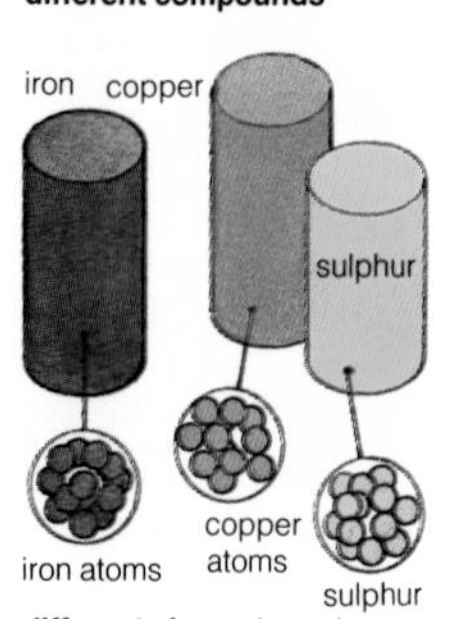

different elements

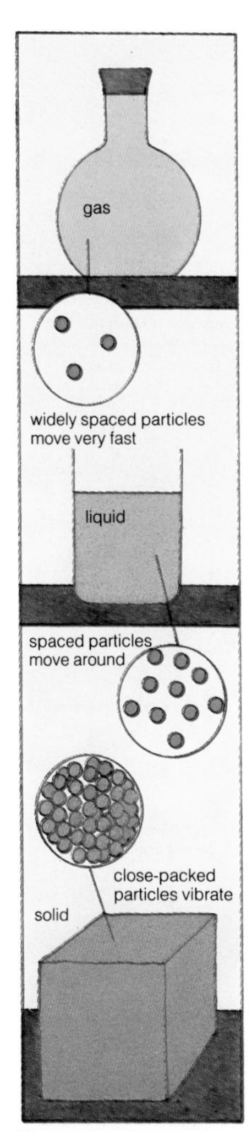

three states of matter

property (*n*) a property can be seen, heard, smelt, or felt by the senses and it allows one material (↑) or substance (↑) to be recognized as different from another material or substance. All materials and substances have physical properties (↓) and chemical properties (p.19).

physical property a property (↑) which does not depend on the effect of other materials or substances. Examples of physical properties are: shape, colour, odour (p.15), solubility (p.87), melting point (p.12), density (p.12). *See chemical property (p.19).*

extensive property a property which does not depend upon the amount of material or substance; such properties are used to identify (p.225) a material or substance, e.g. colour, odour, density, boiling point (p.12).

intensive property a property which depends upon the amount of material or substance; such properties are used to identify different specimens (p.43) of the same material or substance, e.g. mass, volume, concentration (p.81).

characteristic (*adj*) describes a property which readily distinguishes (p.224) an object, material, substance, pattern (p.93) from all other similar things. A characteristic property provides an easy means of recognition, e.g. copper has a characteristic reddish-brown colour, which readily distinguishes it from other metals. **characteristic** (*n*).

feature (*n*) a distinctive property common to a group of materials or substance.

description (*n*) a list of the properties of an object, material, substance, pattern, form of energy, or collection, or a list of the events in a process. **describe** (*v*), **descriptive** (*adj*).

state of matter solid, liquid, and gas are the three states of matter. Any material or substance is a solid, a liquid or a gas.

change of state a physical change (p.13) of a material or substance from one state of matter (↑) to another, e.g. from solid to liquid; from liquid to gas. A change of state is commonly caused by heating or by cooling.

solid (*n*) one of the states of matter (p.9). A solid has a definite (p.226) mass, a definite volume and a definite shape. For example, iron is a solid at room temperature. **solid** (*adj*), **solidify** (*v*), **solidification** (*n*).

melt (*v*) to change a solid (↑) to a liquid (↓) by heating; a solid to change to a liquid when heated. For example, heat melts ice; ice melts when heated. Only one material or substance is concerned in the action. Compare *dissolve* (p.30) which concerns two or more substances. **molten** (*adj*).

molten (*adj*) describes a material or substance in the liquid state. The material or substance is solid at room temperature.

solidify (*v*) to change a liquid (↓) to a solid (↑) on cooling. Solidifying is the opposite action to melting, only one material or substance is concerned in the action. The word is used for materials and substances which are normally solid at room temperature, e.g. molten iron solidifies at about 1500°C.

set (*v.i.*) of suspensions (p.86) in liquids, to form a solid as the liquid evaporates.

freeze (*v*) to change a liquid (↓) to a solid (↑) on cooling below room temperature. The word is used for substances which are normally liquid at room temperature, e.g. water freezes to form ice. Freezing is the opposite action to melting. **freezing** (*adj*), **frozen** (*adj*).

liquid (*n*) one of the states of matter. A liquid has a definite (p.226) mass, a definite volume, but no definite shape, e.g. water and kerosene are liquids at room temperature. A liquid takes the shape of its containing vessel (p.25). **liquefy** (*v*), **liquefaction** (*n*), **liquid** (*adj*).

boil (*v*) to change a liquid (↑) into a gas (↓) by heating. Bubbles (p.40) of gas are formed all over the liquid and the gas is given off (p.41). The temperature of the liquid remains constant (p.106) during boiling. **boiling** (*adj*), **boiled** (*adj*), **boil** (*n*).

boiled (*adj*) describes water that has been boiled for some time. The water no longer contains dissolved (p.30) air.

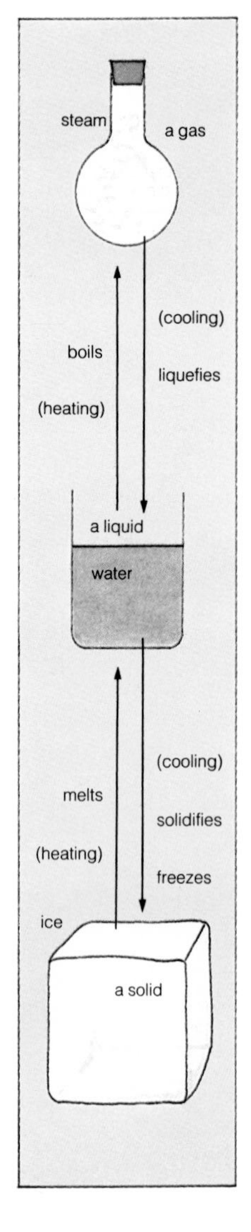

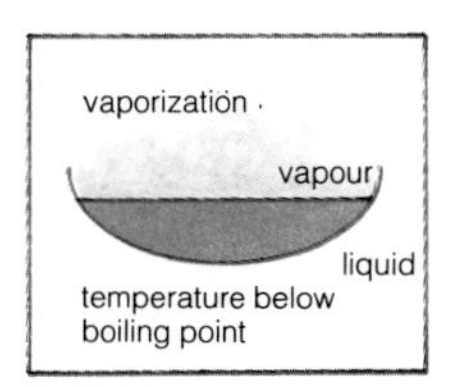

vaporization

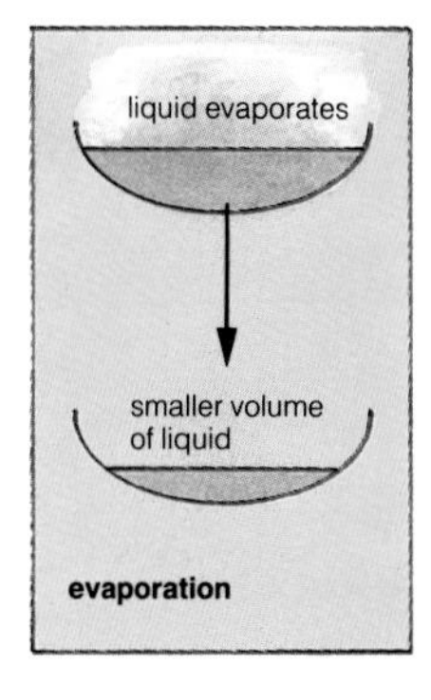

evaporation

liquefy (*v*) to change a solid (↑) or a gas (↓) to a liquid (↑). A gas is liquefied by cooling. **liquefaction** (*n*).

gas (*n*) one of the states of matter (p.9). A gas has a definite (p.226) mass but no definite volume and no definite shape, e.g. air is a gas at room temperature. A gas expands to fill the volume of its containing vessel (p.25). **gaseous** (*adj*).

gaseous (*adj*) describes a substance in the state of a gas (↑), or a chemical reaction (p.62) between gases.

vapour (*n*) a substance in the gaseous (↑) state. A vapour can be changed to a liquid (↑) by increasing the pressure (p.102). A gas is called a vapour below its *critical temperature* (p.102). To compare *gas* with *vapour*: both are in the gaseous state, but a gaseous substance above its critical temperature is a *gas* and cannot be liquefied (↑) however great the pressure, while a gaseous substance below its critical temperature is a *vapour* and can be liquefied by a sufficient increase in pressure. **vaporize** (*v*), **vaporization** (*n*).

vaporize (*v*) to change a liquid (↑) to a vapour (↑) at a temperature lower than that at which it boils. Some solids, e.g. naphthalene, also vaporize in air. **vaporization** (*n*).

evaporate[1] (*v*) to change a liquid (↑) to a vapour (↑) and so to cause the volume of the liquid slowly to become less. The important fact is the volume of liquid becoming less. *See evaporate*[2] *(p.32).* **evaporation** (*n*), **evaporated** (*adj*).

condense (*v*) to change a vapour (↑) to a liquid (↑) by cooling, or by increasing the pressure, or by both; change of a vapour to a liquid because it cools or because the pressure is increased. The word is used for materials and substances which are liquid at room temperature and the usual method of condensation is by cooling. **condensed** (*adj*), **condensation** (*n*).

condensation[1] (*n*) the formation of a liquid from its vapour, e.g. the condensation of steam to water.

fluid (*n*) any liquid or gas is a fluid. A fluid is a substance that flows. **fluid** (*adj*), **fluidity** (*n*).

boiling point the temperature at which a liquid (p.10) boils (p.10). At the boiling point the *vapour pressure* (p.103) of the liquid is equal to the *atmospheric pressure* (p.102). The lower the atmospheric pressure, the lower is the boiling point of the liquid. The boiling point of water is 100°C at *standard atmospheric pressure* (p.102).

melting point the temperature at which a solid substance melts; at the melting point both solid and liquid substance exist (p.213) together. The melting point of a solid varies slightly with the ambient (p.103) pressure (p.102). The term melting point is used for substances which are solid at room temperature. *See freezing point* (↓).

freezing point the temperature at which a solid substance melts. The term freezing point is used for substances which are liquid at room temperature, e.g. the freezing point of water is 0°C but the melting point of naphthalene is 80°C.

mass (*n*) the property of a material or substance that causes it to be attracted by the earth. The *force* attracting an object, or any substance, to the earth is its *weight*. Mass is measured in *kilograms*, weight is measured in *newtons*.

weight (*n*) see mass (↑). **weigh** (*v*).

volume (*n*) the amount of space taken up by an object in three dimensions.

density (*n*) the mass (↑) of 1 m^3 of a material or substance (p.80). Density = mass ÷ volume for any specimen (p.43) of a material or substance. Density is an extensive property (p.9) used in identification (p.225) of material and substances. Density is measured in kg/m^3. **dense** (*adj*).

relative density the density of a material or substance divided by the density of water. Relative density has no units, it is a pure number.

relative vapour density the density of a gas or vapour divided by the density of hydrogen measured at the same temperature and pressure (p.102). Relative vapour density has no units, it is a pure number, and is independent of temperature and pressure. The relative vapour density of a substance is numerically equal to half its *relative molecular mass* (p.114).

vapour density = relative vapour density (↑).

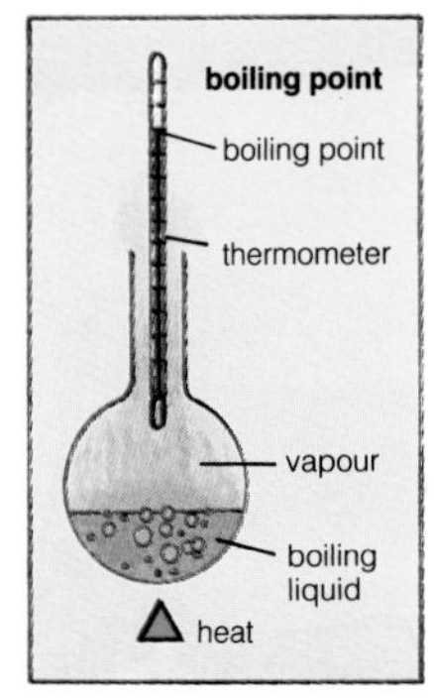

melting point

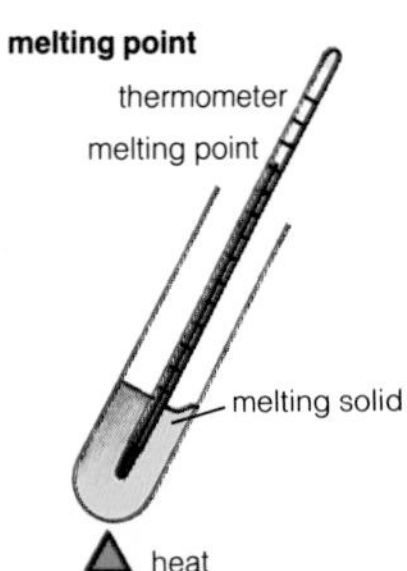

different densities

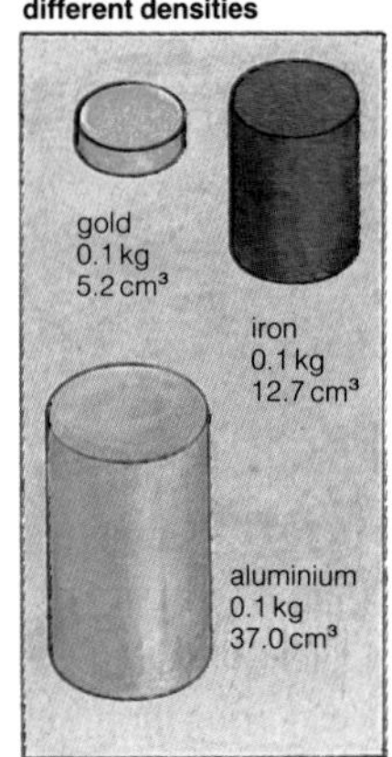

physical change a change in which no new materials or substances are formed. In a physical change, a material or substance may change its state, or some of its physical properties (p.9) may change, e.g. the change from water to steam is a physical change.

state of division a measure of the size of small particles (↓) into which a large piece of a solid has been divided, e.g. marble can be in lumps (↓), chips (↓), or powder (↓), three different states of division.

particle[1] (*n*) a very small piece of solid material or substance (p.8).

lump (*n*) a large piece of a solid material or substance with an irregular (p.93) shape. **lumpy** (*adj*).

chip (*n*) a small piece of a solid material or substance, broken off from a large piece. A chip is smaller than a lump, but bigger than a granule.

flake (*n*) a small, flat lump (↑) of solid material or substance. A flake is similar in size to a chip (↑).

granule (*n*) a small piece of a solid material or substance made up of several grains (↓). **granular** (*adj*).

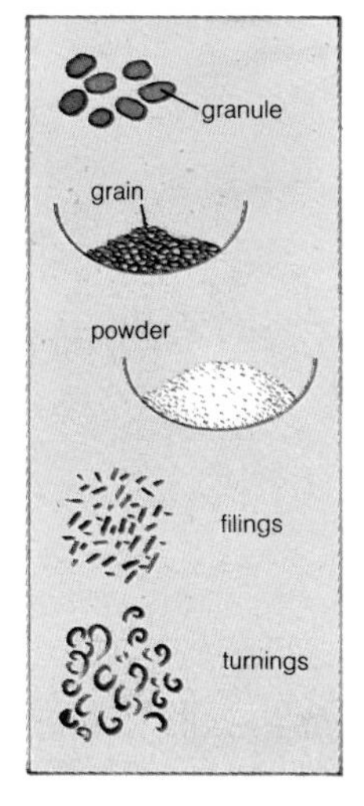

grain (*n*) a very small piece of solid material or substance; a particle (↑) that can be seen by the naked eye. Sand and salt consist of grains.

powder (*n*) a solid material or substance consisting of particles (↑) so small that they cannot be seen by the naked eye. **powdered** (*adj*), **powdery** (*adj*).

filings (*n.pl.*) small particles formed by rubbing a metal; they are similar in size to grains or granules (↑), but long and thin.

turnings (*n.pl.*) particles formed by cutting a metal; they are much larger than filings (↑).

fine (*adj*) describes powders (↑) and filings (↑) in which the state of division (↑) is very small. **fineness** (*n*).

coarse (*adj*) describes powders and filings which are bigger than those described as fine (↑). **coarseness** (*n*).

finely divided (*adj*) describes a solid in powder form with very small particles in the powder, i.e. a fine powder.

texture (*n*) the nature (p.19) of the surface of a solid, i.e. whether it is rough or smooth, is the texture of the surface. The texture of a powder, granules or grains, depends upon the fineness (p.13) or coarseness (p.13) of the particles, e.g. a surface can have a smooth texture or a powder a coarse texture.

massive (*adj*) describes a solid, particularly a metal, which consists of large pieces, including lumps, e.g. massive zinc consists of large pieces of zinc. Massive is the exact (p.79) opposite of finely divided (p.13).

elastic (*adj*) describes a solid which can have its shape changed by a force and which returns to its original (p.220) shape when the force is removed, e.g. a piece of rubber is elastic. The property of such a solid is its **elasticity** (*n*).

plastic[1] (*adj*) describes a solid material or substance which can have its shape changed by a force but which does not return to its original (p.220) shape when the force is removed, e.g. clay is plastic. The property of such a solid is its **plasticity** (*n*).

brittle (*adj*) describes a solid material or substance which breaks into small pieces under a force, e.g. glass is brittle, it breaks into small pieces when hit. The property of such a solid is its **brittleness** (*n*).

ductile (*adj*) describes a solid material or substance which can be drawn out to form a thin wire. Metals and alloys (p.55) are ductile. The property of such a solid is its **ductility** (*n*).

malleable (*adj*) describes a solid material or substance which can have its shape changed to a thin sheet by beating with a hammer, e.g. iron is malleable. Metals and alloys (p.55) are malleable. The property of such a solid is its **malleability** (*n*).

abrasive (*adj*) describes a material which wears away the surface of another material. **abrasion** (*n*).

refractory (*adj*) describes a solid material or substance which can be heated to a high temperature without changing its properties, e.g. some kinds of bricks are refractory. **refractoriness** (*n*).

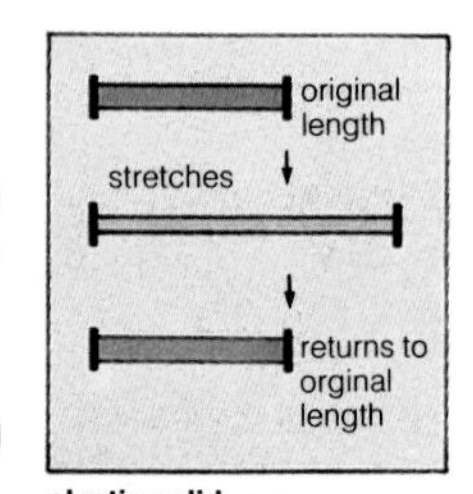

elastic solid

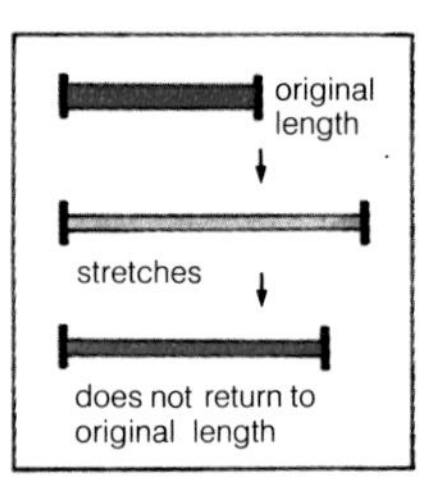

plastic solid

ductile solid

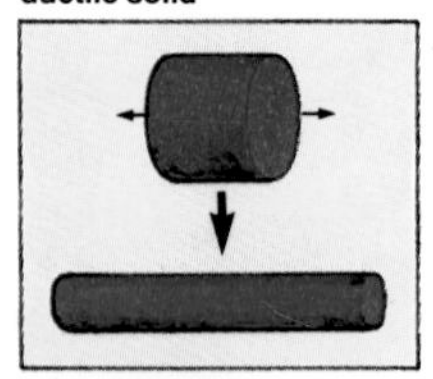

malleable solid

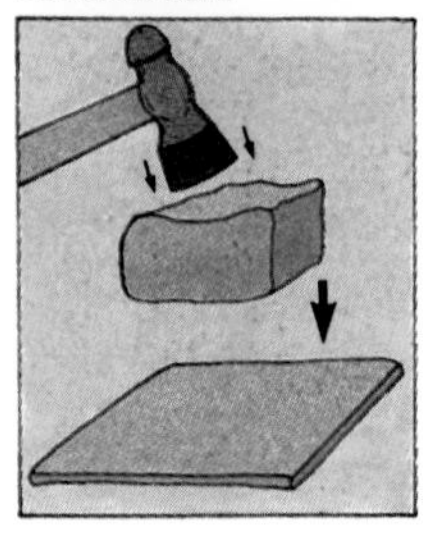

clear (*adj*) describes a liquid which is transparent, e.g. water is a clear liquid. A clear liquid can be coloured or colourless, e.g. tea is a clear, brown liquid; kerosene is a clear, colourless liquid. **clarity** (*n*).

soluble (*adj*) describes a solid, or gaseous (p.11), substance (p.8) which can be dissolved (p.30) in a liquid; the liquid is usually water. A substance can be described as very soluble, slightly (↓) soluble, sparingly (↓) soluble, insoluble (↓) or soluble. For example, sugar is soluble in water (sugar can be dissolved in water); lime is slightly soluble in water. **solubility** (*n*).

insoluble (*adj*) describes a solid, or gaseous (p.11), substance (p.8) that does not dissolve in a liquid. It is the opposite of soluble. Very few substances are completely insoluble.

slightly (*adj*) describes **soluble** (↑) if only a small amount of substance dissolves in a liquid, e.g. lime is slightly soluble in water, that is, only a small amount of lime will dissolve in water.

sparingly (*adj*) describes **soluble** (↑) if an amount which dissolves is very small, even less than dissolves for a slightly soluble (↑) substance, e.g. air is sparingly soluble in water.

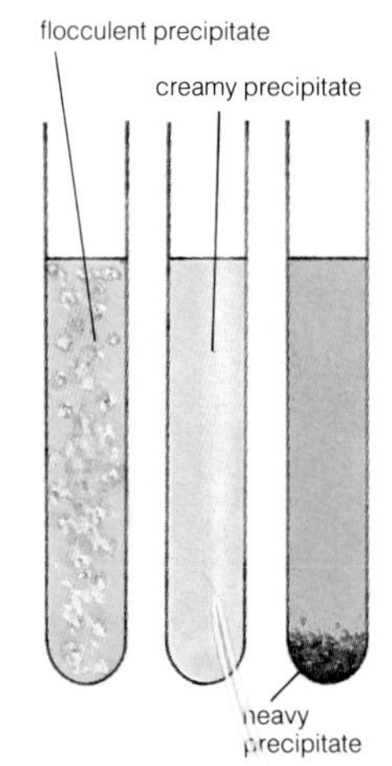

flocculent (*adj*) describes a precipitate (p.30) which has the appearance of wool floating in the liquid, e.g. a precipitate of aluminium hydroxide is flocculent.

milky (*adj*) describes a liquid (p.10) with a white precipitate (p.30) which gives the liquid the appearance of milk. The precipitate is very light, e.g. when carbon dioxide is passed into lime water, a light precipitate of calcium carbonate turns the lime water into a milky liquid.

creamy (*adj*) describes a white precipitate which is heavier than the precipitate forming a milky (↑) liquid; the precipitate still floats in the liquid, e.g. silver chloride forms a creamy precipitate.

heavy (*adj*) describes a precipitate (p.30) which sinks to the bottom of the liquid, e.g. barium sulphate forms a heavy precipitate.

miscible (*adj*) describes liquids that can mix in all proportions (p.76); the result looks like a single liquid, e.g. water and alcohol can mix completely and look like one liquid.

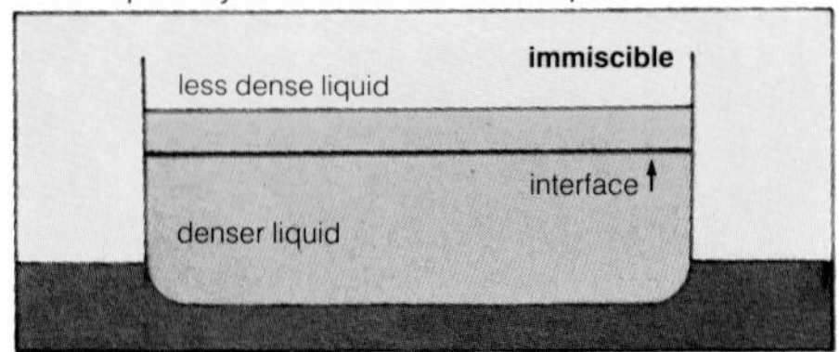

immiscible (*adj*) describes liquids that do not mix at all, e.g. oil and water form two layers (↓) of liquid as oil and water are immiscible.

layer (*n*) a spread of material over the surface (p.16) of another material. A layer can be thick or thin, e.g. a layer of skin covers an orange; a sandwich has 3 layers – bread, meat, bread.

film (*n*) a thin layer (↑) of a substance. It can be a thin layer of a liquid, or a vapour, or a solid; a thin layer of one liquid on another liquid; a thin layer of a solid on another solid. For example, a thin film of oil on water; a thin film of oxide (p.48) on a metal.

interface (*n*) the surface between two layers of liquid, e.g. oil floats on water, where the oil and water meet is an interface.

mobile (*adj*) (1) describes a liquid that flows easily, e.g. kerosene is a mobile liquid. (2) describes an object that can be moved easily, e.g. a bicycle is mobile. For liquids, mobile is the opposite of viscous (↓). **mobility** (*n*).

viscous

viscous (*adj*) describes a liquid that does not flow quickly, e.g. engine oil is viscous. Viscous is the opposite of mobile (↑). **viscosity** (*n*).

volatile (*adj*) describes a liquid that vaporizes (p.11) easily, e.g. petrol is a highly volatile liquid. **volatility** (*n*).

viscosity (*n*) the property of a liquid which prevents it from flowing readily, e.g. olive oil has a high viscosity; water has a very low viscosity.

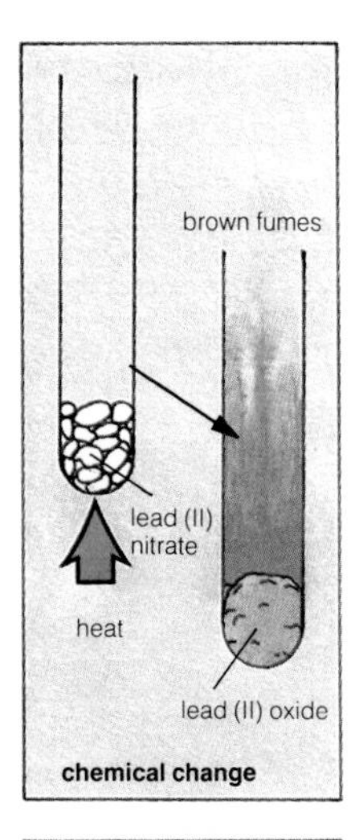

chemical change

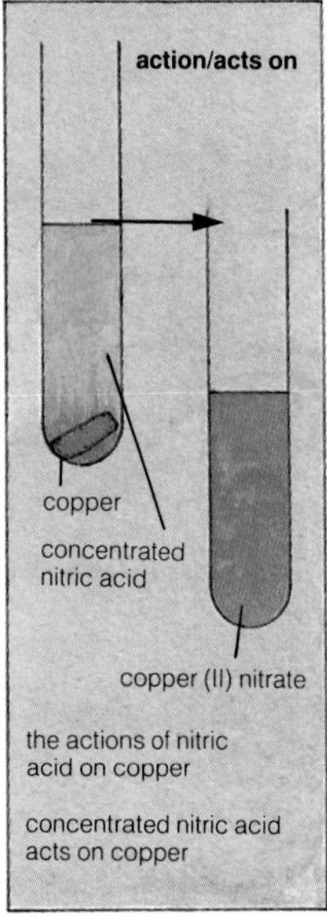

the actions of nitric acid on copper

concentrated nitric acid acts on copper

chemical change (*n*) a change in which new materials or new substances (p.8) are formed, e.g. when chalk is heated, lime and carbon dioxide are formed, this is a chemical change. When sulphur dioxide is passed into a solution of sodium hydroxide, sodium sulphite is formed, this is a chemical change.

chemical property a property which depends on the effect of other substances, e.g. a chemical property of sulphuric acid is that alkalis (p.45) neutralize the acid; a chemical property of water is that it is decomposed (p.65) by sodium.

nature (*n*) the essential (p.226) properties of a material or substance (p.8) form its nature, e.g. the nature of alkalis: they neutralize acids, and have a soapy feel when dissolved in water.

action (*n*) the effect of a substance on another material or substance, or the effect of heat, or an electric current, on a substance (p.8). For example, the action of sulphuric acid on calcium carbonate forms carbon dioxide and calcium sulphate; the action of heat decomposes (p.65) calcium carbonate.

act on to produce a chemical action (↑), e.g. hydrochloric acid acts on calcium carbonate, an action takes place (p.63), carbon dioxide is given off (p.41) and calcium chloride formed.

active (*adj*) (1) describes a substance (p.8) with many chemical properties, or with chemical properties that produce a strong effect, e.g. sulphuric acid is a very active substance, as many actions (↑) occur with other substances, and they are strong actions. (2) describes the part of a mixture (p.54) or molecule (p.77) which has the particular properties exhibited by the whole mixture or molecule, e.g. the hydroxyl group (p.185) of an alcohol (p.175) is the active part of the molecule.

inactive (*adj*) describes a substance (p.8) with few chemical properties, e.g. the alkanes (p.172) are inactive substances. *Compare stable (p.74).*

inert (*adj*) describes: (1) a substance (p.8) which has no chemical properties, e.g. krypton is an inert gas; (2) an atmosphere preventing oxidation. *Compare stable (p.74).*

pure (*adj*) describes a substance (p.8) which has no other substances mixed with it, e.g. pure silver has no other substances mixed with it. **purify** (*v*), **purification** (*n*).

impure (*adj*) describes a substance (p.8) which is not pure (↑) as it has other substances mixed with it; these other substances are *impurities*. The amount of the impurities is usually small, e.g. iron usually contains a small amount of carbon, and is therefore impure iron. **impurity** (*n*).

trace (*n*) a very small amount of an impurity (↑). It is also a very small amount of any substance (p.8) present (p.217) in a mixture or in the earth, e.g. a trace of arsenic in sulphur is a very small amount of impurity; a trace element in the earth is an element (p.8) present in very small amounts.

contaminate (*v*) to make a substance impure (↑) with a small amount of an unwanted impurity (↑), e.g. drinking water contaminated with small amounts of dissolved lead; aluminium contaminated with silicon. **contamination** (*n*), **contaminated** (*adj*).

chemical[1] (*adj*) describes a property or action concerned with the making of new materials or substances (p.8).

chemical[2] (*n*) an element (p.8) or compound (p.8) which takes part in a chemical change, e.g. sodium hydroxide is a chemical; sulphuric acid is a chemical.

fine chemical a chemical which is pure (↑). Fine chemicals are used in analysis (p.82) and for other special purposes. Only small quantities are made of some fine chemicals.

pharmaceutical chemical a pure substance used for medicinal purposes.

technical chemical a reasonably pure substance, not as pure as a fine chemical, but purer than a coarse chemical. Technical chemicals are used for most industrial (p.157) purposes. *See heavy chemical (p.171).*

coarse chemical an impure substance used in industry and agriculture. Suitable for the purpose used and cheap to manufacture.

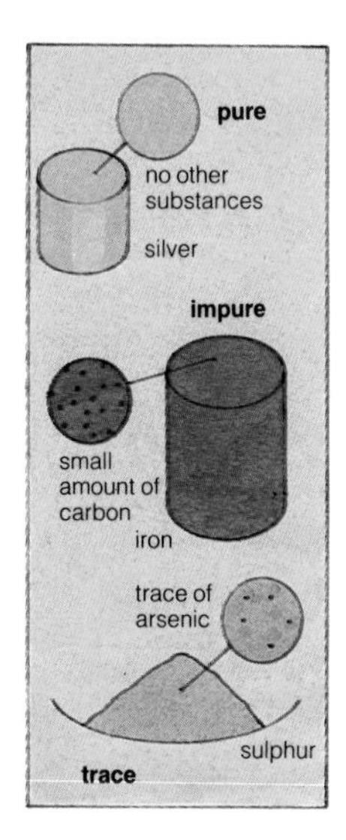

pharmaceutical chemical

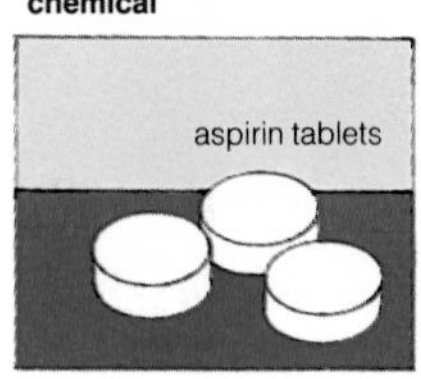

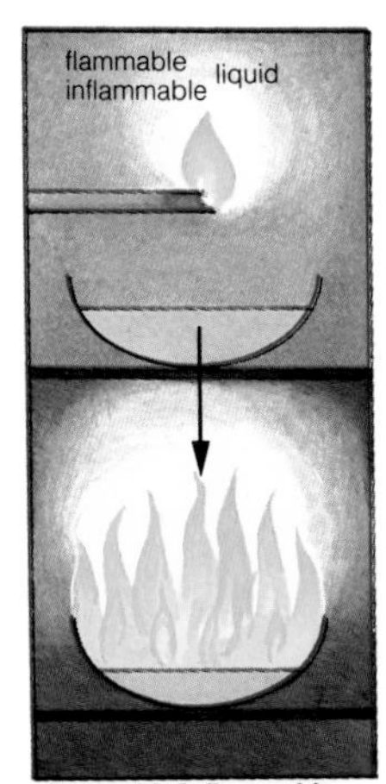

flammable/inflammable

non-inflammable

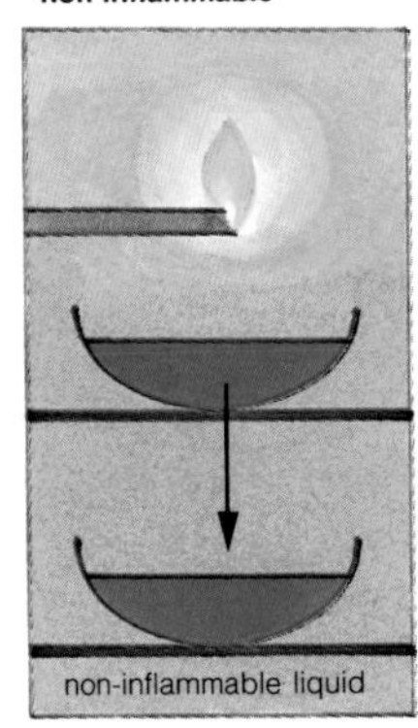

corrosive (*adj*) describes any chemical (↑) which attacks the surface (p.16) of solids and of living things, and destroys these surfaces. Examples of corrosive chemicals are strong mineral acids (p.55), such as concentrated sulphuric acid, which destroy skin. *Compare corrosion (p.61).*

caustic (*adj*) describes a chemical (↑) which attacks and destroys the surface of living things. Strong alkalis (p.45) are described as caustic, e.g. sodium hydroxide is called caustic soda and it burns skin when it comes into contact with it.

mild (*adj*) describes an alkali which is not caustic (↑), but is stronger than a weak alkali. *Mild* describes a condition which is between strong and weak, e.g. mild steel is less strong than hard steel, but harder than iron.

bland (*adj*) describes any chemical (↑) which does not irritate (p.22) nor cause discomfort to people. Bland usually describes food and pharmaceutical chemicals (↑), e.g. olive oil is bland.

passive (*adj*) describes the surface of a substance (p.8) which has been made inactive (p.19). A passive substance may also be inactive because its surface is covered with a thin film (p.18) of oxide (p.48), as on some metals. Usually the surface is made inactive by the attack of a corrosive (↑) chemical, e.g. iron is made passive by the action of concentrated nitric acid. Aluminium is passive because its surface is covered with a thin film of oxide. **passivity** (*n*).

activated (*adj*) describes a substance (p.8) which has been made active (p.19), e.g. activated charcoal has had its power to absorb (p.35) made greater than that of ordinary charcoal.

flammable (*adj*) describes a substance (p.8) which readily bursts into flames.

inflammable (*adj*) another word for flammable (↑).

non-inflammable (*adj*) describes a substance which does not burst into flames, and does not burn. The opposite of flammable (↑).

odoriferous (*adj*) describes a material or substance which produces an odour (p.15), e.g. onions are odoriferous.

irritate (*v*) to make parts of animals painful; to be unpleasant to the senses, particularly touch. For example, wood smoke irritates the eyes and the nose of a person; the eyes become painful, and the nose has an unpleasant feeling. **irritant** (*n*), **irritating** (*adj*).

pungent (*adj*) describes an odour (p.15) which has a strong effect on the nose or the tongue, e.g. vinegar has a pungent odour and taste. The odour is neither pleasant nor unpleasant.

acrid (*adj*) describes an odour (p.15) which is like the smell of wood smoke. It is an irritating (↑) odour.

choking (*adj*) describes an odour which is more irritating (↑) than an acrid (↑) odour; it is felt at the back of the throat and is very unpleasant, e.g. the smell of hydrogen chloride gas is a choking odour.

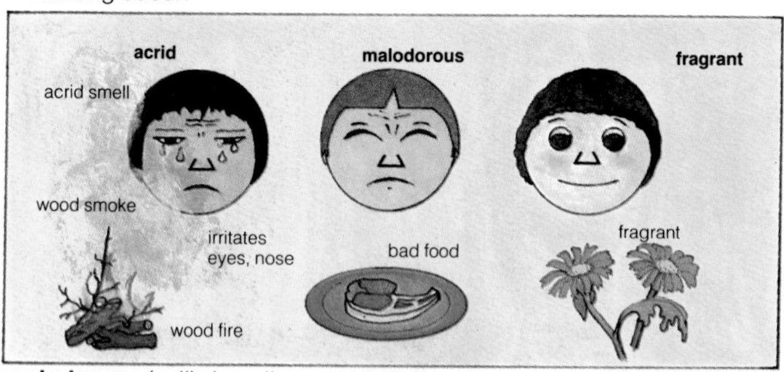

malodorous (*adj*) describes a material or substance (p.8) which has a very unpleasant odour, e.g. bad food is malodorous.

fragrant (*adj*) describes an odour which is pleasant to the senses, e.g. the odours of fruit and flowers are often fragrant.

dense (*adj*) (1) describes fumes (p.33) which are opaque (p.16), e.g. a fire can give off dense smoke and it is not possible to see through the smoke. (2) describes any material or substance (p.8) with a high density (p.12).

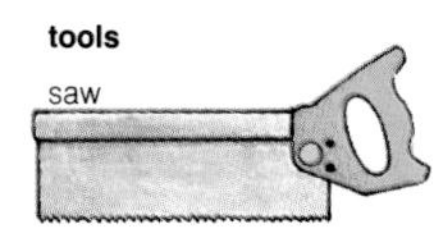

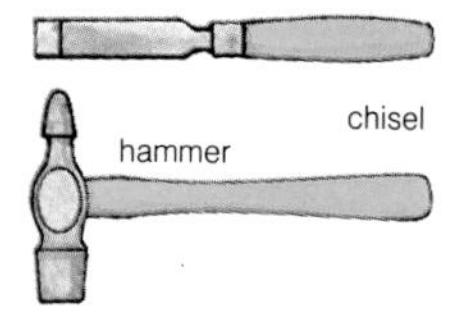

device (*n*) an object made for a special purpose, e.g. a telescope is a device for seeing distant objects; a thermostat is a device for keeping a constant (p.106) temperature.

tool (*n*) a device (↑) which is held in the hand and used to help a person who is working with his hands, e.g. a saw, a spanner, a screwdriver.

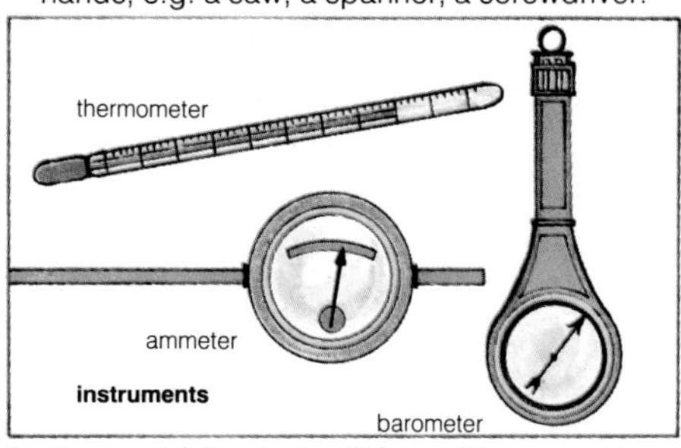

instrument (*n*) a device (↑) used for measuring, recording (p.39) or detecting (p.225), e.g. a thermometer, a barometer, an ammeter, a spectroscope. The work done with an instrument is more accurate (p.227) than the work done with a tool (↑) and needs more knowledge.

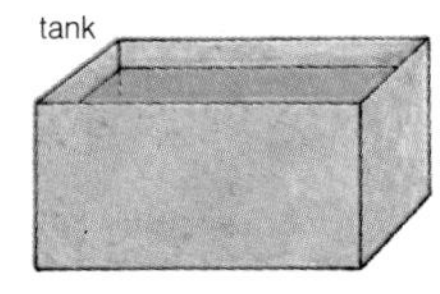

apparatus (*n*) (*apparatus n.pl.*) all the objects, devices (↑), tools (↑) and instruments (↑) used for work in chemistry, or science, e.g. thermometers, beakers, ammeters, thermostats, supports.

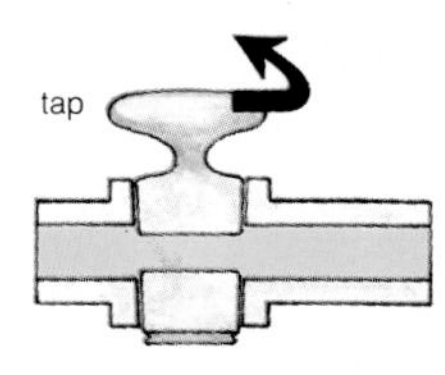

laboratory (*n*) a room in which scientific experiments (p.42) are carried out.

tank (*n*) a large container for liquids.

tap (*n*) a device (↑) which controls the flow of a liquid or gas from a pipe.

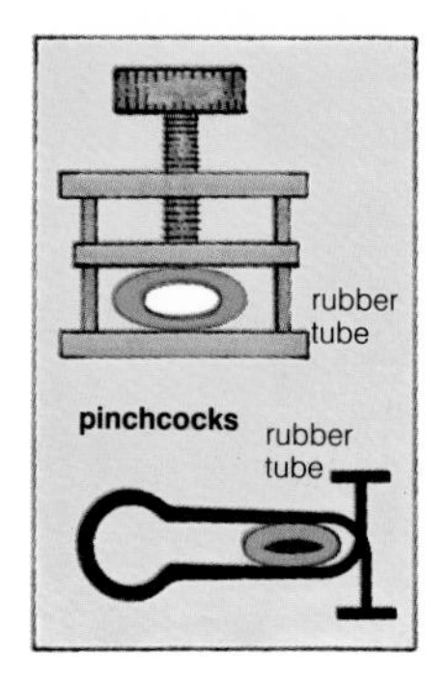

pinchcock (*n*) a device (↑) which closes a tube (p.29) or a pipe.

stopcock (*n*) a kind of tap (↑) used with glass tubes (p.29).

practical (*adj*) describes work with apparatus on experiments, which is opposite to work with writing and ideas.

theoretical (*adj*) describes work with ideas, usually recorded (p.39) in writing. Theoretical work is the opposite of practical (↑) work.

stopper (*n*) an object which closes a hole in a pipe, a bottle, or a flask (↓), e.g. a glass stopper in a bottle.
cork (*n*) a stopper (↑) made from a wood called cork.
bung (*n*) a stopper (↑) made of rubber, or a large stopper made of wood.
delivery tube a glass tube (p.29) connecting different pieces of apparatus (p.23); it conducts fluids (p.11) from one piece of apparatus to another.

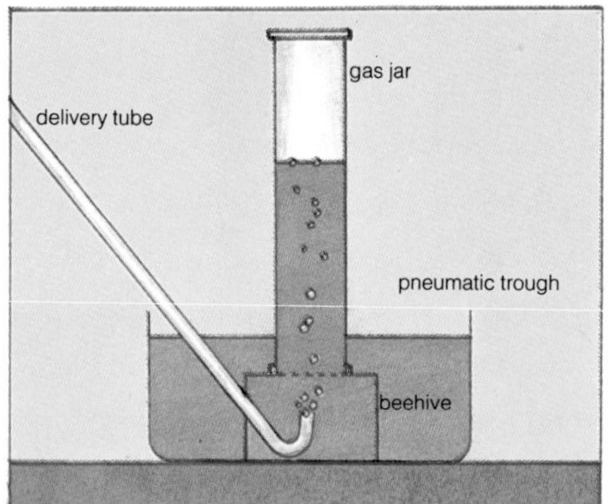

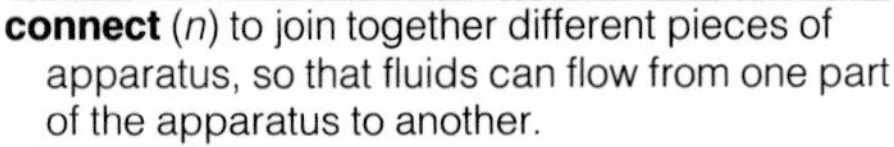

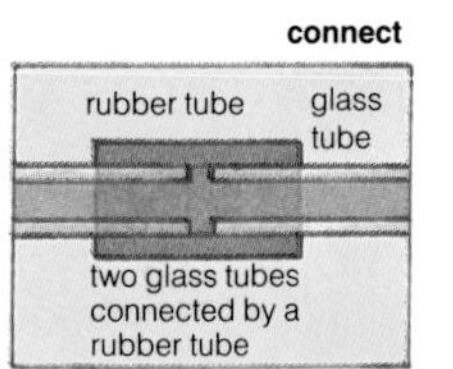

connect (*n*) to join together different pieces of apparatus, so that fluids can flow from one part of the apparatus to another.
disconnect (*n*) to separate different pieces of apparatus so that they are not connected (↑). Disconnect is the opposite of connect.
trough (*n*) (1) a flat container for liquids. (2) the hollow between two waves.
pneumatic trough a flat container used in the collection of gases.
gas-jar (*n*) a tall vessel (↓) for the collection of gases.
beehive (*n*) the stand for a gas-jar which is put in a pneumatic trough (↑).
aspirator (*n*) a large vessel used to supply air or water to apparatus.
eudiometer (*n*) a glass tube (p.29), with a scale, used to measure the volume of gases.

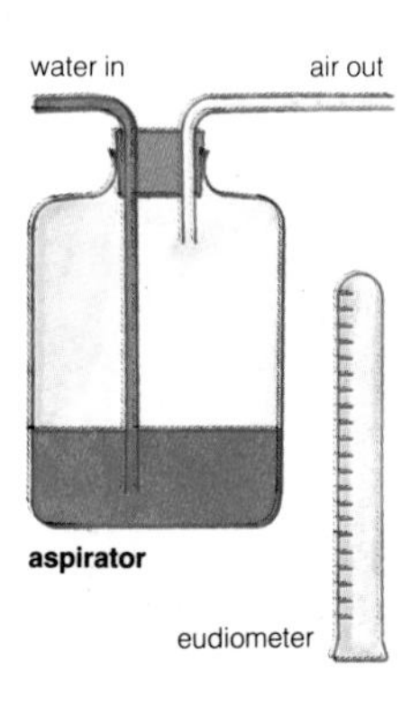

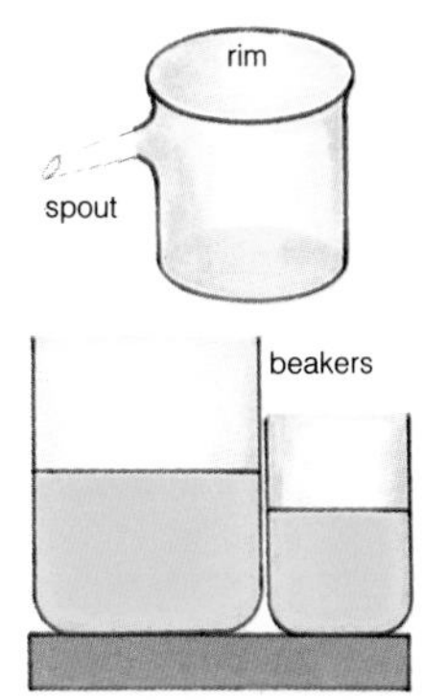

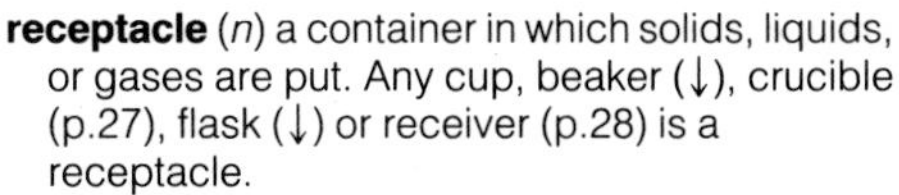

receptacle (*n*) a container in which solids, liquids, or gases are put. Any cup, beaker (↓), crucible (p.27), flask (↓) or receiver (p.28) is a receptacle.

vessel (*n*) a receptacle (↑) for containing liquids.

rim (*n*) the outside edge of the opening to a receptacle (↑).

spout (*n*) a pipe leading from a vessel (↑), through which a liquid passes when leaving the vessel.

leak (*v*) to lose fluids from a receptacle or pipe through a small hole in the receptacle or pipe, e.g. water leaks from a hole in a water pipe.

beaker (*n*) an upright vessel for containing liquids and also in which liquids can be heated.

flask (*n*) a round-shaped vessel or bottle with a long narrow neck; it is used to contain liquids for experiments (p.42). Flasks can have round bottoms or flat bottoms.

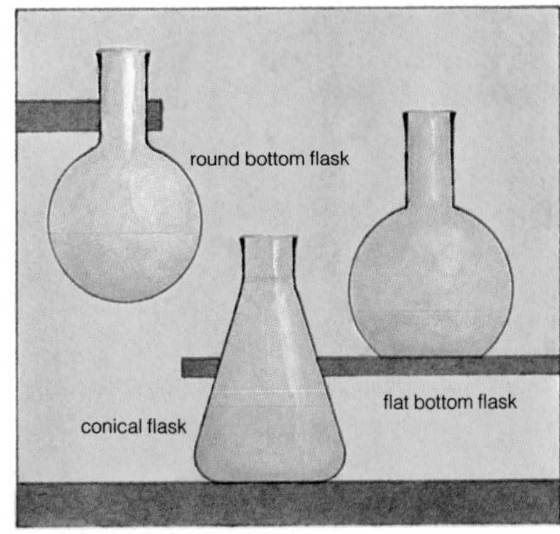

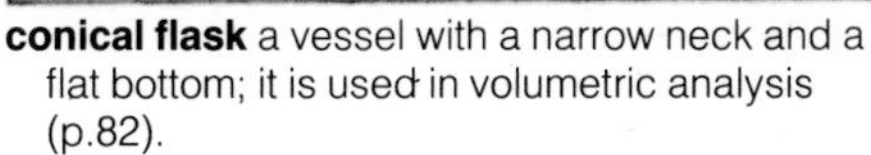

conical flask a vessel with a narrow neck and a flat bottom; it is used in volumetric analysis (p.82).

Erlenmeyer flask a large conical flask (↑) containing more than 500 cm^3 liquid.

U-tube (*n*) a glass tube (p.29) in the shape of the letter U.

Woulfe bottle a glass bottle with two necks; it is used to pass gases through a liquid.

graduation (*n*) a mark on an instrument, or measuring vessel, which is equally spaced from other marks to show a scale (↓) for measurement. **graduated** (*adj*).

scale (*n*) a set of marks, with numbers beside them, rising from a low value to a high value, e.g. the scale on a thermometer rising from 0°C to 100°C. Each mark is a graduation (↑) on the scale.

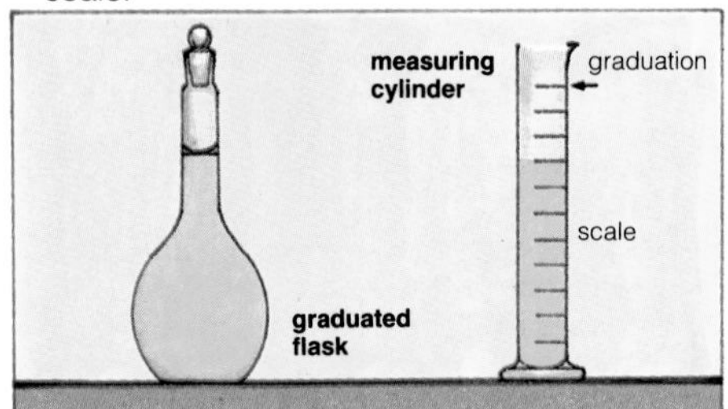

measuring cylinder a tall, narrow vessel, with a scale (↑) for volume; it is used to measure the volume of liquids. Measuring cylinders can measure volumes of 100 cm^3, 1 dm^3, etc.

burette (*n*) a long, narrow vessel with a spout and stopcock; used for measuring the volumes of a liquid which is allowed to run out of the burette. Burettes are used in volumetric analysis (p.82).

pipette (*n*) a vessel of a particular shape which measures a fixed amount of liquid. Common sizes of pipettes are: 10 cm^3, 20 cm^3, 25 cm^3, and 50 cm^3.

graduated flask a flask with a very long and very thin neck. It has a graduation (↑) on the neck and measures the volume of a liquid very accurately. Graduated flasks commonly have sizes of 100 cm^3, 250 cm^3, 500 cm^3 and 1 dm^3. They are used in volumetric analysis (p.82).

calibrate (*v*) to make a scale (↑) on a device, instrument, or other measuring apparatus, so that it measures accurately, e.g. to pass a known electric current through an ammeter (p.123) and to mark the value of the current on a scale, and then to complete the scale by marking other graduations. **calibration** (*n*).

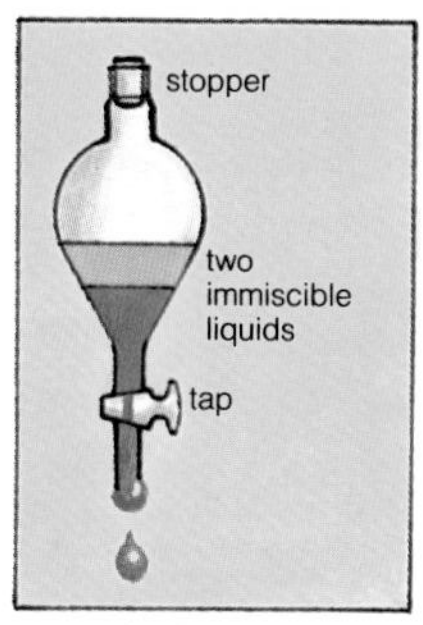

separating funnel

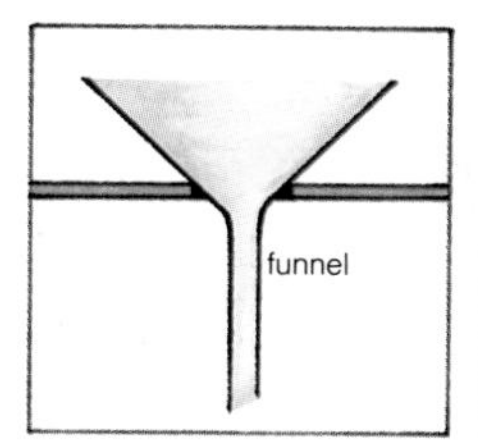

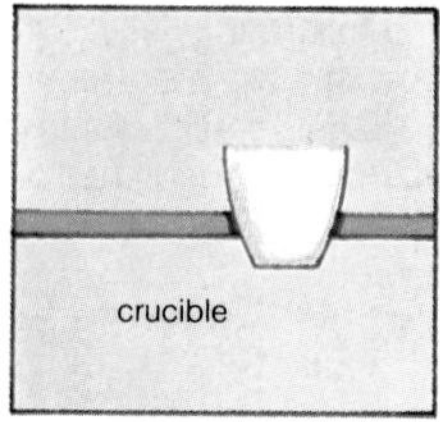

funnel (*n*) a piece of apparatus with a wide mouth and a thin pipe; it is used to put liquids into bottles, flasks and other vessels.

separating funnel a kind of funnel (↑) used to separate immiscible (p.18) liquids.

porous pot (*n*) a vessel (p.25) with porous (p.15) sides; it is used with gases and liquids, and these fluids (p.11) can pass through the sides.

crucible (*n*) a receptacle (p.25) used for heating solids to a high temperature.

generator (*n*) any piece of apparatus which gives a supply of a required gas, e.g. a generator of hydrogen.

Kipp's apparatus a generator used to give a supply of a gas.

thermostat (*n*) a device used to keep a liquid at a constant (p.106) temperature.

balance[1] (*n*) any piece of apparatus used to weigh solids, and liquids contained in vessels. The *mass* (p.12) of a solid is measured by a beam balance or a lever balance. The *weight* of a solid is measured by a spring balance.

a balance

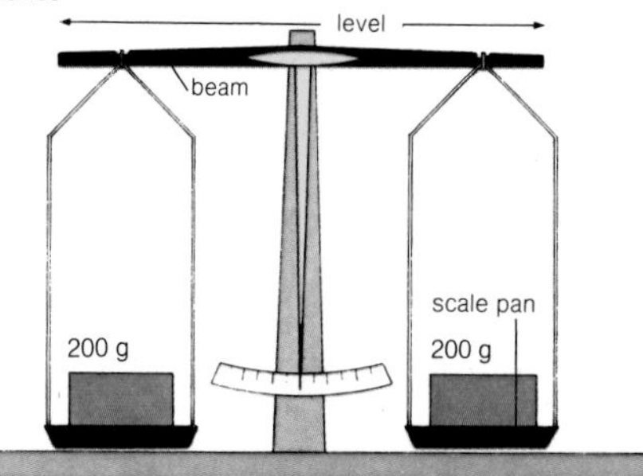

top loading balance

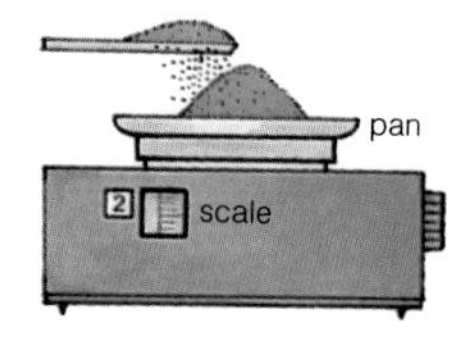

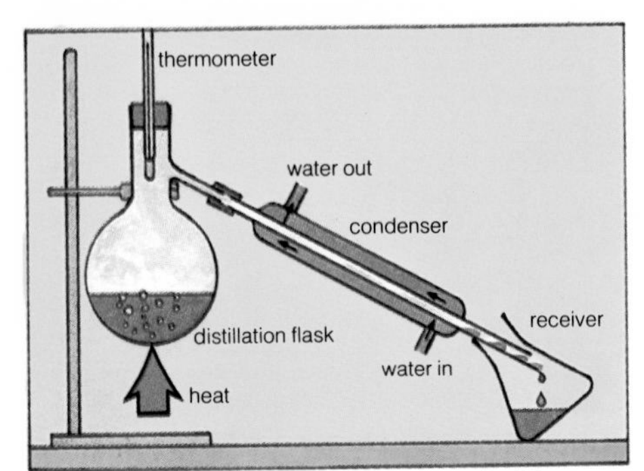

distillation flask a flask used when distilling (p.33) liquids; it has a side-arm on its neck.

condenser (*n*) a long glass tube (↓) which has a larger glass tube around it. Vapour passes down the inner tube and cold water passes through the outer tube. The cold water condenses (p.11) the hot vapour.

receiver (*n*) a vessel (p.25) put at the end of a condenser (↑) to receive the condensed (p.11) liquid.

support (*n*) a wooden or iron rod which takes the weight of a piece of apparatus and prevents the apparatus from falling. **support** (*v*).

thermometer (*n*) an instrument for measuring temperature (p.102). Most thermometers use mercury to measure the temperature on the Celsius scale. On this scale, water freezes at 0°C and boils at 100°C.

lag (*v*) to put a material (p.8) round a pipe, or piece of apparatus, to prevent heat escaping. **lagging** (*n*).

retort (*n*) a vessel (p.25) with a long neck, used in distillation (p.33).

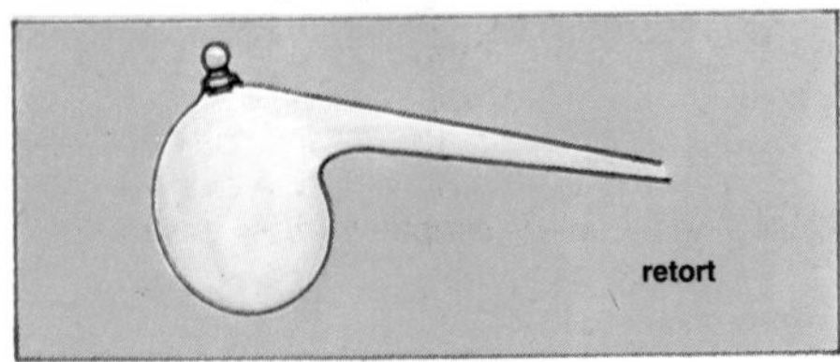

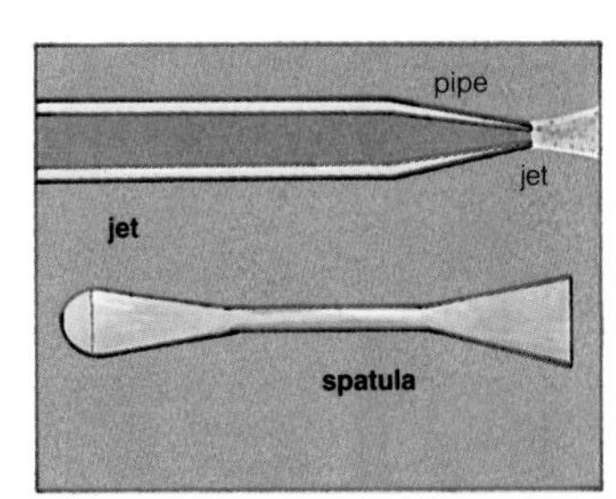

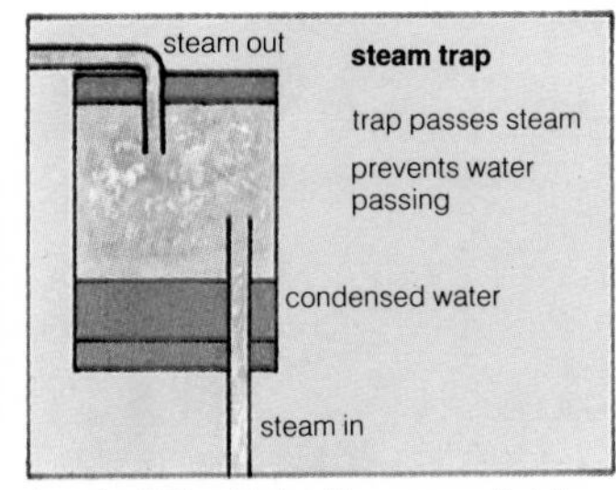

trap (*n*) a device (p.23) which allows one fluid (p.11) to pass through, but prevents another kind of fluid from passing through, e.g. a steam trap allows steam to pass through, but prevents water passing through.

jet (*n*) (1) a very small hole at the end of a pipe. (2) a current (p.122) of fluid which comes out of a jet, e.g. a jet of water.

blowpipe (*n*) a piece of apparatus like a pipe, ending in a jet (↑). It can be used to blow a small, very hot flame onto an object.

spatula (*n*) a tool shaped like a spoon, or which has a flat surface, used for removing small quantities of solids.

tongs (*n.pl.*) an article used to hold hot objects.

tube (*n*) a long, hollow article, through which fluids can flow. Chemical apparatus uses glass and rubber tubes.

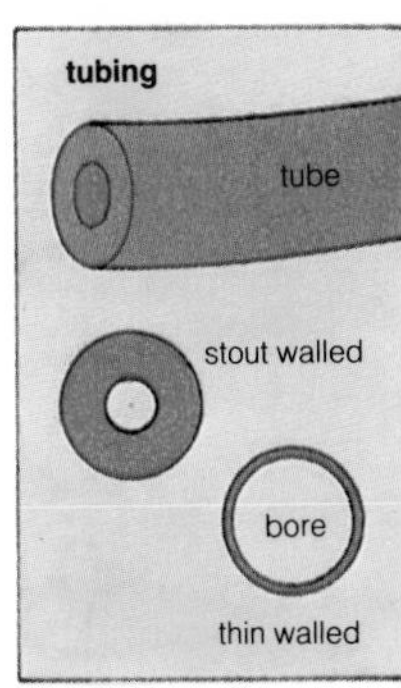

tubing (*n*) different kinds of tubes, either glass or rubber, may be of different bores (↓).

stout-walled (*adj*) describes tubes (↑) with thick walls.

thin-walled (*adj*) describes tubes (↑) with thin walls.

bore (*n*) the size of the hole in a tube (↑), e.g. a bore of 5 mm.

ground glass (*n*) glass which has been rubbed to give a very smooth surface.

quick-fit (*adj*) describes apparatus which uses ground-glass (↑) stopper (p.24) and joints.

diagram (*n*) a drawing, using lines, to show how apparatus (p.23) is connected. A diagram is used because it is simpler to draw a diagram than to draw a picture.

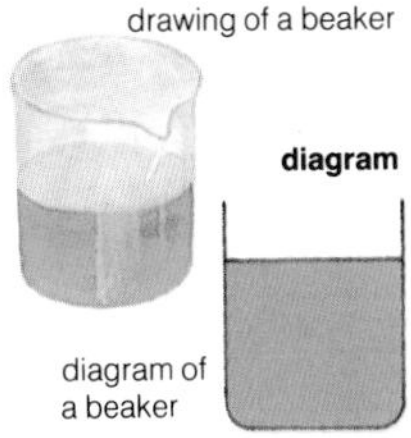

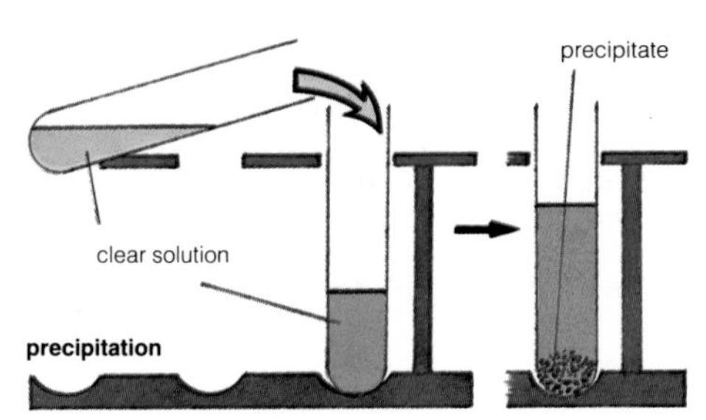

dissolve (*v*) (1) to make a solid (p.10) or a gas (p.11) disappear into a liquid. The result is a solution (p.86), e.g. sugar is dissolved in water, air is dissolved in water. (2) to disappear into a liquid, e.g. solids and gases dissolve in water; sugar dissolves in water. The solid or gas is soluble (p.17) if it dissolves.

precipitate (*n*) a solid (p.10) which appears in a solution when two solutions (p.86) are mixed. It is the result of a chemical reaction (p.62), e.g. when a solution of sulphuric acid is added to a solution of barium chloride, a precipitate appears. Precipitates are described as flocculent, creamy or heavy (*see p.17*). **precipitate** (*v*), **precipitated** (*adj*).

filter (*v*) to separate an insoluble (p.17) solid from a liquid by pouring through a filter in a funnel. The filter can be filter paper or glass wool. **filter** (*n*), **filtered** (*adj*), **filtrate** (*n*), **filtration** (*n*).

filtrate (*n*) the liquid that passes through a filter (↑).

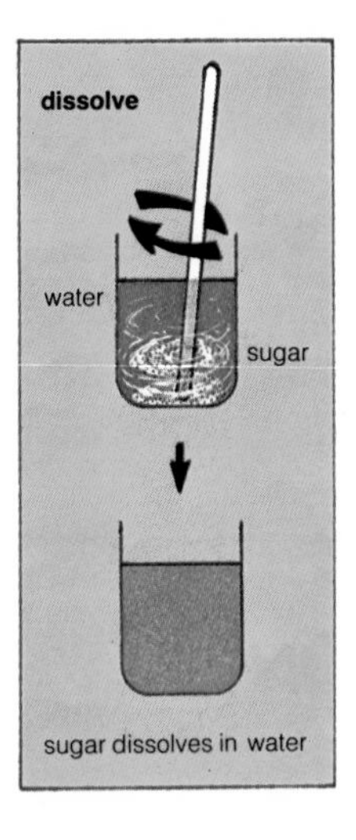

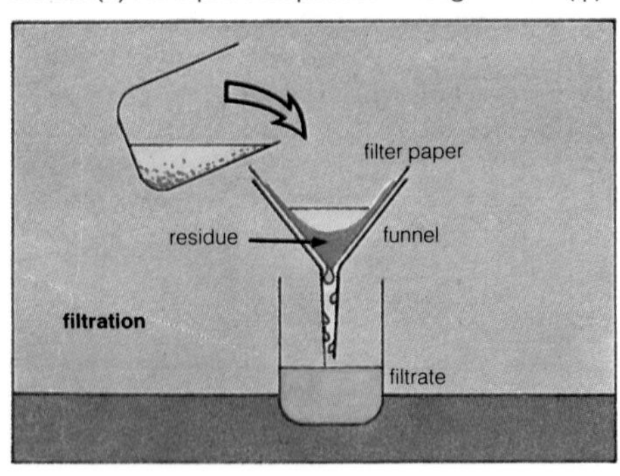

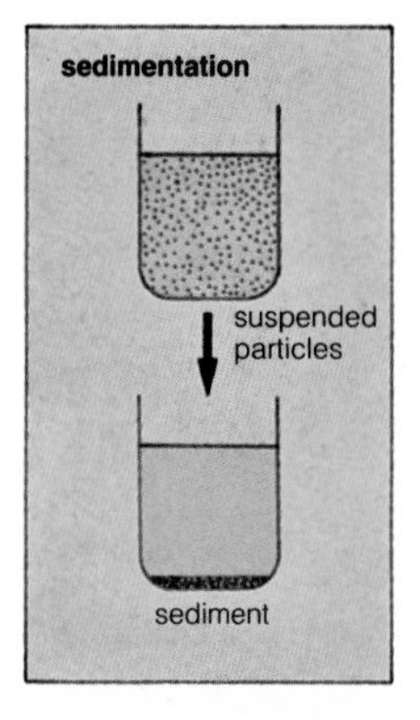

residue (*n*) (1) the solid that does not pass through a filter (↑). A precipitate (↑) is collected as a residue during filtration. (2) the material (p.8) or substances (p.8) left behind after any process, e.g. the residue left in a flask after distillation (p.33); the residue left in an evaporating basin after evaporation (p.11).

suspended (*adj*) describes light particles (p.13) of an insoluble (p.17) solid which float at all levels in a liquid or gas (p.11).

settle (*v*) to fall slowly through a fluid; particles (p.13) of a solid fall slowly to the bottom of a fluid when they settle. For example, insoluble particles fall to the bottom of a liquid; dust settles from the air onto a surface (p.16).

sediment (*n*) solid particles (p.13) that settle (↑)to the bottom of a vessel (p.25) form sediment. **sedimentation** (*n*).

decant (*v*) to pour off a clear (p.16) liquid, leaving any sediment (↑) at the bottom of the vessel (p.25). Before decanting liquid, any suspended (↑) material is allowed to settle as a sediment. *See supernatant (p.90).*

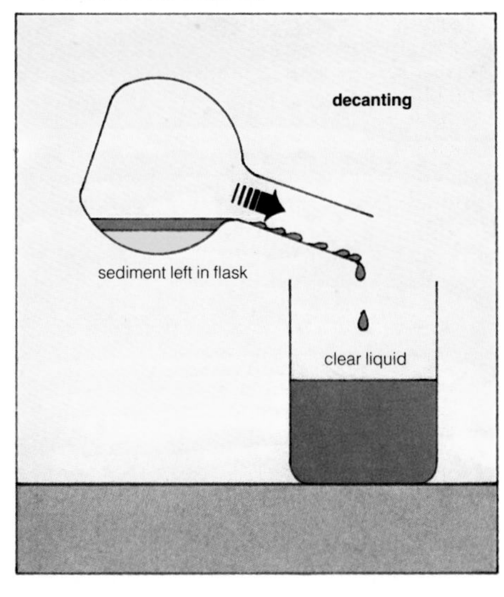

evaporate[2] (*v*) to heat a liquid so that it boils and vapour is given off. This makes the volume of the liquid become less. A liquid can be evaporated to dryness, with all the liquid changed to vapour and dissolved (p.30) solids left as a residue (p.31). **evaporation** (*n*), **evaporated** (*adj*).

evaporating basin a vessel (p.25) used for evaporating (↑) liquids.

concentrate (*v*) to boil away liquid from a solution (p.86) so that the same amount of solid is dissolved (p.30) in less water. The concentration (p.81) of the solution is then increased. **concentration** (*n*).

digest (*v*) to make a solid dissolve (p.30) in a liquid by adding the solid to the hot liquid. The hot liquid is usually stirred (↓). To contrast *digest* and *dissolve*: when a solid is *digested* a chemical action (p.19) takes place between the solid and the liquid; when a solid is *dissolved* a physical change (p.13) takes place, and there are no new substances formed. For example, copper (II) oxide is digested in hot sulphuric acid to make copper (II) sulphate. **digestion** (*n*).

stir (*v*) to move a glass rod, or other article, in a circular path in a liquid or powder (p.13) in order to mix the constituents (p.54) together. **stirrer** (*n*).

fuse (*v*) to melt a powdered solid so that it forms one solid mass. **fusible** (*adj*), **fusion** (*n*), **fusibility** (*n*).

calcine (*v*) (1) to heat a solid to a high temperature to drive off volatile (p.18) substances. (2) to heat a metal to a high temperature to form the oxide (p.48) of the metal. **calcined** (*adj*).

ignite (*v*) to set a material or substance on fire so that it burns (p.59). The ignition temperature of a substance is the lowest temperature at which it will catch fire. **ignition** (*n*), **ignited** (*adj*).

deflagration (*n*) the bursting into flame of a substance caused by chemical action, e.g. the deflagration of phosphorus in chlorine. **deflagrate** (*v*).

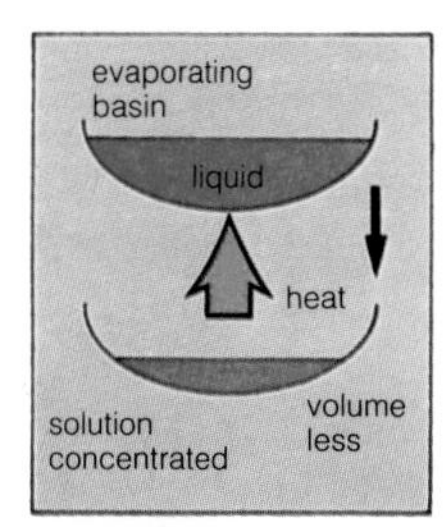

evaporation

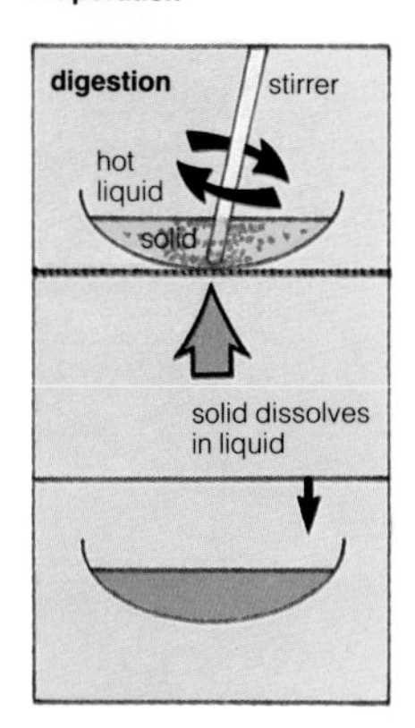

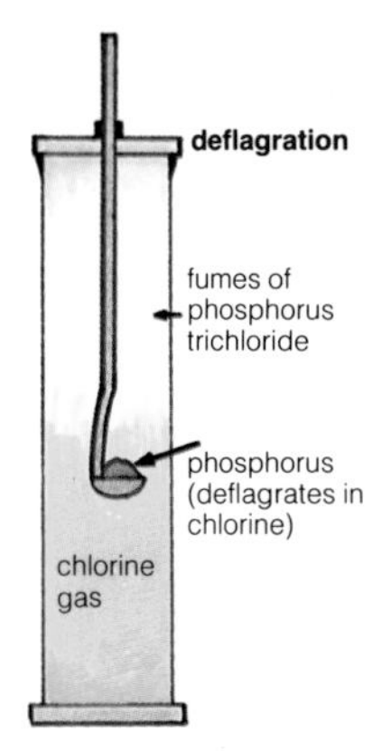

pyrolysis (*n*) the decomposition (p.65) of a chemical compound (p.8) by heat. Pyrolysis is usually used for organic (p.55) compounds; the organic compound is decomposed to simpler compounds.

distil (*v*) to change a liquid to a vapour (p.11) by heating and then to condense (p.11) the vapour back to a liquid. **distillation** (*n*).

distillation (*n*) the process (p.157) of distilling a liquid. Distillation is used to separate two or more liquids which have different boiling points. It is also used to purify (p.43) liquids.

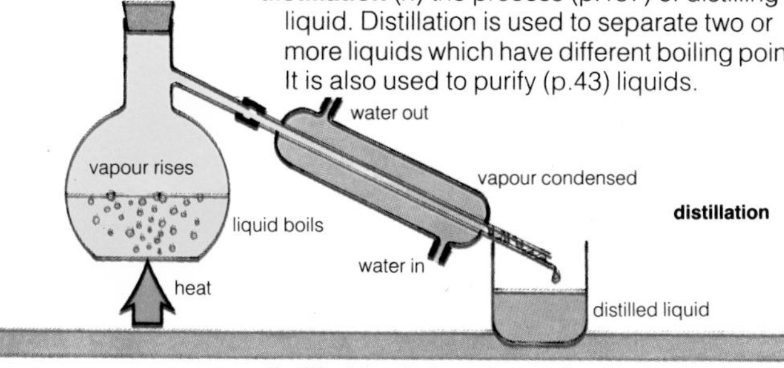

distillation

distilled (*adj*) describes a liquid which has been purified (p.43) by distillation, e.g. distilled water is very pure water.

bumping (*n*) the forming of very large bubbles (p.40) when a liquid boils. As a bubble rises, the vessel containing the liquid jumps up and down; this event is called boiling by bumping. To prevent bumping, small pieces of porous pot (p.27) are put in the liquid.

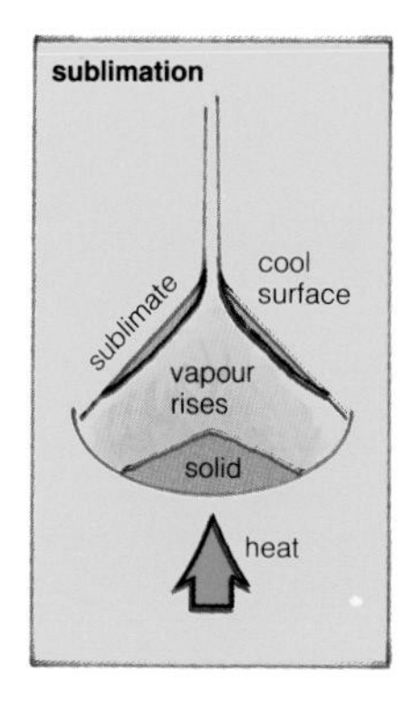

sublimation

sublime (*v*) to change a solid to a vapour (p.11) by heating, and then to cool the vapour so that it changes directly back to a solid, e.g. to sublime sulphur. **sublimation** (*n*), **sublimed** (*adj*).

sublimation (*n*) the process (p.157) of subliming a solid. Sublimation is used to purify solids. Not many solids sublime.

sublimate (*n*) a solid substance that has been formed (p.41) by sublimation.

fumes (*n.pl.*) (1) small particles (p.13) in the air, looking like smoke, are fumes. (2) vapour given off by an acid and combining with water vapour in the air to give the appearance of smoke. (3) any visible (p.42) vapour, especially a vapour which irritates (p.22) the nose and eyes. **fumes** (*v*).

separation (*n*) the way in which substances (p.8) are separated from each other, especially liquids. Immiscible (p.18) liquids are separated by using a separating funnel (p.27). Miscible (p.18) liquids are separated by distillation (p.33). Different methods are used for the separation of solid mixtures (p.54). **separate** (*v*), **separable** (*adj*).

extraction (*n*) (1) the process (p.157) of taking one substance from a mixture (p.54) of substances, e.g. the extraction of a solid by using a solvent (p.86) such as the extraction of iodine from its solution in water by using tetrachloromethane as a solvent. (2) the process of obtaining an element (p.8) from the earth; in a few cases the element can be extracted directly; in most cases an ore (p.154) is taken from the earth and the element extracted from the ore. **extract** (*v*).

dialysis (*n*) a process (p.157) for the separation (↑) of a colloid (p.98) from a crystalloid (p.91). A mixture of a colloid and a crystalloid is put in a receptacle (p.25) made from a membrane (p.99), usually parchment. Water is passed round the membrane; the crystalloid passes through the membrane and is taken away by the water. The colloid remains in the membrane. **dialyze** (*v*).

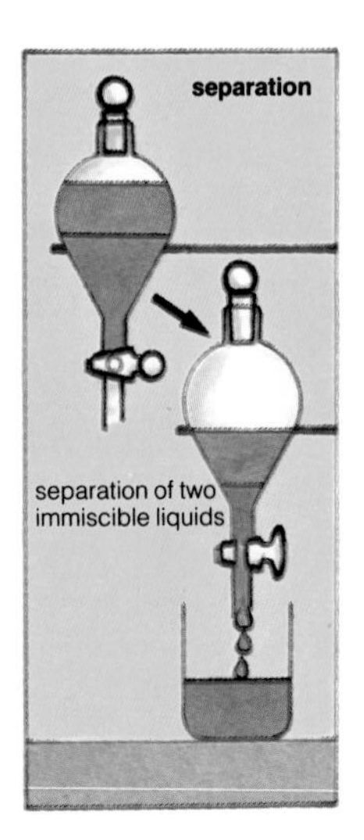

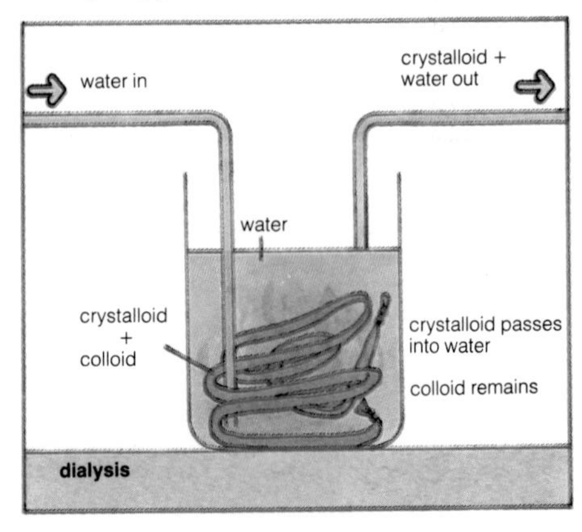

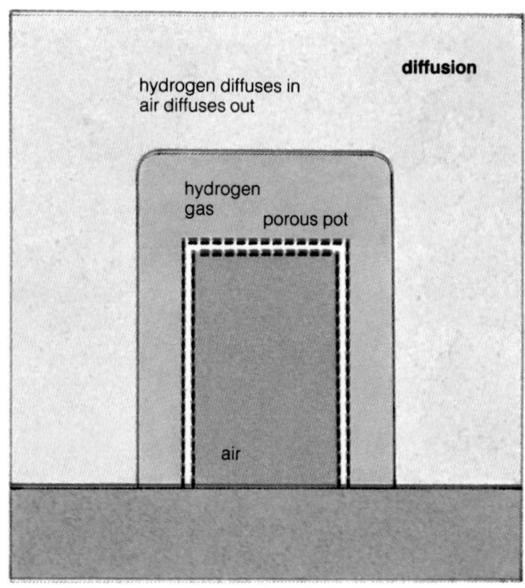

diffusion (*n*) (1) the process (p.157) of one gas spreading through another gas, e.g. petrol vapour spreads through the air in a room by diffusion. (2) when two miscible (p.18) liquids are put together, they spread through each other by diffusion. (3) when a soluble solid is put in a liquid, the solid dissolves and spreads through the liquid by diffusion. (4) gases pass through a porous (p.15) membrane (p.99) by diffusion if the pressure on each side of the membrane is the same. **diffuse** (*v*).

effusion (*n*) the process (p.157) of a gas passing through a porous (p.15) membrane (p.99) or through a small hole when the gas goes from a higher pressure to a lower pressure. Compare diffusion (↑) where the pressure is the same on both sides of the membrane or hole. **effuse** (*v*).

absorption (*n*) (1) the process (p.157) of a solid taking in a liquid or a gas, e.g. the absorption of a gas by charcoal, the absorption of water by a gel (p.100). (2) the process of a liquid taking in a gas, e.g. sodium hydroxide solution absorbs carbon dioxide from the air.

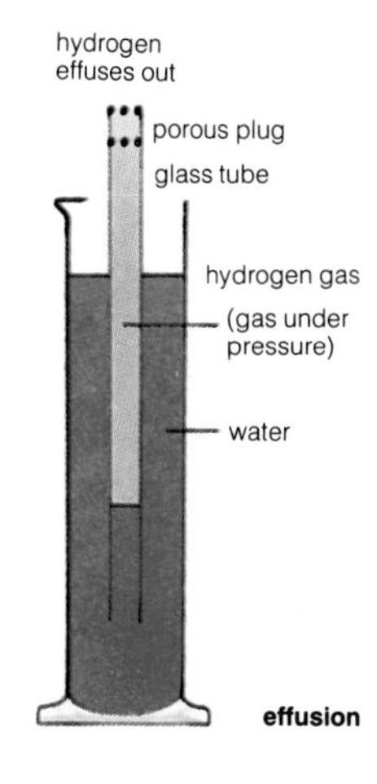

chromatography (*n*) a method of separating a mixture of solutes, by using a solvent and a separating medium. The solvent moves through the *separating medium*, which can be paper (↓) or a column (↓) of an inert (p.19) solid. In gas chromatography volatile constituents (p.54) of a mixture are passed through a column of a porous solid (the separating medium) by a current of an inert gas. **chromatographic** (*adj*).

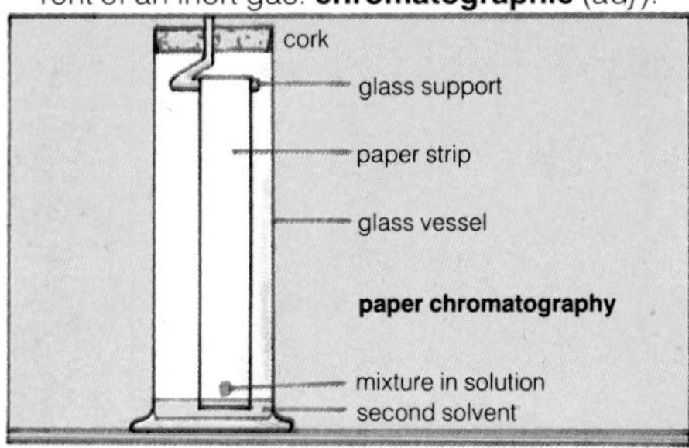

paper chromatography a solid mixture is dissolved in a suitable solvent and drops of this solution are placed on marks on a strip of filter paper. The paper strip is put in a corked container with one end in a second solvent, *see diagram*. The second solvent moves up the paper strip and takes with it the constituents (p.54) of the mixture. Different constituents rise to different heights.

development (*n*) if the constituents of a mixture used in paper chromatography (↑) are colourless, then the paper is sprayed with a dilute solution of a reagent which forms a coloured derivative (p.200) of a constituent. This action, to produce different coloured derivatives, is called *development*. **develop** (*v*).

solvent front the greatest height reached by a solvent in paper chromatography (↑). The constituents of a mixture never reach this height.

chromatogram (*n*) a paper strip, or a column, in chromatography (↑) with the individual constituents marked by coloured spots, using development (↑) if necessary, after separation by a suitable solvent.

chromatogram

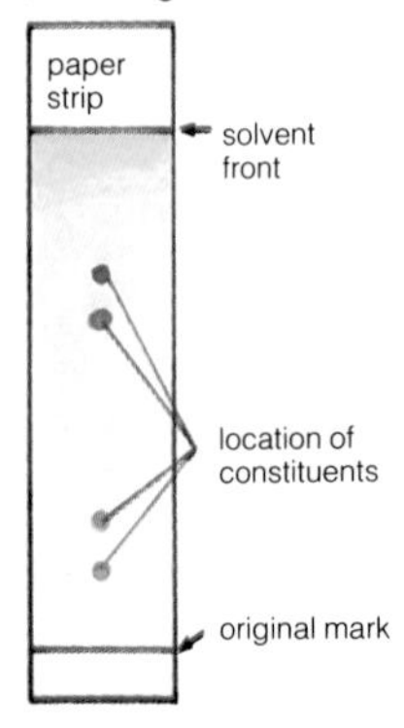

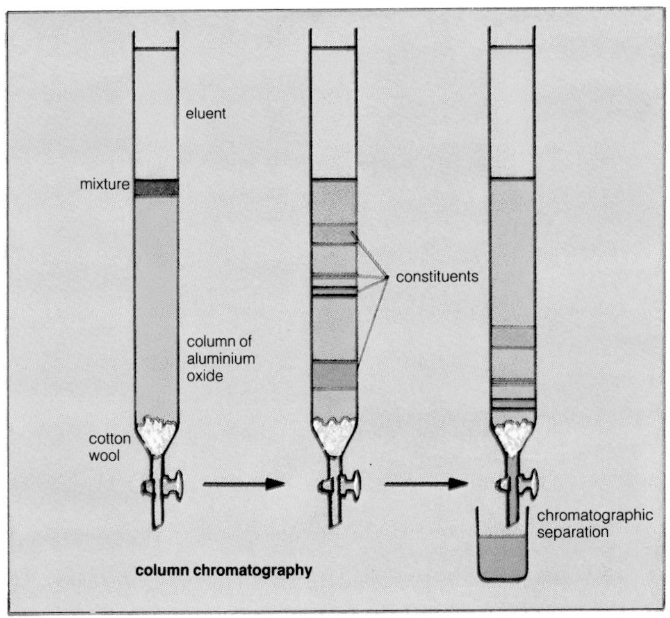

column chromatography

column chromatography a glass tube is packed with an inert (p.19) solid, usually aluminium oxide, forming a column. A mixture is placed on the top of the column. A suitable solvent is added above the mixture, *see diagram*. The solvent flows slowly down the column taking the constituents (p.54) of the mixture with it. Different constituents travel different distances. By continually adding solvent, all the constituents will eventually pass out of the column and can be separately collected.

eluent (*n*) the solvent used to separate a mixture in column chromatography (↑). This method of separation is called *elution*.

location (*n*) the process of finding the positions of constituents (p.54) in a chromatogram (↑) or a column (↑). For colourless constituents, development (↑) or ultra-violet light is used to produce colours.

grind (*v*) to change large lumps of a solid into a powder by rubbing in a pestle and mortar (↓). **ground** (*adj*).
pestle (*n*) an article with a handle and a rounded glass or stone end, used to grind (↑) substances in a mortar (↓).
mortar (*n*) a receptacle (p.25) for holding substances to be ground by a pestle.
triturate (*v*) to mix by grinding (↑). Two or more solids, or a solid and a liquid can be triturated. **trituration** (*n*).
treat (*v*) to add a reagent (p.63) or any chemical to a material or substance to cause a chemical change; e.g. to treat iron with concentrated nitric acid to make the iron passive; to treat cotton with sodium hydroxide solution to give the cotton a shiny surface (p.16). **treatment** (*n*).
acidify (*v*) to add acid (p.45) to a solution (p.86) so that there is excess (p.230) acid present. **acidified** (*adj*).

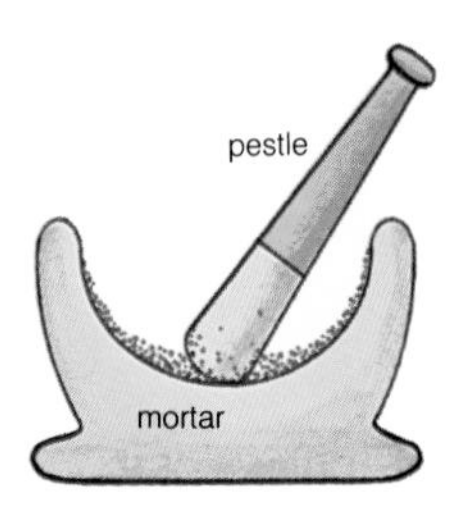

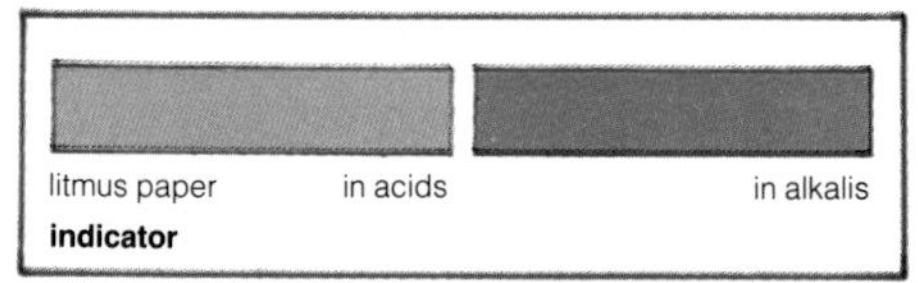

indicator (*n*) a chemical substance (p.8) that shows whether a solution is acidic (p.45), alkaline (p.45) or neutral (p.45). **indicate** (*v*).
indicate (*v*) to show clearly, or to make a sign.
pH (*abbr*) a way of describing how acid or how alkaline a solution is.
pH value (*n*) a number on a scale of 1 to 14 showing the strength or concentration of an acid or alkali. A value of pH = 1 indicates (↑) a strong acid. A value of pH = 14 indicates a strong alkali. A value of pH = 7 indicates a neutral solution.

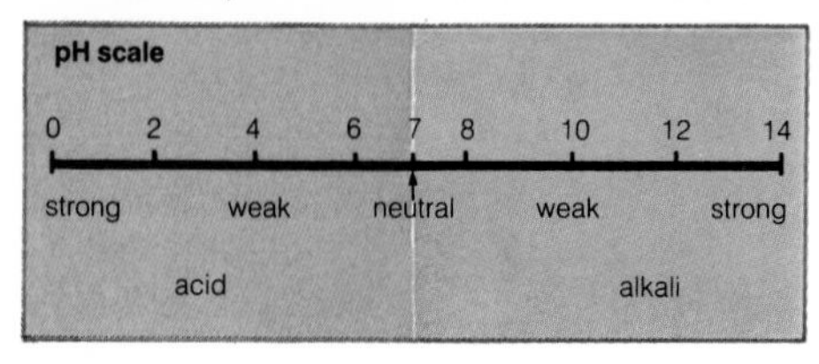

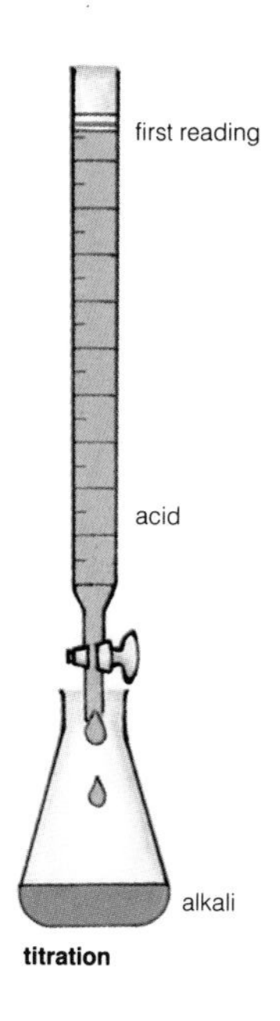

titration

titration (*n*) the process (p.157) of letting a solution (p.86) flow from a burette (p.26) into another solution held in a conical flask (p.25) until a chemical reaction (p.62) is complete. A common titration is between an acid (p.45) and an alkali (p.45) using an indicator (p.38). **titrate** (*v*).

end point the sign that a titration (↑) is complete as the chemical reaction has been completed. In an acid and alkali titration the end point is shown by a change in the colour of an indicator.

titre (*n*) the volume of a solution from a burette needed to reach the end point in a titration (↑).

reading (*n*) the value on a scale of an instrument (p.23) or a piece of apparatus (p.23) which is taken as a measurement, e.g. the reading of the level of the solution in a burette from the scale of volume on the burette; mercury in a thermometer shows the temperature on the thermometer scale, this is the reading on the thermometer.

record (*v*) to write down a reading (↑) or an observed (p.42) effect. **record** (*n*).

result (*n*) a change which is measured or observed (p.42). To contrast *result* and *effect*: when iron is exposed to air and water it rusts (p.61) forming iron oxide; rusting is the effect, and iron oxide is the result. **result in** (*v*).

tabulate (*v*) to write down results (↑) in a table. **tabulation** (*n*).

table of results

	cm^3	cm^3	cm^3
2nd reading	24.1	46.7	26.3
1st reading	1.4	24.1	3.6
titre	22.7	22.6	22.7

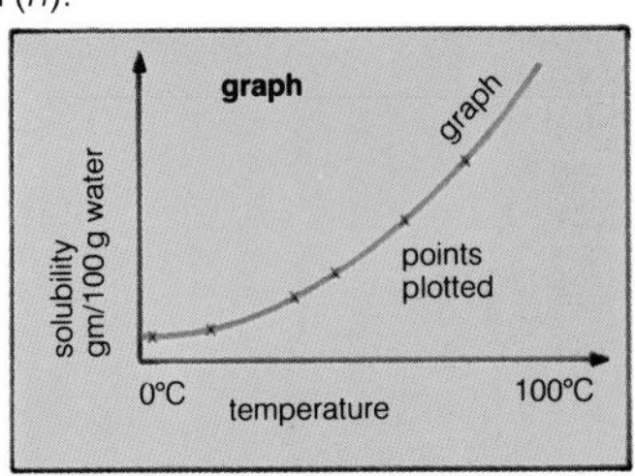

graph (*n*) a line drawn to show the relation between two changing quantities, e.g. pressure and volume of a gas, solubility (p.87) of a substance and temperature. **graphical** (*adj*).

plot (*v*) to make a graph (↑) by putting marks for results (↑) and connecting the marks by a line.

evolve (*v*) to form bubbles (↓) of gas and give off (p.41) the gas. To contrast *form*, *give off*, and *evolve* gases: *form* and *give off* can be used with both physical and chemical changes, *form* describes a weaker effect than *give off*; *evolve* is used only with chemical changes and describes a stronger effect than *give off* and the gas or vapour evolved can be collected. Evolution of gases are described as *steady*, *brisk* and *rapid* as the quantity of gas evolved is increased. **evolution** (*n*).

effervesce (*v*) to evolve (↑) a gas rapidly with the formation of many bubbles (↓) at a liquid surface (p.16), e.g. when a dilute acid is added to a carbonate, the mixture effervesces. **effervescence** (*n*).

bubble[1] (*n*) a small quantity of a gas or vapour with liquid around it. The liquid can be a thin film (p.18) round the bubble or the liquid can be in a vessel with bubbles in the liquid.

bubble[2] (*v*) to cause bubbles (↑) of gas to go through a liquid.

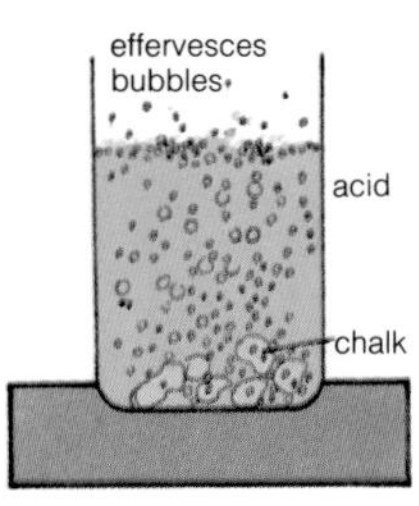

effervescence

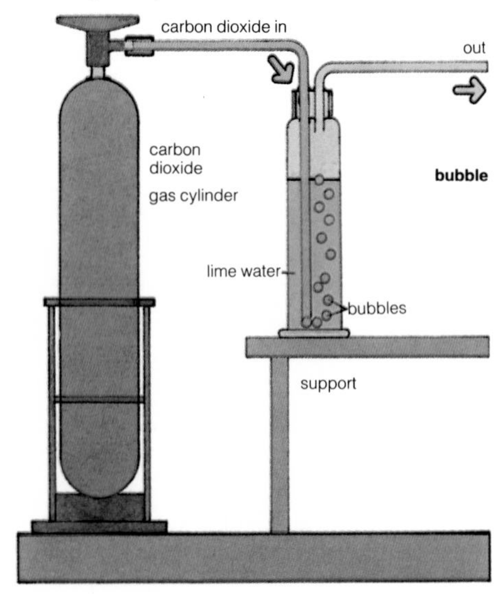

bubble

form (*v*) to cause to come into being, e.g. when two solutions (p.86) are mixed, a precipitate (p.30) is formed; kerosene forms a vapour; hydrogen is formed when calcium metal reacts with water. **formation** (*n*).

give off of an object, living thing, or chemical action, to cause a gas, vapour (p.11), or odour (p.15) to come from itself. For example, when iron is added to dilute sulphuric acid, hydrogen (a gas) is given off by the reaction; when water is boiled, steam is given off by the water; a flower gives off an odour. *See liberate (p.69).*

generate (*v*) to produce (p.62) large quantities of a gas for a particular purpose. *See generator (p.27).* **generator** (*n*).

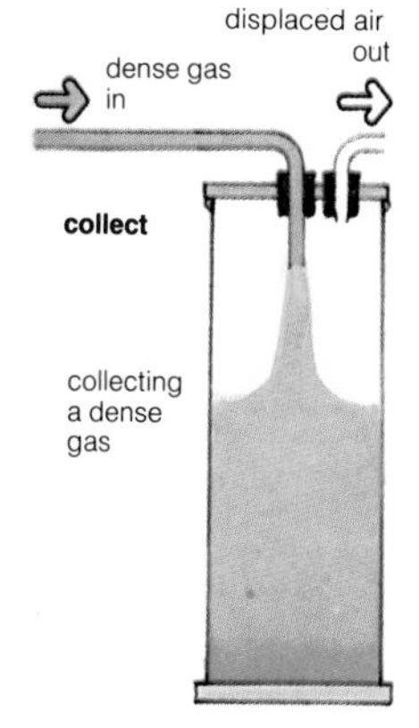

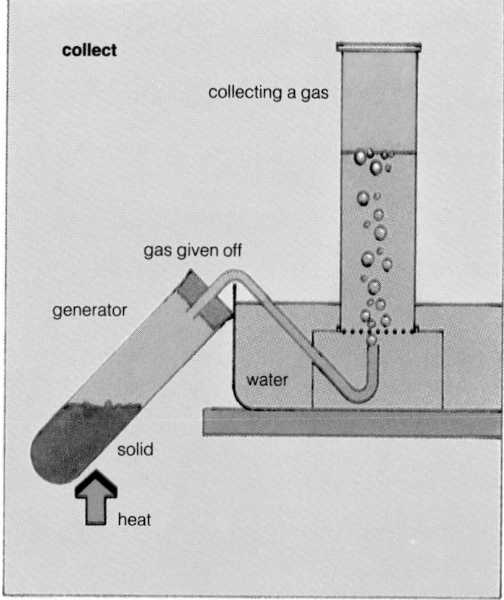

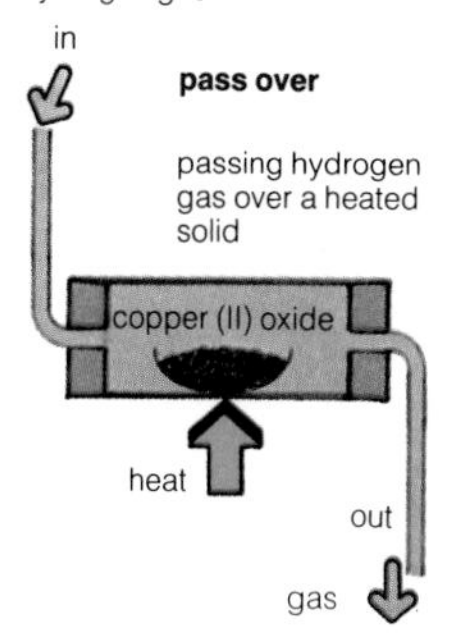

collect (*v*) (1) to obtain a gas in a vessel (p.25), usually a gas-jar (p.24). (2) to obtain a distilled liquid in a receiver (p.28). (3) to obtain a specimen (p.43) of crystals (p.91).

pass over to cause a gas to flow over a solid, e.g. hydrogen is passed over heated copper (II) oxide.

investigate (*v*) to study carefully by means of experiments (↓), tests (↓) and recorded (p.39) facts, e.g. to investigate the properties (p.9) of sulphuric acid is to find and record the properties by making a careful study of them. **investigation** (*n*).

experiment (*n*) to work with instruments and apparatus in a practical (p.23) investigation (↑) on the behaviour and nature of substances.

visible (*adj*) describes anything that can be seen. *See perceptible* (↓).

observation (*n*) the use of the senses for a particular purpose, e.g. when a chemical reaction (p.62) takes place, a person makes observations on any changes of state, colour, or odour, and on any new substances formed. Observations are recorded as results (p.39). **observe** (*v*), **observable** (*adj*).

perceptible (*adj*) describes any changes, usually physical changes, which can be observed (↑) by the senses, although such changes are small. Observable changes can be noticed more easily than perceptible changes. Changes can be *readily, barely* or *hardly* perceptible, showing increasing difficulty of perception. **perception** (*n*), **perceive** (*v*).

test (*n*) the use of a chemical reagent (p.63) to identify (p.225) a substance, a metal, a radical (p.45) or any group of substances. The test gives a result and from the result an inference (↓) is made. Some tests give a definite (p.226) result so the identification is made without doubt. **test** (*v*).

confirmatory (*adj*) describes a test which makes an identification without doubt. **confirm** (*v*), **confirmation** (*n*).

test paper (*n*) a piece of paper with a reagent (p.63) on it; the paper changes colour when testing particular substances and thus identifies (p.225) these substances, e.g. litmus paper.

demonstrate (*v*) to show clearly by practical examples, e.g. to demonstrate the properties of chlorine gas by showing in an experiment (↑) the action of various substances on the gas. **demonstration** (*n*).

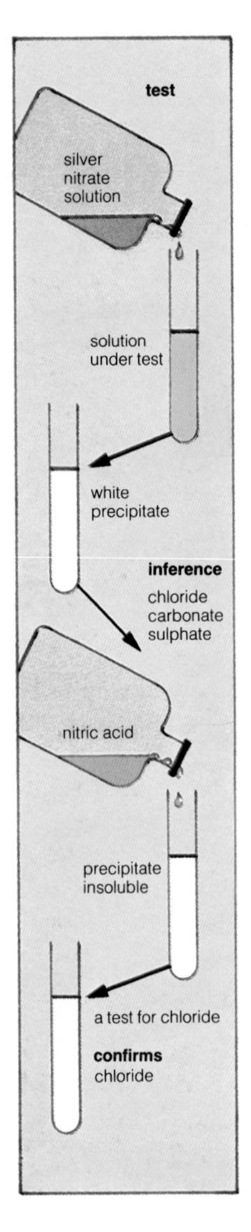

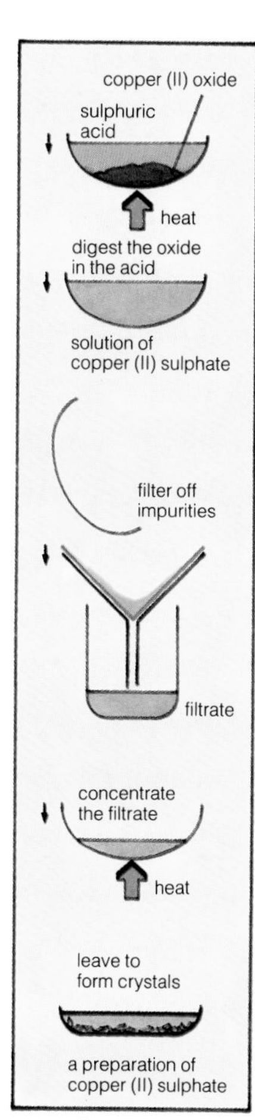

preparation

inference (*n*) the use of results (p.39) to decide on the identification (p.225) of a substance. An inference is only an opinion, a confirmatory (↑) test has to be made to be sure. **infer** (*v*).

conclusion (*n*) (1) the use of results (p.39) to decide on the identification (p.225) of a substance without doubt, e.g. inferences (↑) from tests together with a confirmatory (↑) test allow a conclusion to be made. (2) the use of results to decide on the relation (p.232) between observations (↑), e.g. a set of results on pressure and volume allows a conclusion to be drawn on the relation between the pressure and the volume of a gas. **conclude** (*v*).

technique (*n*) an accepted way of carrying out (p.157) a process (p.157) which needs practical (p.23) skill and a knowledge of chemistry, e.g. if a person knows the technique of distillation (p.33), he can connect suitable apparatus, heat to the correct temperature and collect a pure distillate.

isolate (*v*) to obtain a pure substance from a mixture (p.54) of substances. The pure substance can be a compound (p.8) or an element (p.8), e.g. to isolate bromine from sea water.

preparation (*n*) (1) a substance made in a laboratory (p.23) for a particular purpose, e.g. a student makes crystals of copper (II) sulphate, the substance is a preparation. (2) to make ready the apparatus and any solutions (p.86) needed in an experiment (↑). **prepare** (*v*).

specimen (*n*) (1) a small quantity of a material or substance (p.8) which is isolated (↑) from a mixture. (2) a quantity of a substance which is used as an example of that substance, i.e. the properties of the specimen are the properties of all quantities of the substance. For example, tests carried out (p.157) on a specimen of copper demonstrate (↑) the properties of the specimen and thus the properties of copper.

purification (*n*) a process (p.157) or processes carried out (p.157) to remove impurities (p.20) from a substance, e.g. the purification of silver by which impurities such as lead are removed (p.215). **purify** (*v*).

trivial name	quick lime	blue vitriol	oil of vitriol
traditional name	calcium oxide	cupric sulphate	sulphuric acid
systematic name	calcium oxide	copper (II) sulphate	sulphuric acid

trivial name	green vitriol	common salt	—
traditional name	ferrous sulphate	sodium chloride	potassium permanganate
systematic name	iron (II) sulphate	sodium chloride	potassium manganate (VII)

chemical nomenclature

nomenclature (*n*) a way of naming chemical substances (p.8).
trivial name a nomenclature (↑) for chemical substances which was used before chemistry was studied properly. Examples of trivial names are alum, blue vitriol, lime, and chalk.
traditional name a nomenclature (↑) for chemical substances which shows their chemical composition (p.82); this method was used before systematic names (↓) were used. Some traditional names are still used, and some are the same as systematic names. Examples of traditional names are cupric sulphate and lead nitrate.
systematic name the modern nomenclature (↑) for chemical substances. For inorganic (p.55) compounds, the oxidation number (p.78) of the metal is given, and the acid radical (↓) is described with an oxidation number for the important element (p.8) of the radical, e.g. copper (II) sulphate, iron (III) sulphate, potassium manganate (VII). For organic (p.55) compounds, the systematic name is taken from a suitable alkane (p.172), e.g. *ethanoic* acid is taken from *ethane*; ethanoic acid is the systematic name of acetic acid (traditional name).
binary compound a compound (p.8) formed from the chemical combination (p.64) of two elements (p.8). The systematic name (↑) for such compounds ends in *-ide*, e.g. lead (II) oxide, calcium carbide, phosphorus trichloride.

radical (*n*) a group of atoms, which are part of a molecule (p.77) or form an ion (p.123); the radical often remains unchanged throughout a chemical reaction (p.62) or a series of reactions. The atoms of a radical which forms an ion are held together by covalent bonds (p.136). Some radicals are also functional groups (p.185). Examples of radicals are the sulphate radical, the nitrate radical, the manganate (VII) radical and the ammonium radical.

acid radical a radical (↑) combined (p.64) with hydrogen in an acid (↓).

acid (*n*) a substance (p.8) which contains (p.55) hydrogen which can be replaced (p.68) by a metal, or by a base (p.46). An acid is a covalent (p.136) substance, which when dissolved in water produces hydrogen ions (p.123) in the solution. The strength of an acid is measured by its pH value (p.38). **acidify** (*v*), **acidic** (*adj*).

alkali (*n*) a soluble base (p.46). The solution of an alkali contains hydroxyl (p.132) ions; it reacts with an acid (↑) to produce a salt (p.46) and water only. **alkaline** (*adj*), **alkalinity** (*n*).

acidic (*adj*) (1) describes a substance having the nature (p.19) of an acid (↑). (2) describes a solution (p.86) containing an acid. (3) describes a compound (p.8) which forms an acid when dissolved in water, e.g. sulphur dioxide is an acidic oxide.

alkaline (*adj*) describes a solution (p.86) with the properties of an alkali, e.g. sodium hydroxide solution is alkaline.

neutral (*adj*) describes a substance or a solution which is neither acidic (↑) nor basic (p.46). A neutral solution has a pH value (p.38) of 7. **neutralize** (*v*).

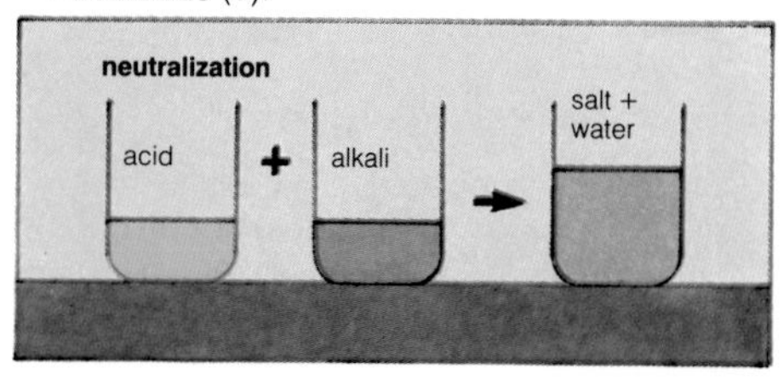

base (*n*) a substance which reacts (p.62) with an acid (p.45) to produce a salt (↓) and water only. Bases are usually the oxides (p.48) or hydroxides (p.48) of metals. Many bases are insoluble. **basic** (*adj*), **basicity** (*n*).

basic (*adj*) describes a substance having the nature of a base (↑), e.g. copper (II) oxide is a basic oxide.

amphoteric (*adj*) describes a substance which has both acidic (p.45) and basic (↑) properties, e.g. aluminium hydroxide reacts with **acids** to form a salt and water and it also reacts with **alkalis** to form a salt and water.

salt (*n*) a compound made by replacing some or all of the hydrogen of an acid (p.45) by a metal. A base or an alkali or a metal reacts with an acid to replace the hydrogen by a metal or by a basic radical such as the ammonium radical. Salts in solution usually form ions (p.123). Examples of salts are copper (II) sulphate (in which copper is combined (p.64) with the sulphate radical (p.45) as it has replaced the hydrogen of sulphuric acid), iron (III) chloride and lead (II) nitrate.

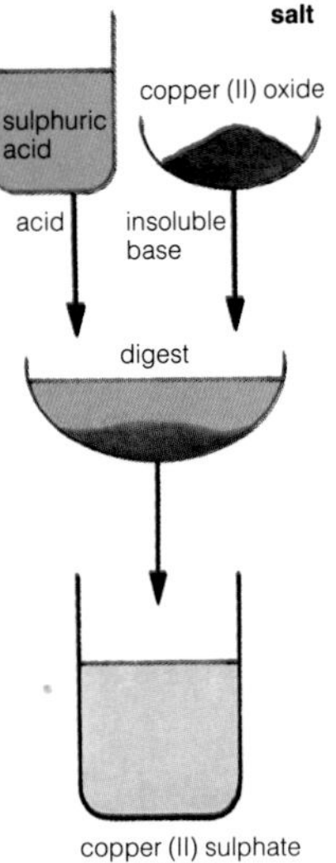

basicity (*n*) the number of hydrogen atoms which can be replaced (p.68) in one molecule of an acid; also the number of hydrogen ions (p.123) formed from one molecule of an acid. For example, hydrochloric acid HCl has one atom of hydrogen that can be replaced in one molecule of the acid, so it has a basicity of one; sulphuric acid H_2SO_4 has two replaceable hydrogen atoms, therefore it has a basicity of two.

monobasic (*adj*) describes an acid with a basicity (↑) of one.

dibasic (*adj*) describes an acid with a basicity (↑) of two.

tribasic (*adj*) describes an acid with a basicity (↑) of three.

normal salt a salt formed when all the hydrogen atoms in an acid (p.45) molecule have been replaced by a metal or a basic radical (p.45), e.g. sodium chloride, copper (II) sulphate, potassium sulphate, sodium carbonate.

acid salt a salt formed when not all the hydrogen atoms in an acid (p.45) molecule have been replaced by a metal or a basic radical (p.45), e.g. potassium hydrogensulphate, calcium hydrogencarbonate. Only dibasic (↑) and tribasic (↑) acids can form acid salts.

basic salt a salt formed when not all the base (↑) reacts with the acid (p.45) in a reaction. The base combined (p.64) with the normal salt (↑) to form an insoluble basic salt, e.g. basic lead carbonate, formed from lead (II) carbonate and lead (II) oxide.

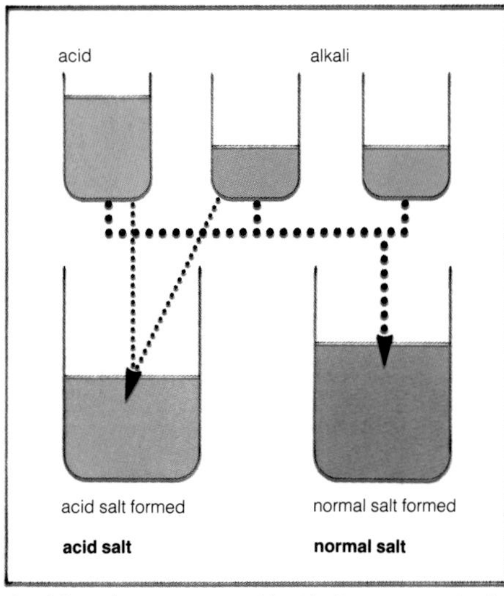

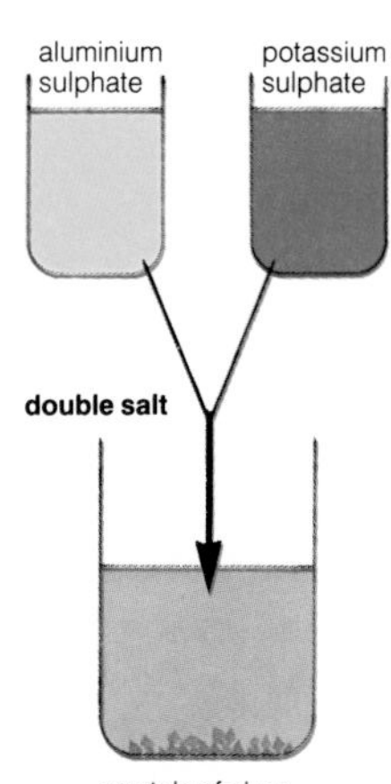

double salt a compound (p.8) of two normal salts (↑) which forms crystals (p.91), e.g. aluminium potassium sulphate, a compound of aluminium sulphate and potassium sulphate, known as alum.

complex salt a compound (p.8) containing a complex ion (p.132); either the metal or the acid radical of a normal salt, or both, can form a complex ion. An example is tetraammine copper (II) sulphate (cuprammonium sulphate); *see also hexacyanoferrate (p.53).*

hydrated (*adj*) describes a salt (p.46) with water chemically combined (p.64) with the compound in its crystals (p.91). The combined water is called *water of crystallization*, e.g. copper (II) sulphate $.5H_2O$, in which five molecules of water are combined with one of the salt in its crystals.

water of crystallization *see hydrated* (↑).

anhydrous (*adj*) describes a salt with no water of crystallization (↑), e.g. sodium chloride crystals have no water of crystallization.

anhydride (*n*) (1) a compound (p.8) formed by removing (p.215) the elements (p.8) of water from a substance, but no water of crystallization (↑), e.g. ethanoic acid (acetic acid) can have the elements of water removed to form ethanoic anhydride. (2) a compound which, when added to water, forms a new chemical compound; the anhydride is named after this new chemical compound, e.g. ethanoic anhydride, when added to water, forms ethanoic acid.

oxide (*n*) a binary compound (p.8) formed by the combination of an element (p.8) with oxygen. The element can be a metal or a non-metal (p.116), e.g. lead (II) oxide, calcium oxide.

hydroxide (*n*) a compound (p.8) formed by the reaction of a basic (p.46) oxide with water and having the radical (p.45) $-OH$. A soluble hydroxide forms hydroxyl (p.132) ions in water, e.g. sodium hydroxide NaOH, calcium hydroxide $Ca(OH)_2$ (formed by the action of water on calcium oxide, a basic oxide).

peroxide (*n*) an oxide which reacts (p.62) with cold dilute sulphuric acid to produce hydrogen peroxide. Peroxides in water form the ion (p.123) $(O-O)^{2-}$. Example: sodium peroxide.

higher oxide (*n*) an oxide with an element showing a higher oxidation number (p.78) than usual, e.g. manganese (VII) oxide is a higher oxide with manganese (IV) oxide the usual oxide.

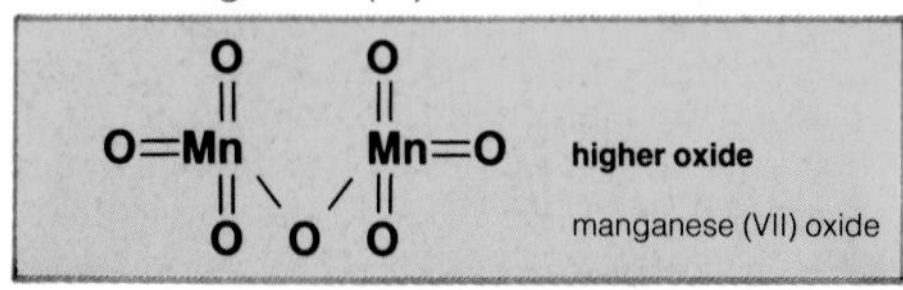

higher oxide
manganese (VII) oxide

oxides

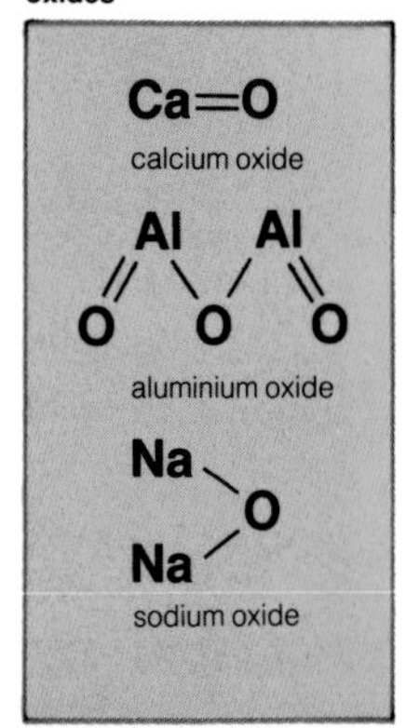

calcium oxide
aluminium oxide
sodium oxide

hydroxide sodium hydroxide

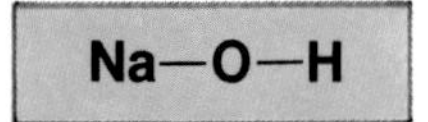

peroxide sodium peroxide

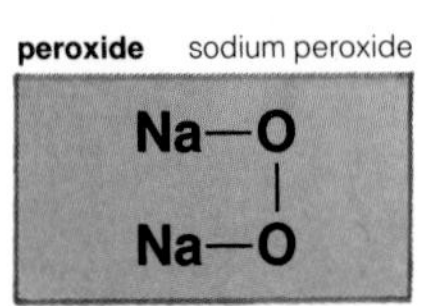

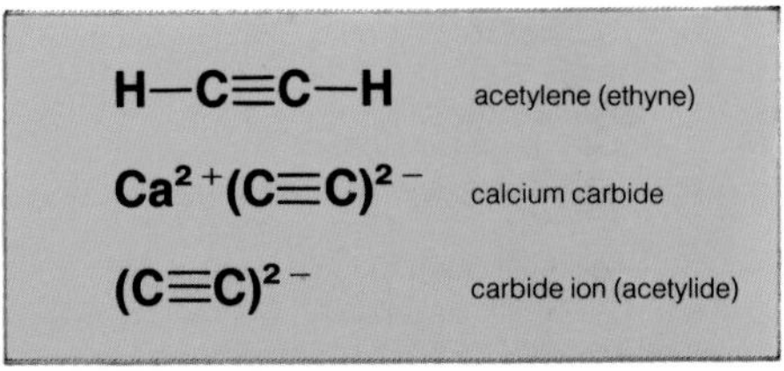

$H-C\equiv C-H$	acetylene (ethyne)
$Ca^{2+}(C\equiv C)^{2-}$	calcium carbide
$(C\equiv C)^{2-}$	carbide ion (acetylide)

carbide (*n*) a binary compound (p.44) containing a metal combined (p.64) with carbon. There are several different kinds of carbides, including the acetylides (↓).

acetylide (*n*) a carbide (↑) which reacts with water to form acetylene (p.174), C_2H_2. The metals of Group I and Group II of the periodic system (p.119) form acetylides. The compounds are ionic (p.123); an acetylide ion with an electrovalency of 2 is formed. Examples of acetylides are Na_2C_2 (sodium acetylide) and CaC_2 (calcium carbide or calcium acetylide).

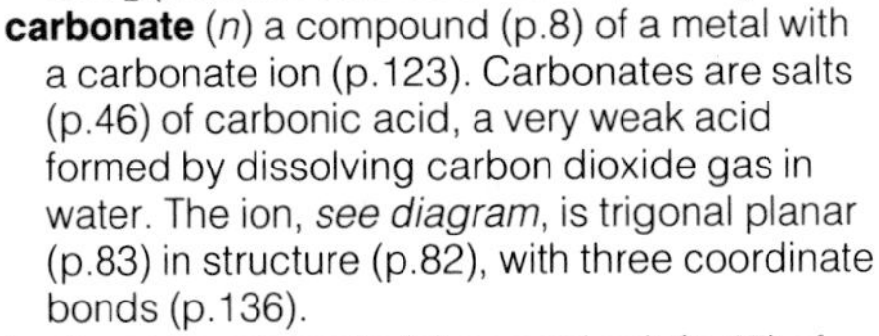

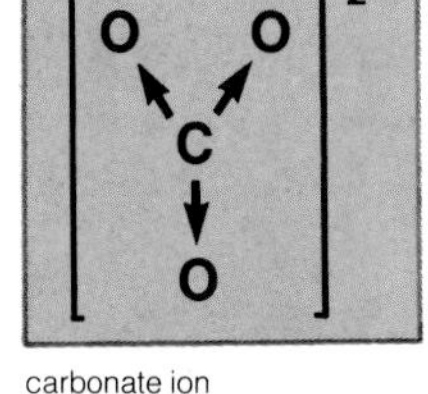

carbonate ion

carbonate (*n*) a compound (p.8) of a metal with a carbonate ion (p.123). Carbonates are salts (p.46) of carbonic acid, a very weak acid formed by dissolving carbon dioxide gas in water. The ion, *see diagram*, is trigonal planar (p.83) in structure (p.82), with three coordinate bonds (p.136).

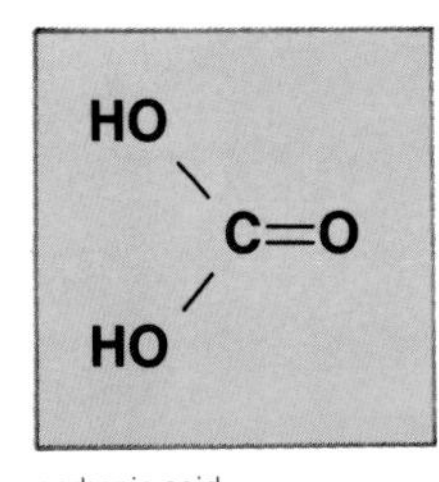

carbonic acid

hydrogen carbonate (*n*) an acid salt (p.47) of carbonic acid; only one hydrogen atom is replaced (p.68) by a metal, e.g. sodium hydrogencarbonate, $NaHCO_3$.

bicarbonate (*n*) the traditional name (p.44) for a hydrogencarbonate.

sodium hydrogen carbonate

$Na^{+}O^{-}$
C=O
O

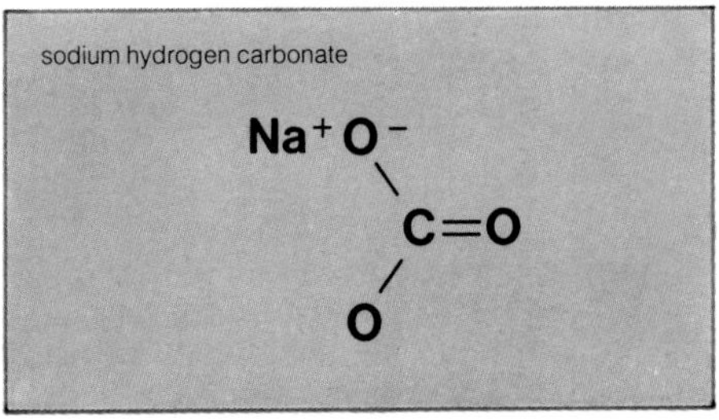

chloride (*n*) a compound (p.8) of an element (p.8) and chlorine. Metals form ionic (p.123) chlorides; non-metals form covalent (p.136) chlorides, e.g. Na^+Cl^- (ionic chloride); CCl_4 (covalent chloride). *See chloro group (p.187)*. Chlorides are salts (p.46) of hydrochloric acid.

bromide (*n*) a compound (p.8) of an element (p.8) and bromine. Metals usually form ionic (p.123) bromides; non-metals form covalent (p.136) bromides. *See bromo group (p.187)*. Bromides are salts (p.46) of hydrobromic acid.

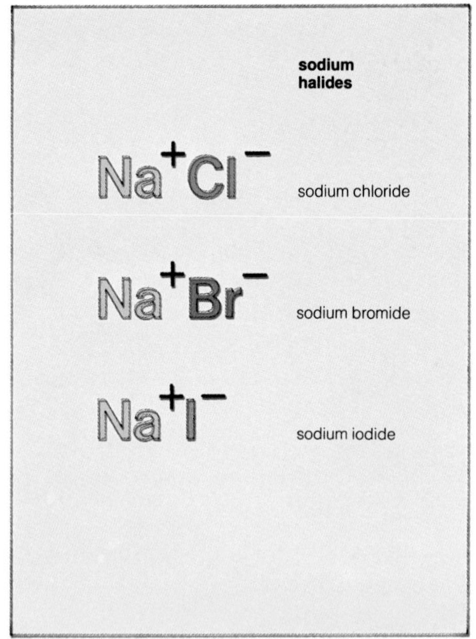

iodide (*n*) a compound (p.8) of an element (p.8) with iodine. Metals usually form ionic (p.123) iodides; non-metals form covalent (p.136) iodides. *See iodo group (p.187).* Iodides are salts (p.46) of iodic acid.

halide (*n*) a compound (p.8) which is a chloride, a bromide, an iodide, or a fluoride. It is formed from an element (p.8) and a halogen (p.117).

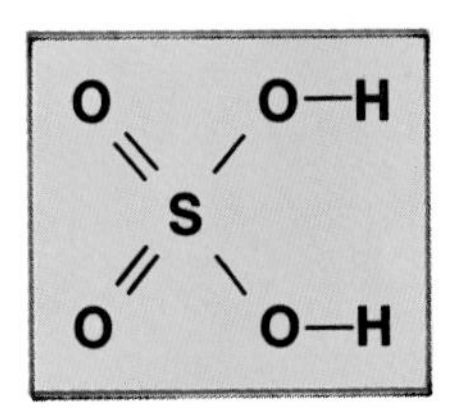

sulphuric acid (covalent)

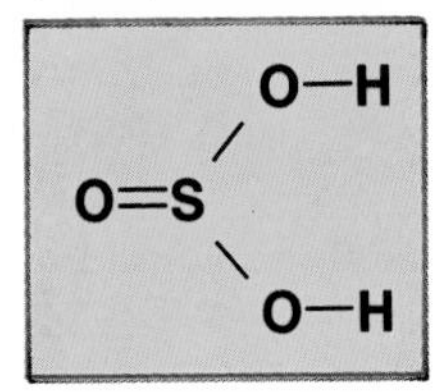

sulphurous acid (covalent)

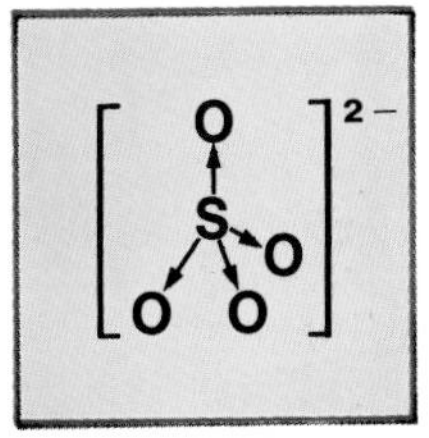

sulphate ion (tetrahedral)

sulphide (*n*) a binary compound (p.44) of an element (p.8) with sulphur. Metals form ionic (p.123) sulphides, most of which are insoluble. Non-metals form covalent (p.136) sulphides. Sulphides are salts (p.46) of the weak acid, hydrogen sulphide, H_2S.

sulphate (*n*) a compound (p.8) of a metal with a sulphate ion (p.123). Sulphates are salts (p.46) of sulphuric acid, a strong acid which is dibasic, forming normal and acid salts. The sulphate ion, *see diagram*, it tetrahedral (p.83) in structure (p.82) with four coordinate bonds (p.136).

hydrogen sulphate (*n*) an acid salt of sulphuric acid, i.e. an acid sulphate (↑). For example, sodium hydrogensulphate $NaHSO_4$ which forms the ion HSO_4^-.

bisulphate (*n*) traditional name (p.44) for hydrogen sulphate.

sulphite (*n*) a compound (p.8) of a metal with a sulphite ion (p.123). Sulphites are salts (p.46) of sulphurous acid, a dibasic acid forming normal and acid salts. The sulphite ion, *see diagram*, is trigonal pyramidal (p.84) in structure (p.82) with three coordinate bonds (p.136) and a lone pair (p.133) of electrons.

hydrogen sulphite (*n*) an acid salt of sulphurous acid, i.e. an acid sulphite, e.g. sodium hydrogensulphite $NaHSO_3$ with the ion HSO_3^-.

bisulphite (*n*) traditional name for hydrogen-sulphite (↑).

thiosulphate (*n*) a compound (p.8) of a metal with a thiosulphate ion (p.123). Thiosulphates are the salts (p.46) of thiosulphuric acid $H_2S_2O_3$. The thiosulphate ion, is tetrahedral (p.83) in structure (p.82) with four coordinate bonds (p.136).

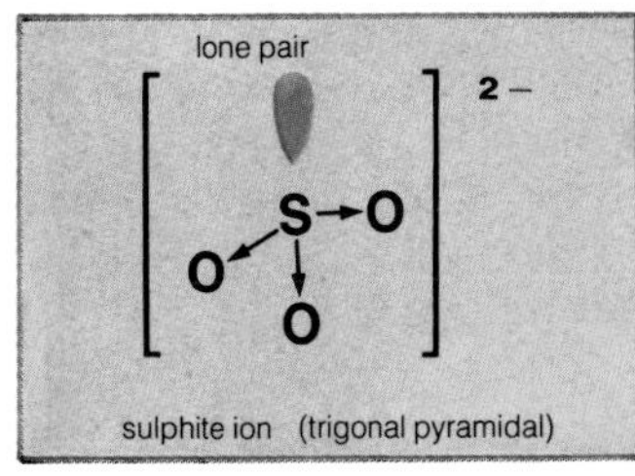

sulphite ion (trigonal pyramidal)

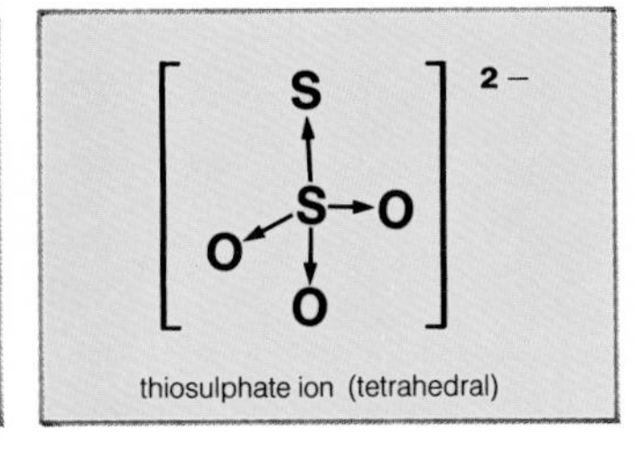

thiosulphate ion (tetrahedral)

nitride (*n*) a binary compound (p.44) of a metal with nitrogen. The common nitrides are formed with the metals of groups I, II and III of the periodic system (p.119). These nitrides react (p.62) with water to form ammonia and the hydroxide of the metal.

nitrate (*n*) a compound (p.8) of a metal with a nitrate ion (p.123). Nitrates are salts (p.46) of nitric acid, a strong acid which is monobasic (p.46). The nitrate ion, *see diagram*, is trigonal planar (p.83) in structure (p.82) with three coordinate bonds (p.136).

nitrite (*n*) a compound (p.8) of a metal with a nitrite ion (p.123). Nitrites are salts (p.46) of nitrous acid, a monobasic acid (p.45). The nitrite ion, *see diagram*, is non-linear (p.83) with two coordinate bonds (p.136).

chromate (VI) (*n*) a compound (p.8) of a metal with the chromate (VI) ion (p.123). Chromic acid, the acid of the salts (p.46), cannot be isolated (p.43). Although the acid is dibasic, only the normal salts (p.46) can be made. The chromate (VI) ion, *see diagram*, is tetrahedral (p.83) in structure (p.82) with four coordinate bonds (p.136).

chromate (*n*) traditional name (p.44) for chromate (VI).

dichromate (VI) (*n*) a compound of a metal with the dichromate (VI) ion (p.123). Dichromic acid, the acid of the salts (p.46), cannot be isolated (p.43). Although the acid is dibasic, only the normal salts (p.46) can be made. The dichromate (VI) ion, *see diagram*, has the structure of two tetrahedra with six coordinate bonds (p.136) and two covalent bonds (p.136).

dichromate (*n*) traditional name (p.44) for dichromate (VI).

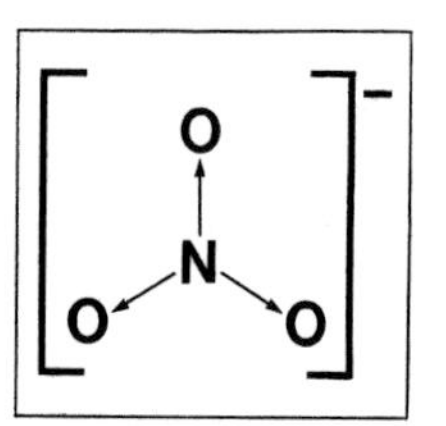

nitrate ion trigonal planar

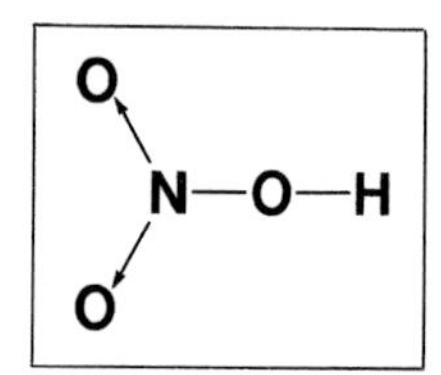

nitric acid

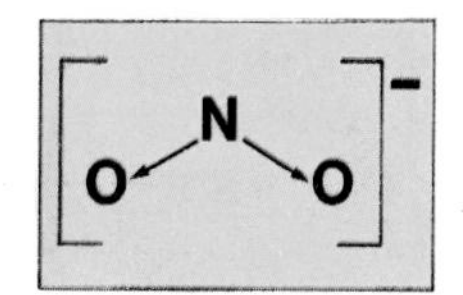

nitrite ion (non-linear)

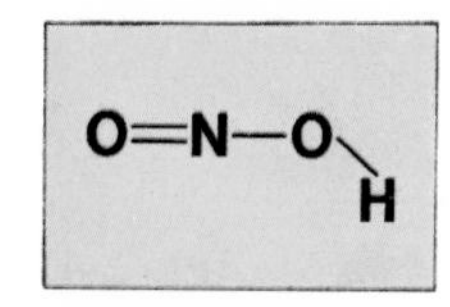

nitrous acid

chromate (VI) ion tetrahedral

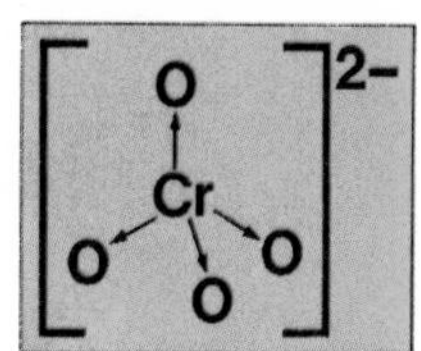

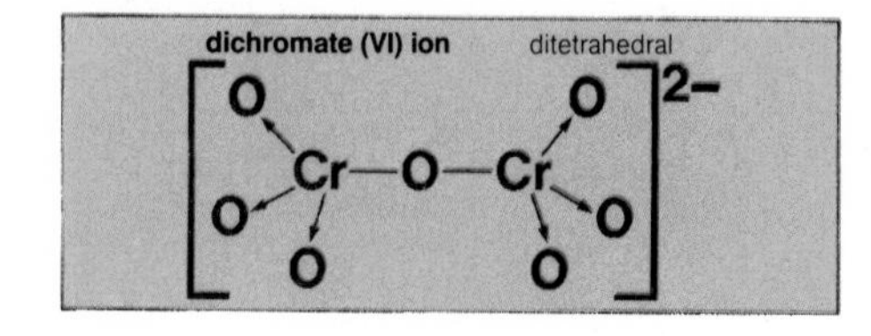

manganate (VI) ion
tetrahedral

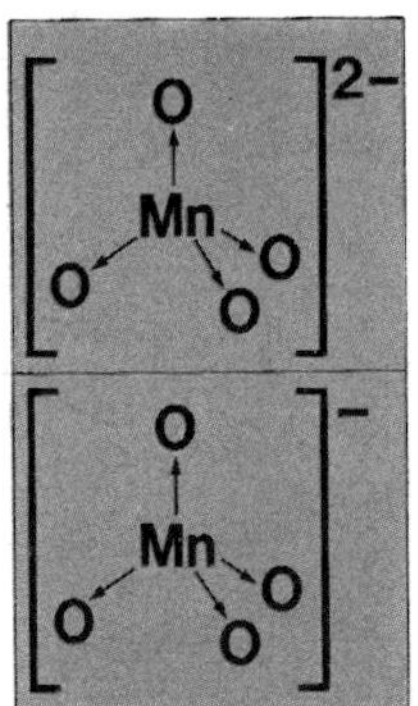

manganate VII ion
tetrahedral

manganate (VI) (*n*) a compound (p.8) of a metal with the manganate (VI) ion (p.123). The acid of these salts (p.46) cannot be isolated (p.43); no acid salts are formed, only normal salts, although the acid is dibasic (p.46). The manganate (VI) ion, *see diagram*, is tetrahedral (p.83) in structure (p.82) with four coordinate bonds (p.136). The manganese atom has an oxidation number (p.78) of 6. An example of a salt is potassium manganate (VI) K_2MnO_4.

manganate (*n*) traditional name (p.44) for manganate (VI).

manganate (VII) (*n*) a compound (p.8) of a metal with the manganate (VII) ion (p.123). The acid of these salts (p.46) cannot be isolated (p.43); it is monobasic (p.46). The manganate (VII) ion, *see diagram*, is tetrahedral (p.83) in structure (p.82) with four coordinate bonds (p.136). The manganese atom has an oxidation number of 7. An example of a salt is potassium manganate (VII) $KMnO_4$.

permanganate (*n*) traditional name (p.44) for manganate (VII).

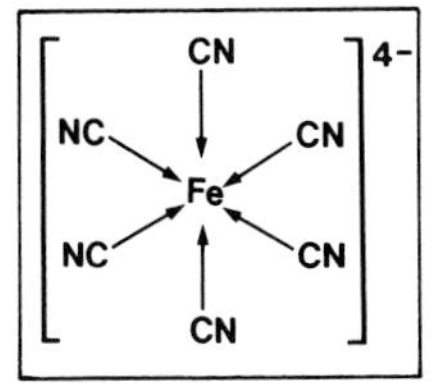

hexacyanoferrate (II) ion

hexacyanoferrate (II) (*n*) a compound (p.8) of a metal and the hexacyanoferrate (II) ion (p.123). There is no acid for these salts (p.46). The hexacyanoferrate (II) ion, *see diagram*, is octahedral (p.83) in structure (p.82) with six coordinate bonds (p.136). The iron atom has an oxidation number of 2. An example of a salt is potassium hexacyanoferrate (II) $K_4Fe(CN)_6$. The hexacyanoferrate ion is a complex ion (p.132).

ferrocyanide (*n*) the traditional name (p.44) for hexacyanoferrate (II).

hexacyanoferrate (III) (*n*) a compound (p.8) of a metal and the hexacyanoferrate (III) ion. There is no acid for these salts (p.46). The hexacyanoferrate (III) ion has the same structure as the hexacyanoferrate (II) ion (↑), but the iron atom has an oxidation number (p.78) of 3; it is a complex ion. An example of a salt is potassium hexacyanoferrate (III) $K_3Fe(CN)_6$.

ferricyanide (*n*) the traditional name (p.44) for hexacyanoferrate (III).

mixture (*n*) different substances (p.8), put together, form a mixture. The substances can be elements (p.8), compounds (p.8) or materials (p.8), e.g. a mixture of charcoal (the element, carbon), sulphur (an element) and potassium nitrate (a compound) forms a mixture called gunpowder. A particular substance can be separated (p.34) from a mixture. Mixtures can be solid, liquid, or gaseous (p.11). **mix** (*v*).

constituent (*n*) (1) a member of a mixture (↑), e.g. a mixture of sulphur and copper has two constituents; charcoal, sulphur and potassium nitrate mixed together form gunpowder, sulphur is one of the constituents of gunpowder. Each constituent keeps its own properties in a mixture. A liquid mixture can be a true solution (p.86), a colloidal solution (p.98) or a suspension (p.86). (2) a part of a compound (p.8), e.g. a molecule of sulphur dioxide consists of (p.55) one atom of sulphur and two atoms of oxygen, it has three constituent atoms. Constituents of a compound are elements (p.8), radicals (p.45), functional groups (p.185) and ions (p.123). **constitute** (*v*), **constitution** (*n*).

ingredient (*n*) a substance (p.8) needed for a mixture (↑) before it is put in the mixture, e.g. charcoal, sulphur and potassium nitrate are ingredients of gunpowder before they are mixed to form gunpowder; after mixing they are constituents (↑).

homogenous (*adj*) describes a material or substance (p.8) which is the same throughout in its properties (p.9) and composition (p.82). Homogenous also describes a chemical reaction (p.62) with all substances in the same state of matter (p.9), e.g. all in the gaseous state, or all in the liquid state. **homogeneity** (*n*), **homogenize** (*v*).

heterogenous (*adj*) describes a material, substance (p.8), or chemical reaction (p.62) which is not the same throughout in its properties (p.9), composition (p.82), or state of matter (p.9). It is the opposite of homogenous (↑). **heterogeneity** (*n*).

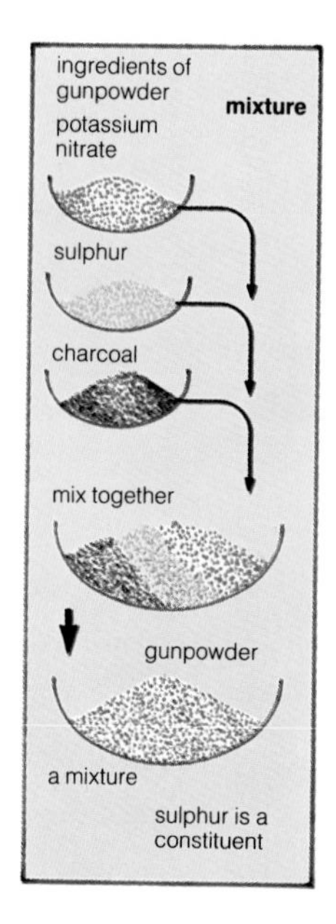

mixture

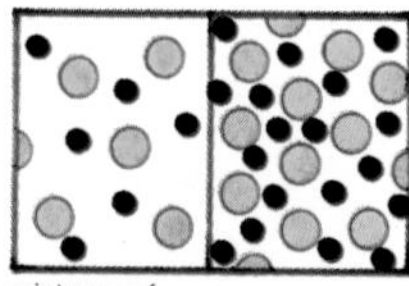

mixtures of iron and sulphur varied proportions

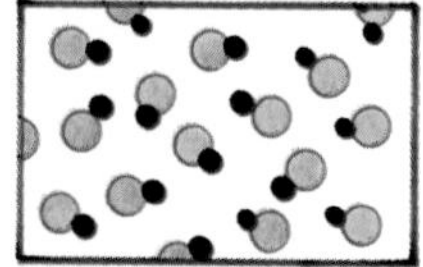

compound of iron and sulphur constant proportions

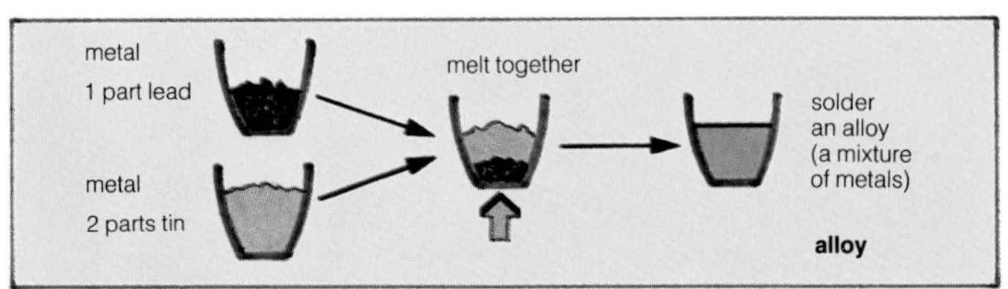

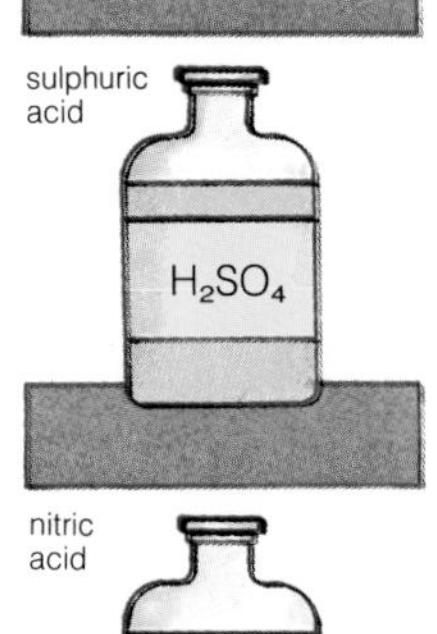

alloy (*n*) a mixture (p.54) of two or more metals, or of a metal and a non-metal. The mixture is homogenous (p.54) and a new material is formed. An alloy makes a particular metal more useful, e.g. silver is soft, but a silver alloy is hard and more useful than pure (p.20) silver. **alloy** (*v*).

brass (*n*) an alloy (↑) of 60–90% copper with zinc; other metals may be added also. Brass is harder than pure copper, and is more useful for a number of purposes.

amalgam (*n*) an alloy (↑) of mercury with other metals. An amalgam is usually soft and may even be liquid. **amalgamate** (*v*).

consist of to have as constituents (↑) when all the constituents are named, e.g. gunpowder *consists of* charcoal, sulphur and potassium nitrate; brass *consists of* copper and zinc.

contain (*v*) to have as constituents (↑) when some, or only one, of the constituents are named but not all, e.g. gunpowder contains sulphur and charcoal; gunpowder contains sulphur.

organic (*adj*) describes substances (p.8) (and the study of these substances in chemistry) which are compounds (p.8) of carbon, but not the oxides or the carbonates of carbon. All substances in living things are organic, but many organic substances are not found in living things. *See hydrocarbon (p.172) and carbohydrate (p.205)*.

inorganic (*adj*) describes substances (p.8) (and the study of these substances in chemistry) which are not organic (↑). Such substances are generally obtained from minerals (p.154).

mineral acid (*n*) an acid obtained from a mineral (p.154) by chemical processes (p.157). The three most important mineral acids are hydrochloric, sulphuric and nitric acids.

air (*n*) a mixture (p.54) of gases which forms the atmosphere and is the cause of atmospheric pressure (p.102). Air contains (p.55) about 20% oxygen, 79% nitrogen, 1% noble gases (↓) and 0.3% carbon dioxide; this is the composition (p.82) of dry air. In addition air always contains water vapour (p.11).

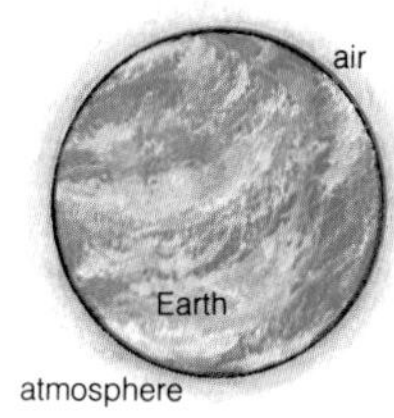

air forms the atmosphere

air

diluent (*n*) a substance added to a solution or to a mixture of solids or gases to reduce (p.219) the concentration of the solution or to reduce the proportion (p.76) of one of the constituents (p.54) of the mixture, e.g. water can be added to a concentrated (p.88) alkali as a diluent; nitrogen in the air is a diluent for oxygen.

noble (*n*) (1) *noble* gases of the air (↑) describes the gases helium, neon, argon, krypton and xenon. The gases are considered inert (p.19). (2) *noble* metals describes gold, platinum and other metals which do not react with the usual mineral acids.

pollution (*n*) undesirable substances (p.8) in the air, water or earth; the surroundings (p.103) become unhealthy or impure, e.g. the pollution of the air by smoke.

water (*n*) a liquid substance formed by the combination (p.64) of hydrogen and oxygen (formula: H_2O).

water vapour pressure all water on the Earth's surface evaporates (p.11) and so air contains water vapour. The vapour exerts (p.106) a pressure, which is part of atmospheric pressure (p.102).

saturated water vapour pressure The water vapour pressure (↑) when the air is saturated (p.87) with water vapour. It is the highest water vapour pressure at a particular temperature. Saturated water vapour pressure increases with a rise in temperature.

water cycle a cycle (p.64) in which water evaporates (p.11) from the Earth's surface and forms clouds, the clouds break to form rain, the rain passes through the Earth to rivers, lakes and finally the sea. Water evaporates from the rivers, lakes and sea. Respiration (p.61) also helps in the cycle.

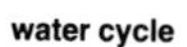

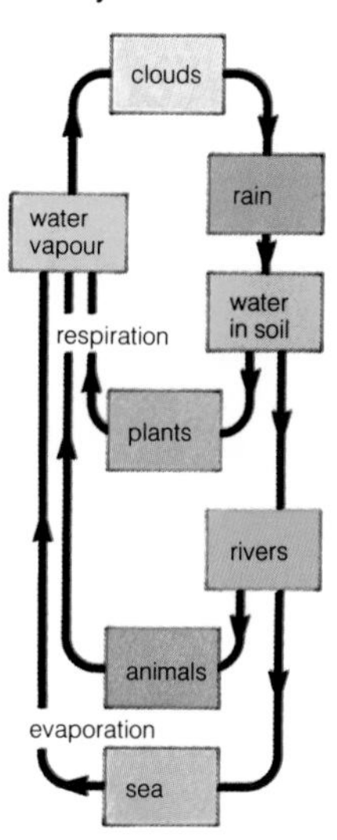

hard water water which does not form a lather (↓) easily with soap. It contains (p.55) salts of calcium and magnesium which form insoluble salts with the soap. **hardness** (*n*).

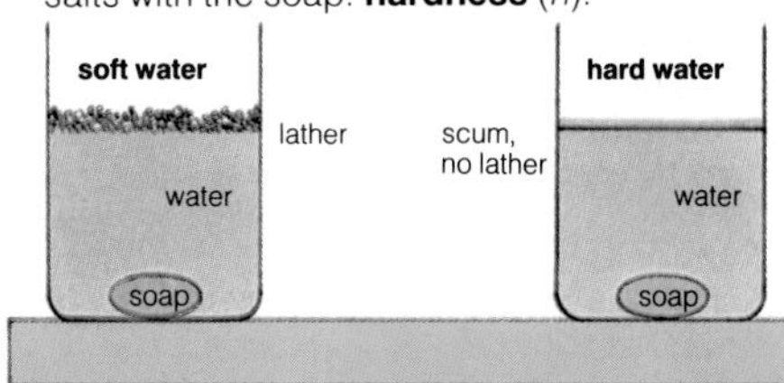

soft water water which forms a lather (↓) easily with soap. It does not contain salts of calcium and magnesium, or else such salts have been removed (p.215). **soften** (*v*).

lather (*n*) a large quantity of very small bubbles (p.40) formed when soap is mixed with water. **lather** (*v*).

temporary hardness a hardness of water (↑) which can be removed (p.215) by boiling the water. It is caused by calcium hydrogen-carbonate being dissolved in the water.

permanent hardness a hardness of water (↑) which cannot be removed (p.215) by boiling. It is removed by (a) adding sodium carbonate, (b) by detergents (p.171), or (c) by zeolite (↓). It is caused by dissolved salts of calcium and magnesium, such as the sulphate.

water softening

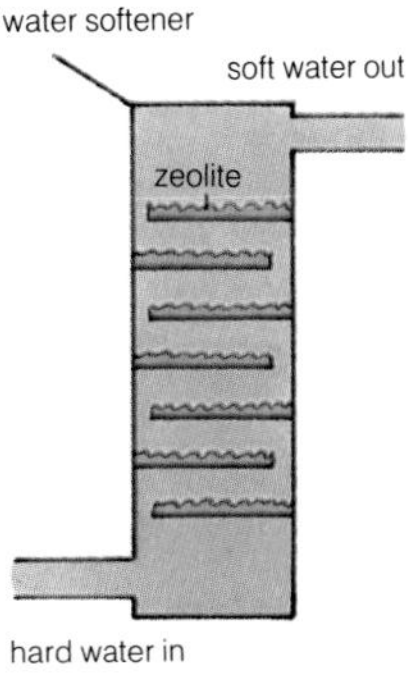

water softener

soft water out

zeolite

hard water in

water softening any process by which hard water (↑), with either temporary or permanent hardness (↑), is changed to soft water (↑).

zeolite process a process (p.157) using minerals (p.154), called zeolites, to soften (↑) water. A zeolite contains sodium ions (p.123) which can be replaced by other metal ions. In water softening, the zeolite removes (p.215) the calcium and magnesium ions from hard water and replaces (p.68) them with sodium ions.

contamination (*n*) the presence in water, food, or any other substance or material (p.8) of causative agents of disease, e.g. the presence of viruses, bacteria, protozoa, etc. which cause disease.

combustion (*n*) any chemical reaction (p.62) in which heat, and usually light, is produced. Combustion commonly is the burning of organic (p.55) substances during which oxygen from the air (p.56) is used to form carbon dioxide and water vapour. **combustible** (*adj*).

rapid combustion combustion (↑) in which heat is produced (p.62) at a high temperature, usually with flames, by the combination (p.64) of substances with oxygen.

slow combustion combustion (↑) in which heat is produced (p.62) at a low temperature, without flames, by the combination (p.64) of substances with oxygen.

explosion (*n*) a chemical reaction which takes place very quickly, forming large volumes of gases and releasing (p.69) energy (p.135). The energy causes a rise in temperature. An explosion produces (p.62) sound and often light. The increase in pressure caused by the large volume of gas causes destruction in the surroundings. **explode** (*v*), **explosive** (*adj*).

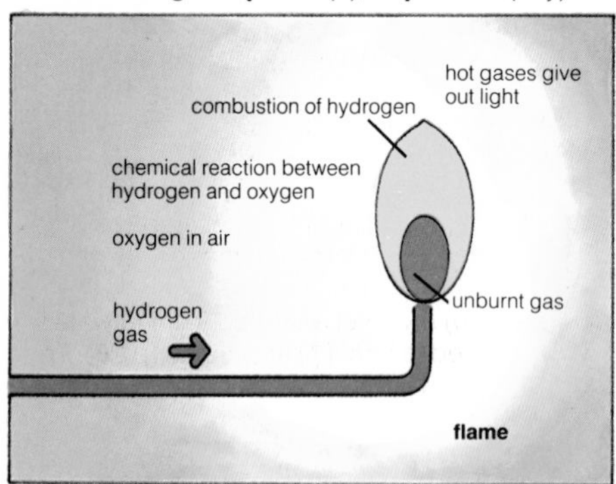

flame (*n*) combustion (↑) of gases produces (p.62) a flame. The hot gases give out light.

spontaneous combustion combustion (↑) which takes place without any apparent (p.223) cause, e.g. the spontaneous combustion of phosphorus in oxygen.

incombustible (*adj*) describes any substance which does not undergo (p.213) combustion (↑).

burn (*v*) to undergo (p.213) combustion (↑) producing (p.62) heat, flame and sometimes smoke, e.g. wood burns with a smoky flame; to burn coal to produce heat.

glow (*v*) to give out light because of a chemical reaction (p.62) or because a substance (p.8) is heated strongly, e.g. a piece of wood glows when there are no flames; a piece of iron glows when heated strongly, **glow** (*n*).

smoulder (*v*) to give out smoke because of a chemical reaction (p.62) taking place at a low temperature without a flame.

warm (*v*) to supply only enough heat to make substances react (p.62) or a process (p.157) to take place. To raise the temperature of a liquid or a solid so that it can still be touched without discomfort. For example, to warm a solution so that crystals dissolve. **warm** (*adj*).

heat (*v*) to supply enough heat to raise the temperature so that a substance is hot enough for a chemical reaction (p.62) to take place or a process (p.157) to take place. The temperature is too high for the substance to be touched with comfort. For example, to heat zinc in a stream of oxygen. **hot** (*adj*), **heat** (*n*), **heater** (*n*).

char (*v*) to change an organic (p.55) substance to carbon by heating it or by using a strong dehydrating (p.66) agent. For example, to char wood by heating it, forming charcoal; to char sugar by adding concentrated sulphuric acid.

water-bath (*n*) a vessel (p.25) containing water, which is used to heat (↑) apparatus (p.23). This way of heating does not allow the temperature to rise above 100°C.

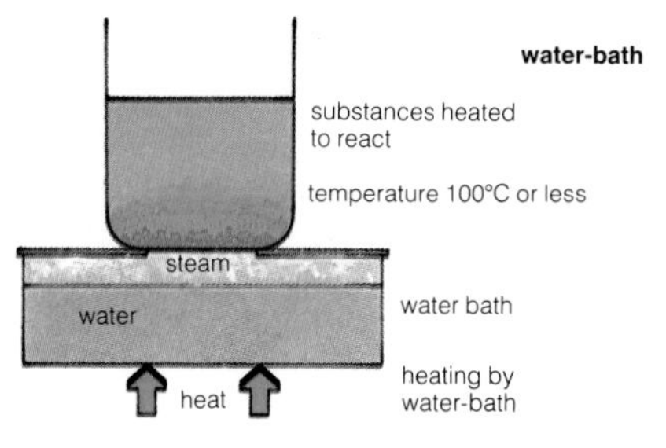

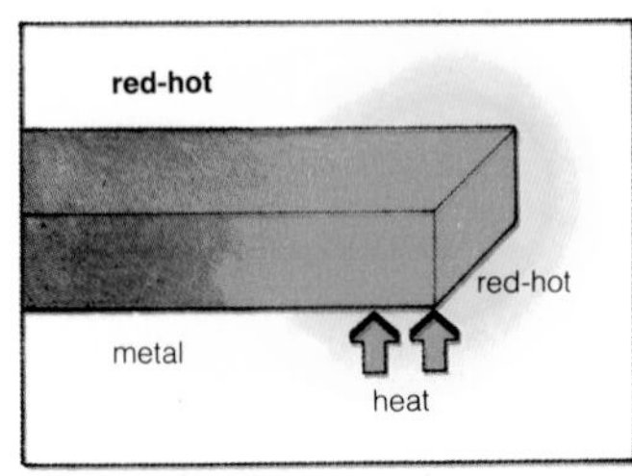

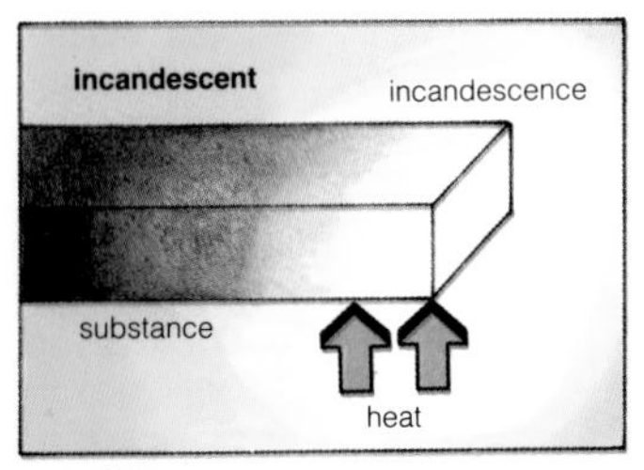

red-hot (*adj*) describes a substance, usually a metal, heated to such a high temperature that it glows (p.59) red.

incandescent (*adj*) describes a substance heated to such a high temperature that it gives out white light. The temperature of an incandescent substance is higher than the temperature of a red-hot (↑) substance. **incandescence** (*n*).

light[1] (*n*) a form of energy whose effect on the eye causes the sense of seeing. White light is composed (p.82) of all colours. Different colours can cause different chemical effects. Some substances, e.g. silver chloride, are decomposed by light falling on them. *See photochemical (p.65).*

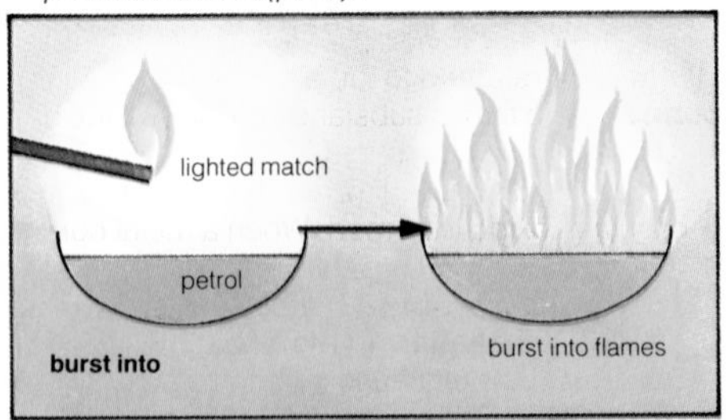

burst into to appear suddenly, as flames do, e.g. when a lighted match is put near petrol, the petrol bursts into flames. *Burst into* is used when a lot of flames appear; with a small flame, the substance *catches fire*.

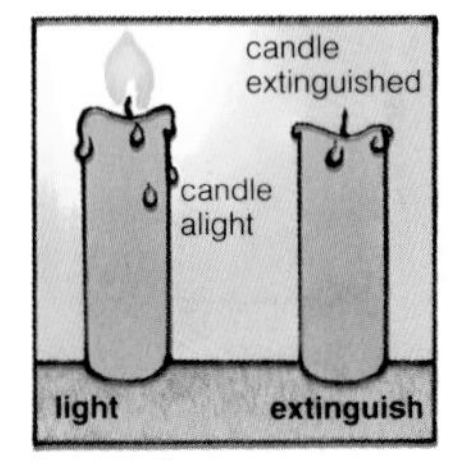

light[2] (*v*) to put a flame to a supply of a gas, so that the gas catches fire, i.e. flames appear, e.g. to light a supply of hydrogen from a jet.

extinguish (*v*) to cause a flame to stop burning.

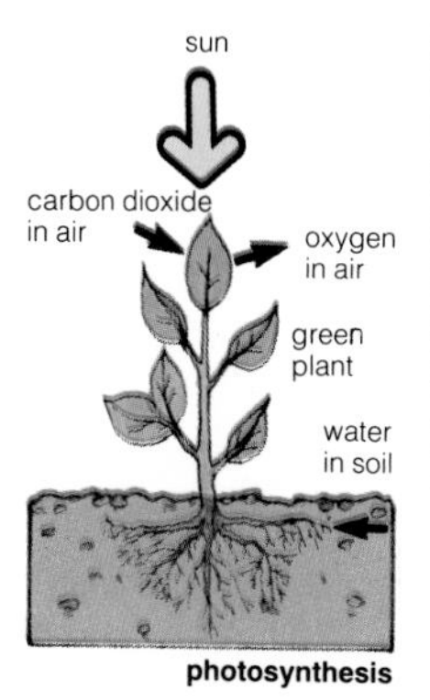

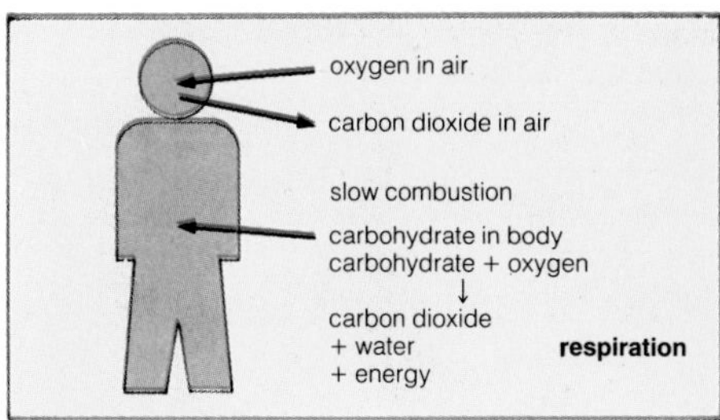

respiration (*n*) a process (p.157) in plants and animals by which oxygen reacts (p.62) with carbohydrates (p.205) to form carbon dioxide and water vapour, with the release (p.69) of energy (p.135). It is a kind of slow combustion (p.58). **respire** (*v*).

photosynthesis (*n*) a process (p.157) in green plants by which carbon dioxide from the air and water from the soil is used to make carbohydrates (p.205). Energy (p.135) from the sun is used in the chemical reaction (p.62). **photosynthesize** (*v*), **photosynthetic** (*adj*).

carbon cycle

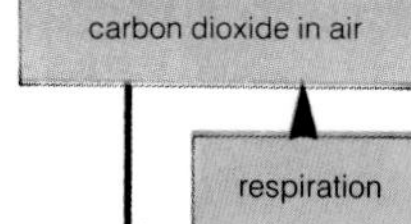

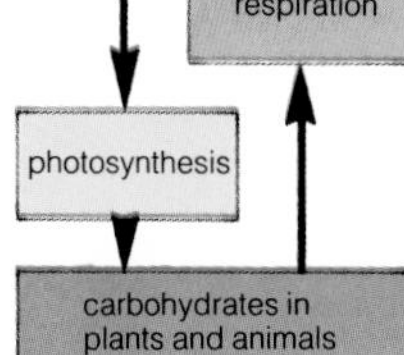

carbon cycle a cycle (p.64) of chemical reactions in which green plants convert (p.73) carbon dioxide to carbohydrates by photosynthesis (↑). The carbohydrates are used by plants and animals in respiration (↑) and the carbon dioxide is returned to the air (p.56).

corrode (*v*) to form a substance on the surface of a metal. The metal reacts (p.62) with particular gases in the air to form the substance; in many cases an oxide is formed. When a metal corrodes, small holes appear on the surface and the strength of the metal becomes less. Oxygen, carbon dioxide, sulphur dioxide or hydrogen sulphide in the air cause a metal to corrode. **corrosion** (*n*), **corrosive** (*adj*).

rust (*n*) corrosion (↑) of iron, forming a red dust which is the rust on the surface of the iron. Rust is an oxide of iron. **rust** (*v*), **rusty** (*adj*).

tarnish (*v*) to corrode if the metal has a shiny surface, e.g. silver tarnishes by the forming of a layer of silver sulphide on the surface; silver sulphide is a corrosion, or tarnish. **tarnish** (*n*).

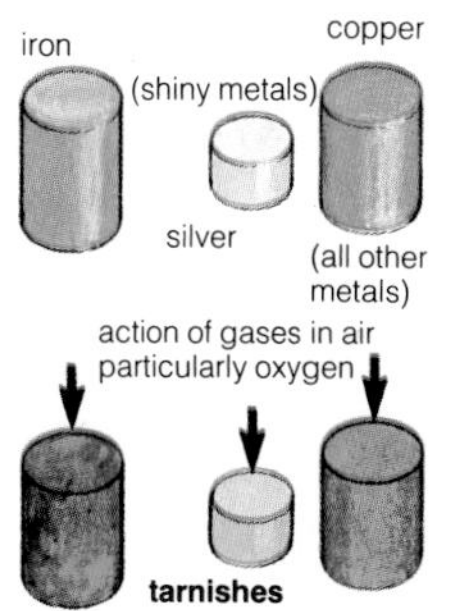

chemical reaction a process (p.157) in which new substances (p.8) are formed, i.e. a chemical change takes place. Energy, usually heat, is needed to make the chemical reaction take place, or the chemical reaction produces (↓) heat.

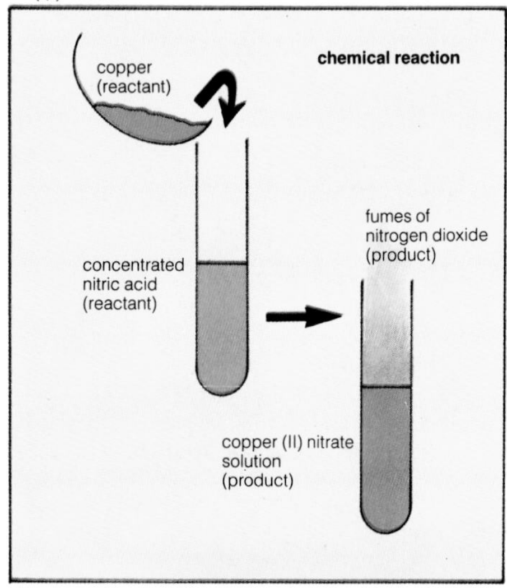

reaction (*n*) in chemistry, a process (p.157) in which two substances (p.8) have an effect on each other and new substances are produced (↓). **react** (*v*).

react (*v*) to behave in such a way that a chemical reaction (↑) takes place, e.g. sodium metal reacts with water, a chemical reaction takes place and sodium hydroxide and hydrogen are formed (sodium and water react).

reactant (*n*) a substance which takes part in a chemical reaction (↑), e.g. sodium and water are reactants when these substances react.

product (*n*) a new substance formed from a chemical reaction (↑), e.g. hydrogen and sodium hydroxide are the products of the reaction between sodium metal and water. **produce** (*v*).

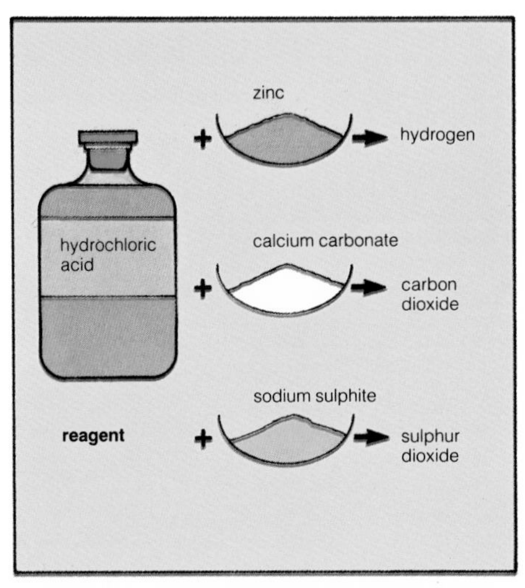

reagent (*n*) a substance which causes a chemical reaction (↑) to take place. The mineral acids (p.55) are common reagents as they have known effects on many inorganic (p.55) substances.

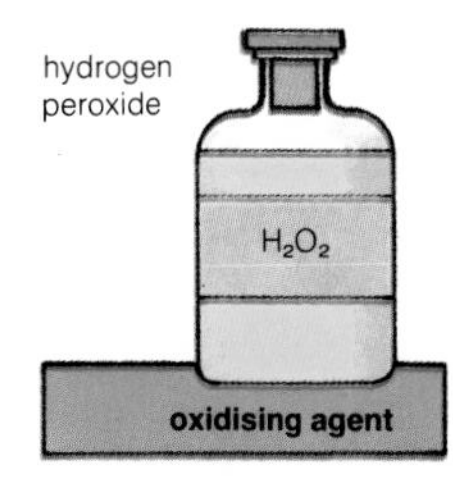

agent (*n*) a substance, or a form of energy, which is used to produce a named effect, e.g. an oxidizing agent causes oxidation (p.70); light causes a photochemical (p.65) effect and so is an agent for the effect.

take place to come into being, when describing an event, i.e. an event takes place. When an event *happens*, it is not expected, when an event *takes place*, it is planned. *See occur*[1] (↓).

occur[1] (*v*) to come into being at a certain time when describing an event, e.g. an event occurs, it is known when it will occur, although a person does not control it. To contrast *take place* and *occur*: a chemical reaction (↑) *takes place* because a person controls it, it cannot *occur* because it has not happened before and it is under the control of a person. A birthday occurs once a year, it does not take place.

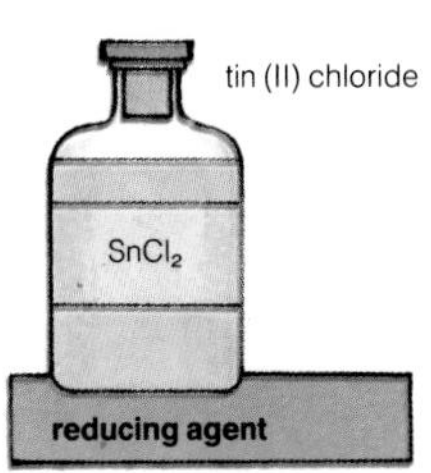

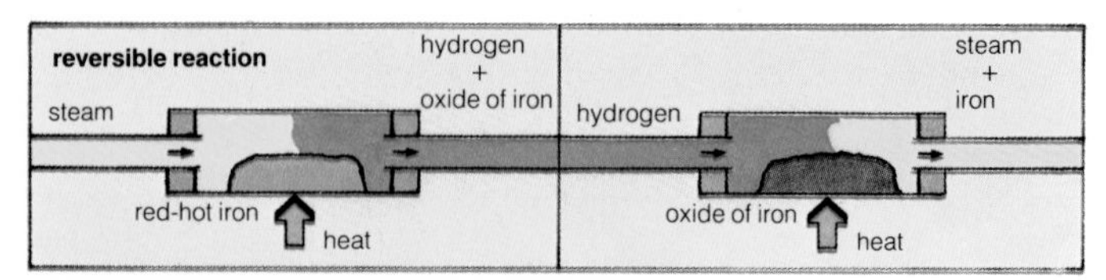

reversible reaction a reaction (p.62) in which reactants (p.62) form products (p.62) and the products can then react to form the reactants. For example, steam and iron form hydrogen and an oxide of iron. The oxide of iron and hydrogen react to form steam and iron. The reaction is written thus:

steam + iron $\rightleftharpoons$ oxide of iron + hydrogen

irreversible reaction a reaction (p.62) in which the products will not react and the reaction stops when the products are formed.

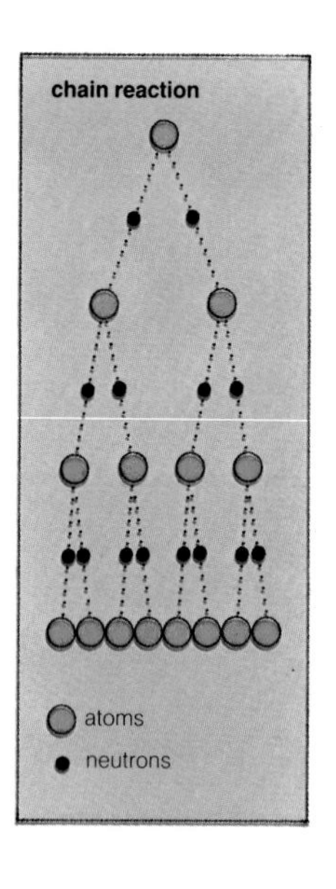

chain reaction (*n*) a reaction (p.62) in which the first reaction between molecules (p.77) or atoms (p.110) forms products which then react with further molecules or atoms so that each reaction becomes stronger until the reaction is explosive.

cycle (*n*) a set of events which is repeated time after time. Each complete set is one cycle, and the cycle is repeated. For example, a carbon atom starts in a molecule of carbon dioxide in the air; it is used in photosynthesis (p.61) in a plant and becomes part of a carbohydrate molecule; the carbohydrate is eaten by an animal and digested; the digested product takes part in respiration (p.61) and the carbon atom becomes part of a carbon dioxide molecule, and is breathed out; the carbon atom is once again in a carbon dioxide molecule in the air, the cycle has been completed, and starts all over again. **cyclic** (*adj*).

combination (*n*) the joining of elements (p.8) by chemical bonds (p.133) to form compounds (p.8). The joining of compounds by addition (p.188) to form new substances, e.g. the combination of iron and oxygen forms iron (III) oxide; the combination of zinc and chlorine forms zinc chloride; the addition of bromine to ethene (p.174) is a combination of the two compounds.

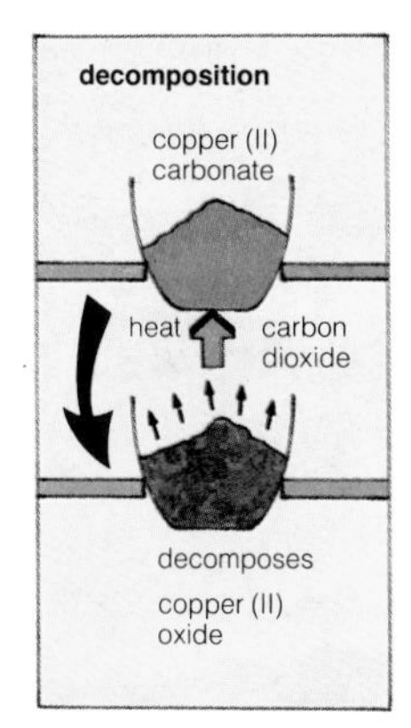

decomposition (*n*) the changing of pure substances into simpler compounds. The process (p.157) is irreversible. Decomposition can be caused by temperature, light, electric current or micro-organisms, e.g. the decomposition of copper (II) carbonate when heated into copper (II) oxide and carbon dioxide; the decomposition of silver bromide by light into silver and bromine; the decomposition of copper (II) sulphate by an electric current. **decompose** (*v*).

dissociation (*n*) the separation of an ionic (p.123) or a covalent (p.136) compound (p.8) into simpler compounds; the process (p.157) is reversible, i.e. the simpler compounds can unite again. For example, sodium chloride, in water, undergoes (p.213) dissociation into ions (the ions can unite to form sodium chloride again); the dissociation of ammonium chloride into ammonia and hydrogen chloride (ammonia and hydrogen chloride can combine to form ammonium chloride). **dissociate** (*v*).

thermal (*adj*) caused by heat.

photochemical (*adj*) describes a chemical reaction caused by light. *See electrochemical (p.128).*

thermal decomposition decomposition (↑) caused by heating a compound.

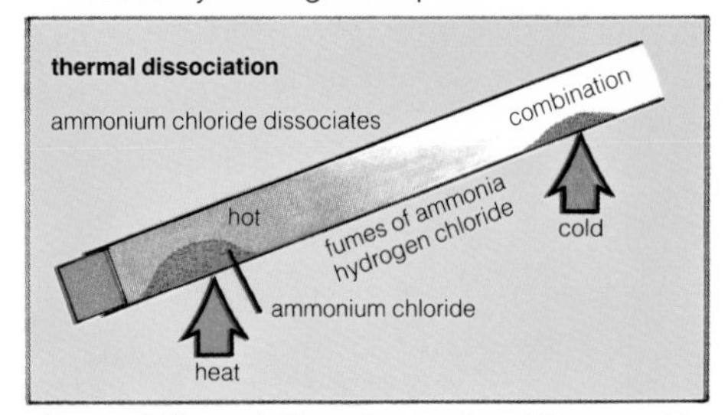

thermal dissociation dissociation (↑) caused by heating a compound, e.g. the thermal dissociation of ammonium chloride.

disintegrate (*v*) to break into small pieces because of physical or chemical action, e.g. when hit hard a mineral (p.154) disintegrates into small pieces. **disintegration** (*n*).

hydrolysis[1] (*n*) decomposition (p.65) caused by the chemical action of water. The salts of weak acids or weak bases undergo (p.213) hydrolysis, e.g. iron (III) chloride in water decomposes into iron (III) hydroxide and hydrochloric acid, as iron (III) hydroxide is a weak base.

dehydrate (*v*) (1) to take away water from a substance, to make it dry, e.g. to take water from ethanol (ethyl alcohol) to make it as dry as possible. (2) to take away the elements (p.8) of water from a compound by a chemical action, e.g. concentrated sulphuric acid removes water from ethanol to form ethene:

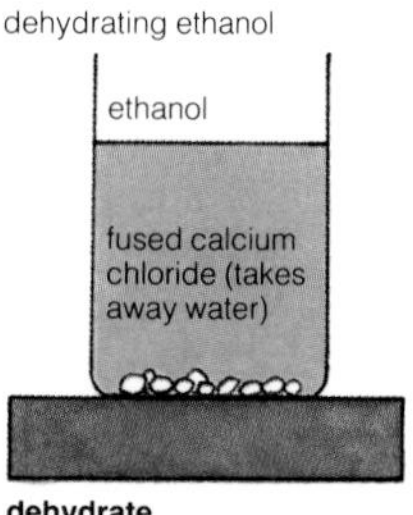

dehydrate

$$C_2H_5OH \rightarrow C_2H_4 + (H_2O \text{ taken away})$$

Dehydrate is the opposite process of *hydrate* (p.90). **dehydration** (*n*).

desiccate (*v*) to take away all traces of water, i.e. a stronger action than dehydrate (↑). **desiccation** (*n*), **desiccator** (*n*), **desiccant** (*n*).

desiccant (*n*) a substance which will desiccate (↑) compounds (p.8).

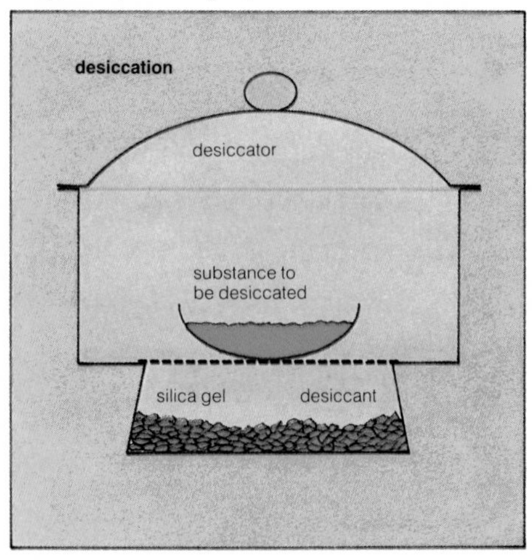

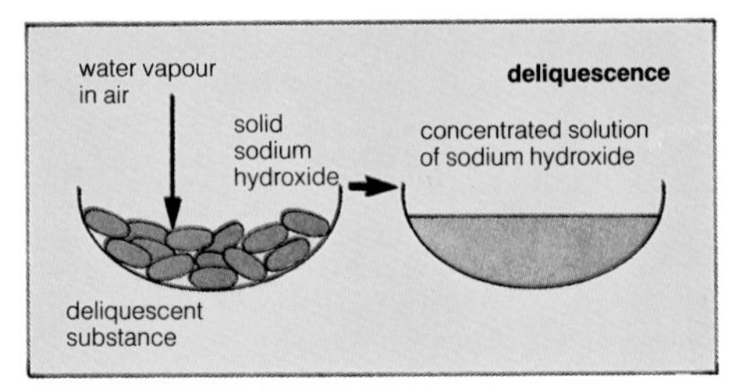

deliquesce (*v*) to absorb (p.35) water from the air by crystals (p.91) so that the crystals dissolve (p.30) and form a concentrated solution, e.g. sodium hydroxide flakes deliquesce in air and form a solution. **deliquescent** (*adj*), **deliquescence** (*n*).

effloresce (*v*) to lose water of crystallization (p.91) from a crystal to the air, e.g. sodium carbonate crystals effloresce in air and form a powder. $Na_2CO_3.10H_2O \rightarrow Na_2CO_3.H_2O$. **efflorescent** (*adj*), **efflorescence** (*n*).

hygroscopic (*adj*) describes crystalline (p.15) and amorphous (p.15) substances which take water from the air and become damp, e.g. sodium chloride is hygroscopic and becomes damp when left in air. **hygroscopicity** (*n*).

neutralization (*n*) the reaction (p.62) between an acid (p.45) and a base (p.46) or alkali, during which they destroy each other's properties and form a salt and water. The solution of the salt is neutral, i.e. neither acidic nor alkaline. For example, the neutralization of hydrochloric acid by sodium hydroxide solution. **neutralize** (*v*), **neutral** (*adj*).

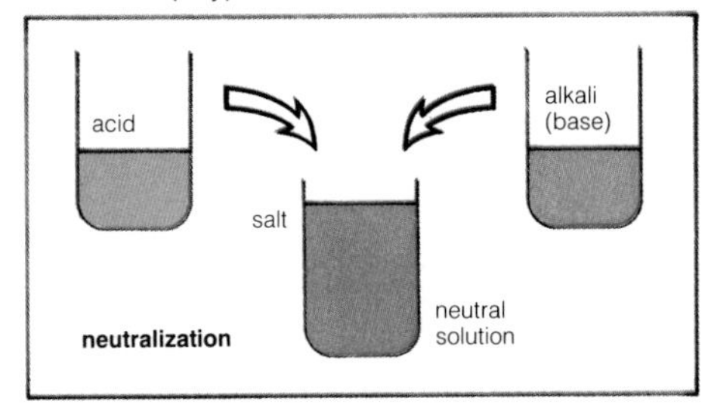

displacement (*n*) a chemical reaction (p.62) in which one element takes the place of another element which is in a compound. For example, iron put into a solution of copper (II) sulphate causes the displacement of copper. The reaction is: $Fe + CuSO_4 \rightarrow Cu + FeSO_4$ Copper metal appears in the solution as the iron disappears. **displace** (*v*).

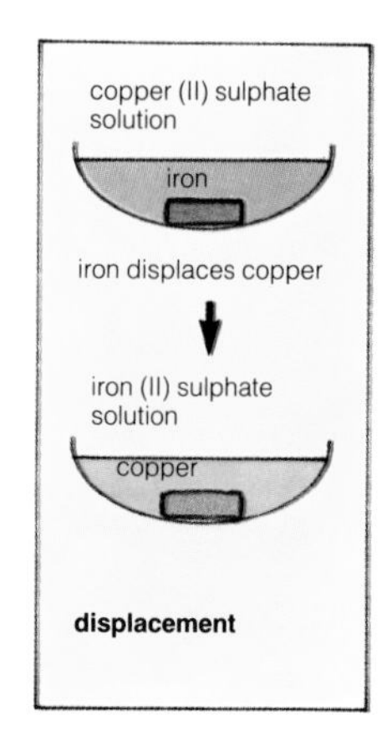

replace (*v*) (1) to put one thing in place of another because the thing replaced is no longer wanted or no longer of use. (2) to put one atom (p.110), a radical, or a functional group (p.185) in place of another atom, radical or group. Usually hydrogen in an acid is replaced by a metal, when the metal acts on (p.19) the acid to form a salt and hydrogen. **replacement** (*n*), **replaceable** (*adj*).

replaceable (*adj*) describes a hydrogen atom in an acid which can be replaced, directly or indirectly, by a metal atom.

base exchange a chemical reaction (p.62) in which two inorganic radicals (p.45) replace (↑) each other, e.g. as in the reaction:
calcium chloride + sodium carbonate → calcium carbonate + sodium chloride
In this example the chloride and carbonate radicals exchange with each other.

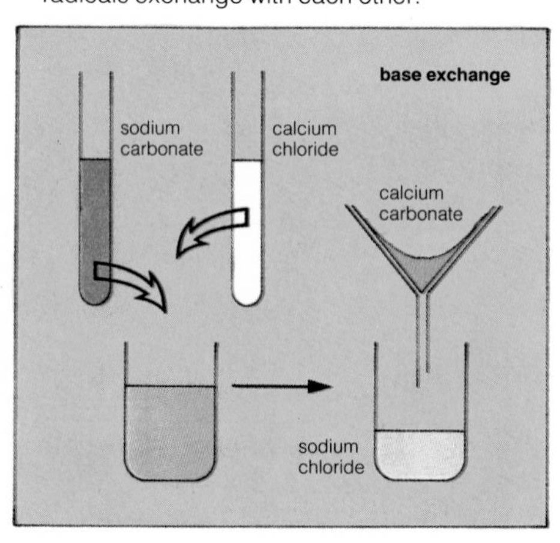

release (*v*) to set free something held physically or to set free energy. Usually a gas in a liquid is released, e.g. when a finely divided (p.13) solid is put into a solution of carbon dioxide, the carbon dioxide gas is released into the air; when a gas tap is opened, gas is released into a burner. **release** (*n*).

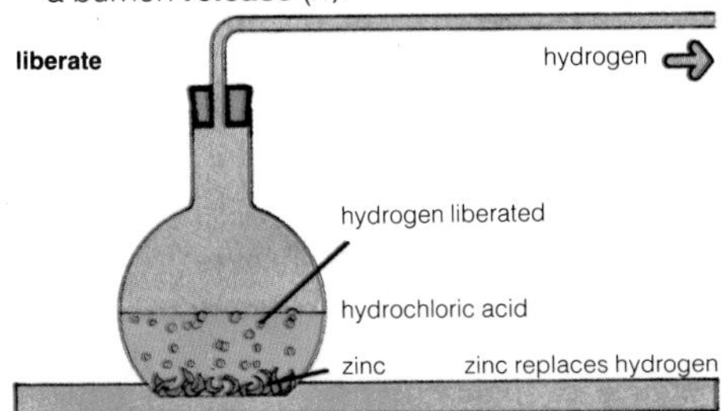

liberate (*v*) to set free a gas by breaking chemical bonds (p.133), e.g. when zinc metal is put in hydrochloric acid, hydrogen is liberated, as bonds in the acid have been broken. To compare *give off* (p.41) with *liberate*: hydrogen is *given off* when zinc is put in hydrochloric acid, this is a practical observation (p.42) as no information is given about the reaction; hydrogen is *liberated* when zinc is put in hydrochloric acid gives information about the chemical reaction. **liberation** (*n*).

affinity (*n*) a measure of the ability of one element or compound to react (p.62) with another compound. The greater the affinity, the stronger is the reaction, e.g. sodium hydroxide has an affinity for carbon dioxide; concentrated sulphuric acid has a great affinity for water, the action is violent.

moderate (*adj*) describes a reaction midway between weak and strong. Many reactions are moderate and so are not usually described as such, only weak, strong, violent (↓) and explosive (p.204) reactions are so described.

violent (*adj*) describes a reaction which is stronger than strong; the strength of the reaction can break apparatus unless care is taken. **violence** (*n*).

oxidation (*n*) (1) the addition of oxygen to an element or compound. (2) the removal (p.215) of hydrogen from a compound. (3) the removal of electrons (p.110) from an atom (p.110) or ion (p.123). (4) an increase in the oxidation number (p.78) of an element. For example: (1) calcium + oxygen forms calcium oxide; (2) hydrogen chloride is oxidized to chlorine by the removal of hydrogen; (3) a Cu atom is oxidized to Cu^{2+}, or a Fe^{2+} atom is oxidized to Fe^{3+}; (4) Mn(VI) is oxidized to Mn(VII). **oxidize** (*v*), **oxidizing** (*adj*), **oxidant** (*n*).

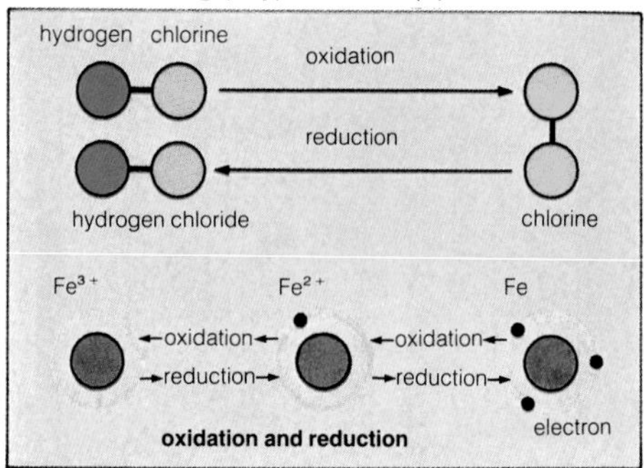

oxidation and reduction

oxidation and reduction

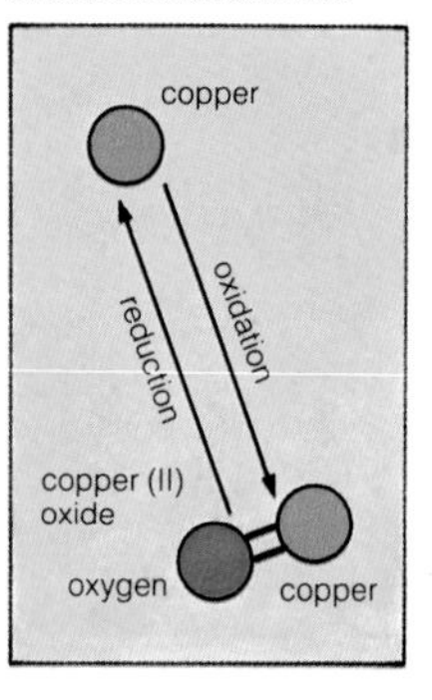

reduction[1] (*n*) (1) the removal (p.215) of oxygen from a compound. (2) the addition of hydrogen to an element or compound. (3) the addition of electrons (p.110) to an atom (p.110) or ion (p.123). (4) a decrease (p.219) in the oxidation number (p.78) of an element. For example: (1) the reduction of zinc oxide to zinc; (2) the reduction of chlorine to hydrogen chloride; (3) the reduction of Cu^{2+} to Cu or of Fe^{3+} to Fe^{2+}; (4) Mn(VII) is reduced to Mn(IV). **reduce** (*v*), **reducing** (*adj*).

redox process if substance A oxidizes (↑) substance B, then substance B reduces (↑) substance A. Oxidation and reduction always occur together, so such a reaction is called a redox process, e.g. hydrogen reduces copper (II) oxide to copper, but copper (II) oxide oxidizes hydrogen to water (hydrogen oxide), the reaction is a redox process.

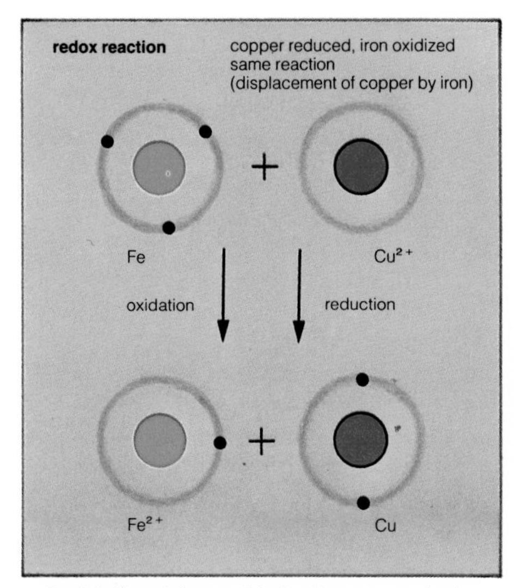

autoxidation (*n*) an oxidation by oxygen in the air at room temperature, e.g. the oxidation of iron in the air to form rust (p.61).

disproportionation (*n*) a process (p.157) in which a substance undergoes (p.213) oxidation (p.70) and reduction (p.70) at the same time, e.g. Cu^{+} undergoes disproportionation into Cu and Cu^{2+}. One atom of Cu^{+} is oxidized to Cu^{2+} and one atom is reduced to Cu.

$$Cu_2{}^{+}SO_4{}^{2-} \rightarrow Cu + Cu^{2+}SO_4{}^{2-}$$

oxidizing agent a substance which oxidizes (↑) elements or compounds, e.g. oxygen is an oxidizing agent, it oxidizes iron to iron (III) oxide; potassium manganate (VII) solution is an oxidizing agent, it oxidizes iron (II) sulphate to iron (III) sulphate.

oxidant (*n*) another name for oxidizing agent (↑).

reducing agent a substance which reduces (↑) compounds, e.g. hydrogen is a reducing agent, it reduces copper (II) oxide to copper; tin (II) chloride solution is a reducing agent, it reduces iron (III) sulphate to iron (II) sulphate.

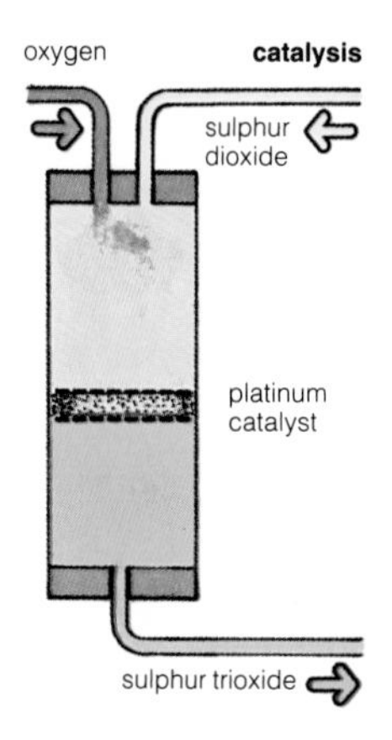

catalyst (*n*) a substance which increases the rate of chemical reaction (p.62) but itself remains chemically unchanged, e.g. platinum is a catalyst for the reaction between sulphur dioxide and oxygen, it increases the rate of reaction. Many finely divided (p.13) metals act as catalysts. **catalyze** (*v*), **catalysis** (*n*).

catalysis (*n*) a process in which a catalyst (↑) increases the rate of reaction (p.149).

negative catalyst a substance which decreases (p.219) the rate of reaction (p.149); it reacts with and destroys the catalyst.

inhibitor (*n*) a substance which slows down a chemical reaction (p.62); it may be a negative catalyst (↑) or a retarder (↓).

retarder (*n*) a substance used to decrease the rate of reaction (p.149) by physical and other means, e.g. the addition of a colloid (p.98) to slow down an ionic (p.123) reaction.

promoter (*n*) a substance used to increase the action of a catalyst (↑), e.g. finely divided (p.13) iron catalyzes the reaction between nitrogen and hydrogen to form ammonia, the addition of iron oxide increases the effect of the catalyst, that is, iron oxide is a promoter.

autocatalysis (*n*) a process (p.157) which takes place when one of the products (p.62) of a chemical reaction (p.62) acts as a catalyst (↑) for the reaction, e.g. potassium manganate (VII) in acid solution oxidizes ethanedioic (oxalic) acid, the Mn^{2+} ions, a product of the reaction, catalyze the reaction so autocatalysis takes place.

enzymatic (*adj*) describes any effect caused by enzymes. An enzyme is a catalyst (↑) produced by living things and which takes part in chemical changes in living things. **enzyme** (*n*).

poison (*n*) a substance which prevents a catalyst (↑) from acting, e.g. arsenic is a poison which prevents platinum from acting as a catalyst. Substances which are poisonous for catalysts are also poisonous for living things as they prevent enzymatic (↑) reactions in living things.

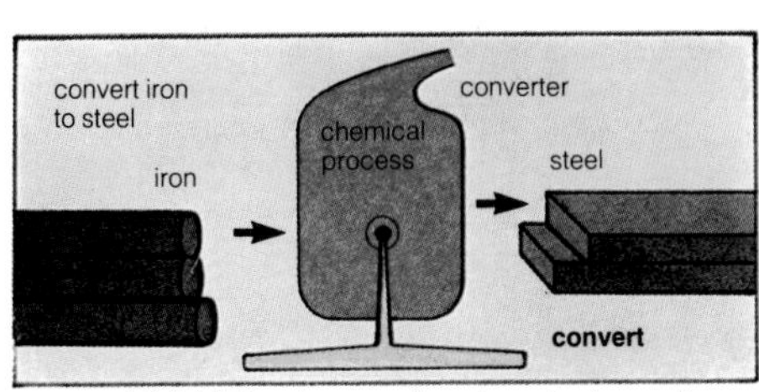

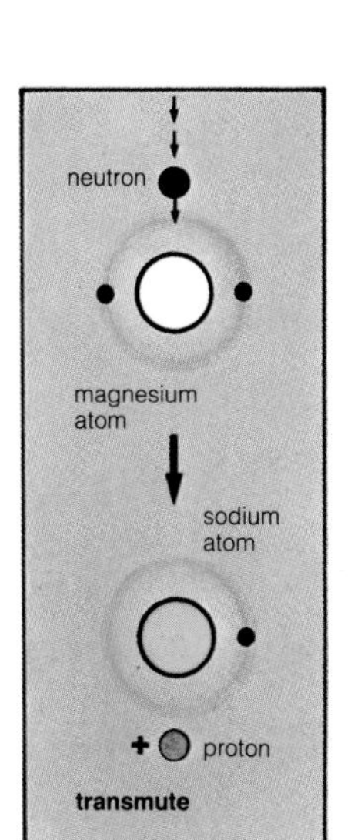

transmutation of magnesium by neutrons

convert (*v*) (1) to change, by chemical reaction (p.62), an element or a compound to another compound that is wanted for a particular purpose. (2) to change the physical nature (p.19) of a material (p.8), a substance, or apparatus. For example to convert nitrogen to ammonia; to convert nitrogen to ammonia; to convert iron to steel; to convert water to steam. *See transform (p.144).* **conversion** (*n*), **converter** (*n*).

transmute (*v*) to change one element (p.8) into another, using a radioactive (p.138) change or using bombardment (p.143) of subatomic particles (p.110). **transmutation** (*n*).

decolorize (*v*) to take away colour so that a coloured substance becomes colourless (p.15), e.g. iodine solution is brown, but when iodine is converted to iodide ions, by adding potassium iodide, the iodine solution is decolorized and becomes colourless. **decolorization** (*n*).

bleach (*v*) to take colour from a coloured substance, usually a pigment (p.162), leaving the substance white, e.g. to bleach cotton with chlorine, the natural yellow colour of cotton is changed to white; to bleach paper using sulphur dioxide. Chlorine and sulphur dioxide are the common bleaches. **bleach** (*n*), **bleaching** (*n*).

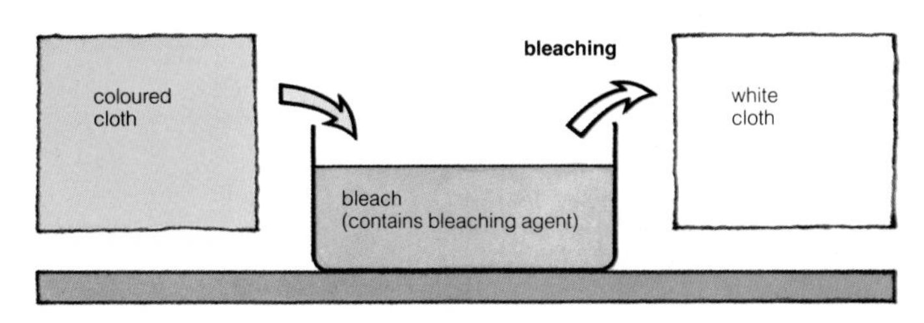

detonate (*v*) to initiate (↓) an explosion (p.58), using an electric spark, or similar agent, called a **detonator** (*n*).

initiate (*v*) to cause a process (p.157) to start when the agent of the cause takes no further part in the process, e.g. chlorine and methane do not react under room conditions, but heating the mixture decomposes (p.65) chlorine molecules (p.77) to chlorine atoms; these atoms react with methane and afterwards the reaction continues without heating, i.e. heat initiates the reaction. **initiation** (*n*).

decrepitate (*v*) to burst with small explosive (p.58) sounds, when crystals (p.91) are heated, e.g. lead (II) nitrate crystals decrepitate when heated and many small explosions can be heard. **decrepitation** (*n*).

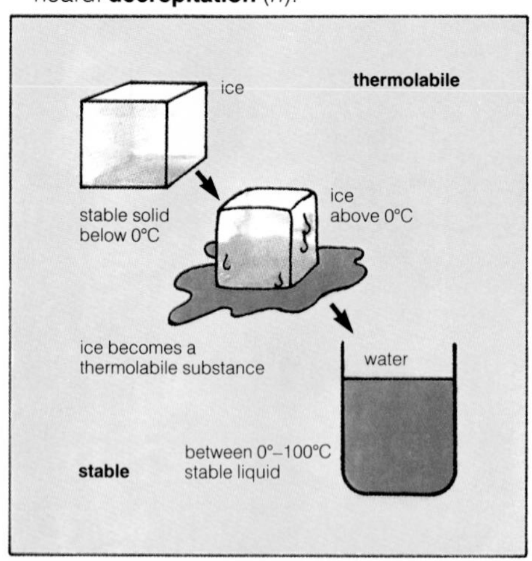

stable (*adj*) describes a substance (p.8) which is not readily changed by heat, chemical reagents (p.63), light, or other forms of energy, e.g. calcium carbonate is stable at all temperatures except very high ones. *Compare inert (p.19), inactive (p.19).* **stability** (*n*), **stabilizer** (*n*), **stabilize** (*v*).

stabile (*adj*) another name for stable (↑).

labile (*adj*) describes a substance (p.8) that readily changes, either physically or chemically, i.e. it decomposes readily. Under normal conditions (p.103), a labile substance is stable (↑), but a change in conditions can easily produce a change in the substance.

unstable (*adj*) describes a substance which tends (p.216) to decompose without apparent (p.223) causes, e.g. nitrogen triiodide is unstable when dry, an insect touching the dry powder can cause it to explode.

thermostable (*adj*) describes a substance which is stable (↑) when heated.

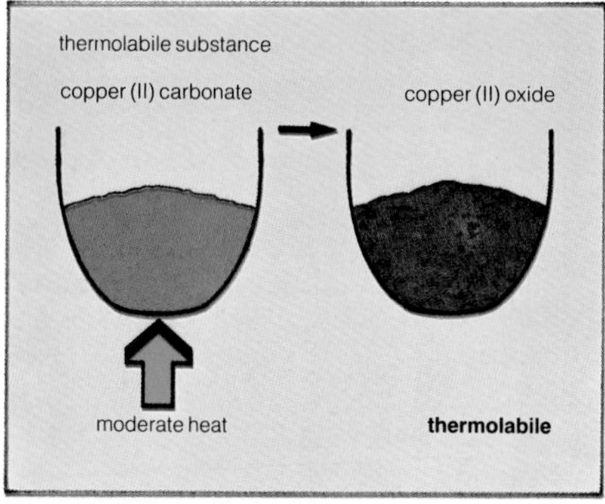

thermolabile (*adj*) describes a substance which loses its nature (p.19) or decomposes when heated.

spontaneous (*adj*) describes an event which takes place without any apparent (p.223) cause, e.g. an unstable (↑) compound can explode spontaneously. **spontaneity** (*n*).

instantaneous (*adj*) describes an event which lasts such a short time that the length of time cannot be measured, e.g. when a match is struck the appearance of the flame is instantaneous. **instant** (*n*).

nascent (*adj*) describes an element at the moment of its coming into being, e.g. when zinc acts on hydrochloric acid, the hydrogen evolved is nascent at the moment of formation.

nascent

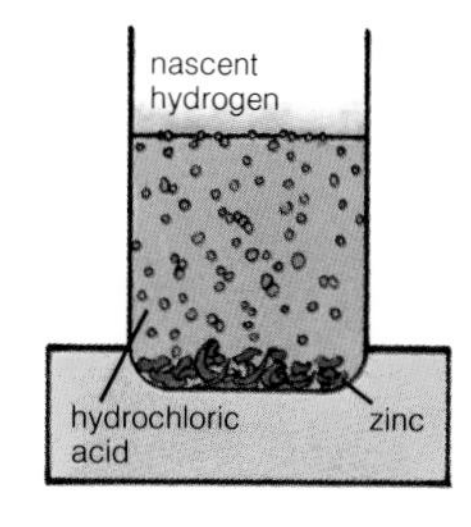

theory (*n*) a description of the causes and effects of natural events and processes (p.157). The theory must rest on experimental observation (p.42). Certain parts of the theory may have to be assumed (p.222) as observation on these parts is not possible, e.g. an atom (p.110) cannot be observed, its properties are assumed. A theory can be used to describe, explain, classify (p.120) and predict (p.85) natural events or processes. **theoretical** (*adj*).

atomic theory the theory that all solids, liquids and gases are composed (p.82) of atoms.

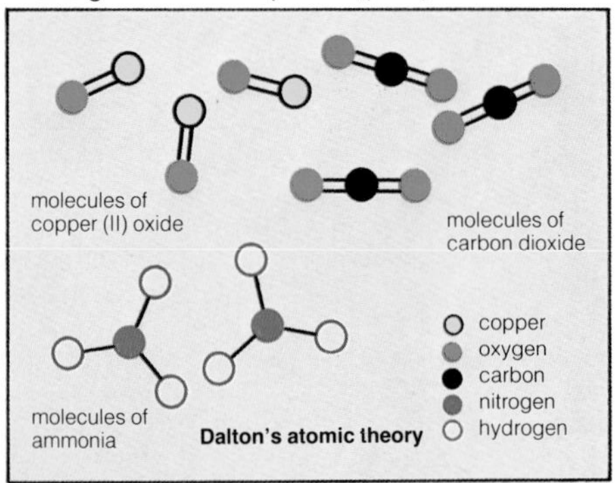

Dalton's atomic theory

Dalton's atomic theory all elements (p.8) are made up of small particles (p.110) called atoms. An atom cannot be made, destroyed or divided. Atoms of the same element are alike. Compounds (p.8) are made by atoms combining chemically to form a molecule. A molecule has a small, whole number of atoms. All molecules of a compound are alike. Chemical change takes place when small, whole numbers of atoms combine or are separated.

law of constant proportions all pure specimens (p.43) of the same chemical compound (p.8) contain the same elements (p.8) combined in the same proportions (↓) by mass.

proportion (*n*) if several ratios (p.79) are equal, then their figures are in proportion, e.g. 3/4 = 6/8, so 3 is to 4 as 6 is to 8, the figures are in proportion. **proportional** (*adj*).

combining weight the weight of an element or radical (p.45) which will combine (p.64) with, or displace (p.68), 1 g of hydrogen or 8 g of oxygen. This name is not much used in modern chemistry.

molecule (*n*) the smallest particle (p.110) of an element or compound (p.8) that can exist (p.213) by itself. A molecule generally consists of a group of atoms (p.110) combined by covalent (p.136) bonds. Ionic (p.123) compounds do not form molecules. **molecular** (*adj*).

symbol (*n*) a letter or sign used to show a chemical element, a quantity, a mathematical operation, or a piece of apparatus, e.g. Na is used for sodium, Cl for chlorine, V for volume, p for pressure. *See diagram*.

symbols in chemistry

Na	sodium	H	hydrogen
K	potassium	C	carbon
Cu	copper	O	oxygen
Fe	iron	N	nitrogen
Mn	manganese	S	sulphur
Ba	barium	Cl	chlorine
Zn	zinc	I	iodine
Mg	magnesium	P	phosphorus
Ca	calcium	Ag	silver

symbols for quantities

t	time	p	pressure
m	mass	T	temperature
l	length	c	concentration
V	volume	n	mole fraction
ρ	density	L	Avogadro constant
R	molar gas constant	F	Faraday constant

formula (*n*) (*formulae n.pl.*) (1) a chemical formula shows the number of atoms (p.110) of each element in a molecule or the ions in a compound. There are other kinds of formulae; *see structural formula (p.181)*. (2) a physical formula shows the relation (p.232) between different quantities (p.81), e.g. pV = constant; this is a formula showing the relation between the pressure and the volume of a mass of gas.

formula weight this name has been replaced by *relative formula mass* (↓).

relative formula mass the mass of one mole (p.80) of molecules (p.77), or ions (p.123) of a compound (p.8). It is calculated from the formula (↑) of a compound and the *relative atomic masses* (p.113) of the elements in the compound.

equation (*n*) the formulae (↑) of compounds or elements are arranged in an equation to show the reactants (p.62) and products (p.62) of a chemical reaction. For example:

$$2KNO_3 \xrightarrow{heat} 2KNO_2 + O_2$$

This shows (1) two molecules of potassium nitrate decompose (p.65) on heating to form two molecules of potassium nitrite and one molecule of oxygen OR (2) two moles (p.80) of potassium nitrate, when heated, decompose to form two moles of potassium nitrite and one mole of oxygen. Ionic (p.123) equations show reactions between ions. For example:

$$Ba^{2+} + SO_4^{2-} \rightarrow Ba^{2+}.SO_4^{2-}$$

oxidation number a number showing the oxidation state (p.135) of a metal in a compound (p.8) with the number written in roman numerals, e.g. lead (II) nitrate, with lead having an oxidation number of II and an oxidation state of + 2; in potassium manganate (VI), maganese in the acid radical (p.45) has an oxidation number of VI and an oxidation state of + 6.

balance[2] (*v*) (1) to keep an object still, under equal and opposite forces. (2) to have equal numbers of atoms of each element on the opposite sides of an equation, e.g. $2KNO_3 = 2KNO_2 + O_2$ is balanced because there are two potassium atoms, two nitrogen atoms and six oxygen atoms on each side of the equation. **balanced** (*adj*).

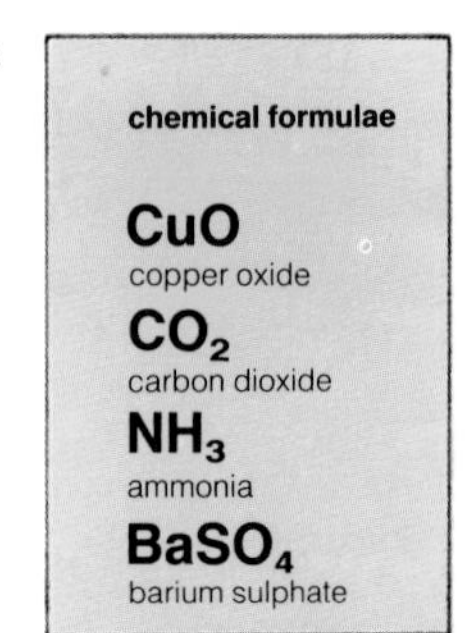

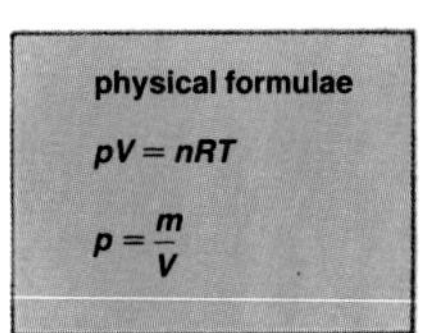

calculation of density
mass of metal = 296.25 g volume of metal = 37.5 cm^3 density of metal = $\frac{296.25\,g}{37.5\,cm^3}$ = 7.9 gcm^{-3}

calculation

calculation
26.7 × 6 160.2 + 31.3 191.5 − 46.7 144.8 a calculation with numbers

calculate (*v*) to get a result (p.39) by using arithmetical processes on numbers or the values of quantities (p.81), e.g. to calculate density (p.12) from measurements on the mass and volume of a specimen (p.43). **calculation** (*n*).

error (*n*) a mistake in a calculation (↑) or in a statement (p.222). The error can be caused by incorrect calculation, readings (p.39), inferences etc.

exact (*adj*) describes a calculation (↑) or statement (p.222) without errors (↑).

approximate (*adj*) describes a measurement which is sufficiently correct to be used in a calculation, e.g. the relative atomic mass of oxygen is 15.99, an approximate value is 16, good enough for most calculations. **approximation** (*n*).

ratio (*n*) the ratio of two numbers, or of two quantities (p.81), is obtained by dividing one by the other and simplifying the fraction, e.g. the ratio of 28 to 40 is 28/40 = 7/10, i.e. the ratio is 7:10.

multiple (*adj*) describes (1) an object made up of two or more like parts. (2) a number which consists of a smaller number multiplied a number of times, e.g. 55 and 44 are multiples of 11.

fixed (*adj*) describes a quantity (p.81) that has been made unchanging instead of allowing it to change or to be changed, e.g. a metal cylinder contains a fixed volume of gas; the volume has been made unchanging because the volume of the cylinder does not change. **fix** (*v*).

arbitrary (*adj*) describes something that has been decided because it is the most suitable or the easiest thing to do and has no relation to any theory (p.76), e.g. the choice of the metre as the standard of length was an arbitrary choice.

mean (*adj*) describes a value of a quantity (p.81) or a number which is equally far from a high and a low value. If there are two numbers x and y, then $(x + y)/2$ is the arithmetic mean, and $\sqrt{x\ y}$ is the geometric mean, e.g. $6\frac{1}{2}$ is the arithmetic mean of 9 and 4; 6 is the geometric mean of 9 and 4. **mean** (*n*).

average (*adj*) describes a value (or a number) obtained by adding together all the values and dividing by the number of values, e.g. the average of 14, 16 and 18 is $(14 + 16 + 18) \div 3 = 16$.

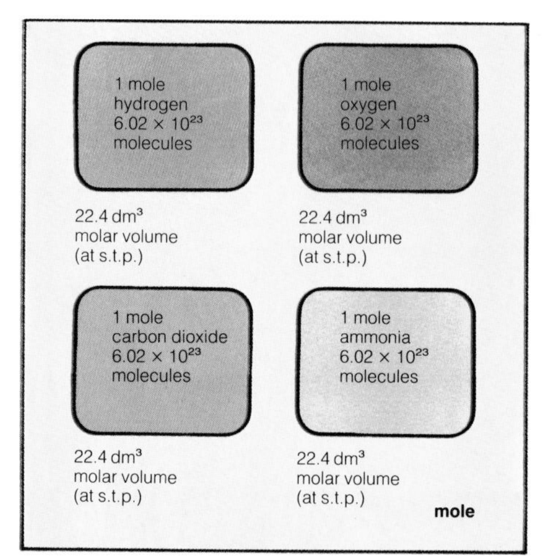

mole (*n*) the standard (p.229) for measurement of an amount (↓) of substance. One mole of a substance is that amount of substance which contains the same number of elementary particles (p.13) as there are atoms in 0.012 kg (12 grams) of carbon-12. (The isotope (p.114) of carbon with a mass number (p.113) of 12). The particles can be ions (p.123), atoms (p.110), molecules (p.77), electrons (p.110) or any other named particle. If the particles are not named, they are assumed (p.222) to be atoms in elements, molecules in covalent (p.136) compounds or ions in an electrovalent (p.134) compound. **molar** (*adj*).

mole fraction the amount of a substance expressed as a fraction of a mole (↑), e.g. 0.2 mole is a mole fraction.

molar volume the volume of one mole (↑) of a substance in a named state of matter (p.9). The molar volume of any gas at s.t.p. (p.102) is always 22.4 dm^3. The symbol V_m is used for molar volume.

Avogadro constant the number of particles in 1 mole (↑). The approximate value is 6.02×10^{23}. The symbol for the Avogadro constant is L.

amount (*n*) the amount of a substance is proportional (p.76) to the number of elementary particles (p.13) it contains. The amount of substance is a physical quantity (↓).

quantity (*n*) (1) any measurement of materials, substances (p.8) or energy (p.135) is a quantity. Examples of quantities are mass, length, time, temperature, amount of matter, atomic number, mass number, wavelength, concentration, heat, density. *Compare quality (p.15).* (2) a measurement of materials or substances that does not give a value, e.g. a quantity of lime; this could be measured in kilograms or moles (↑). **quantitative** (*adj*).

concentration (*n*) the amount (↑) of solute (p.86) dissolved in a solution (p.86). Concentration can be stated as: (a) grams of solute in 1 dm^3 solution; (b) moles of solute in 1 dm^3 solution; (c) as a percentage. For example, 80 g of sodium hydroxide dissolved in 1 dm^3 solution has a concentration of 80 g dm^{-3}, 2 mol dm^{-3} or 8% **concentrate** (*v*), **concentrated** (*adj*).

dilution (*n*) (1) the process (p.157) of adding more solvent (p.86) to a solution (p.86); this lowers the concentration (↑) of the solute (p.86), e.g. the dilution of a concentrated solution of sodium hydroxide to make a dilute solution. (2) the volume in dm^3 of a solution containing one mole of solute, e.g. a solution of sodium hydroxide containing 80 g dm^{-3} has a dilution of 0.5. *See diluent (p.56).* **dilute** (*v*), **dilute** (*adj*).

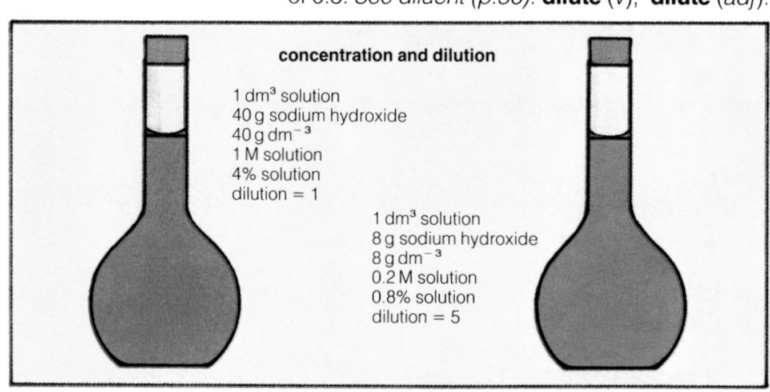

analysis (*n*) a process (p.157) to find the constitution (↓) of a compound (p.8), the composition (↓) of a mixture, the concentration (p.81) of a solution, or the identity (p.225) of a substance, e.g. the analysis of the constitution of ethanoic acid; the analysis of a mixture in order to name the substances present. *See assay (p.155).* **analyse** (*v*), **analytical** (*adj*).

volumetric analysis a way of analysis which uses solutions (p.86) of known concentrations (p.81) in titration (p.39). The reactions of the solutions may be acid-alkali, redox (p.70), or other kinds.

gravimetric analysis a way of analysis which uses reactions (p.62) forming precipitates (p.30). The precipitates are dried and weighed. Other reactions, such as the reduction (p.70) of oxides to metals may be used. In all cases, a solid is prepared (p.43) and weighed.

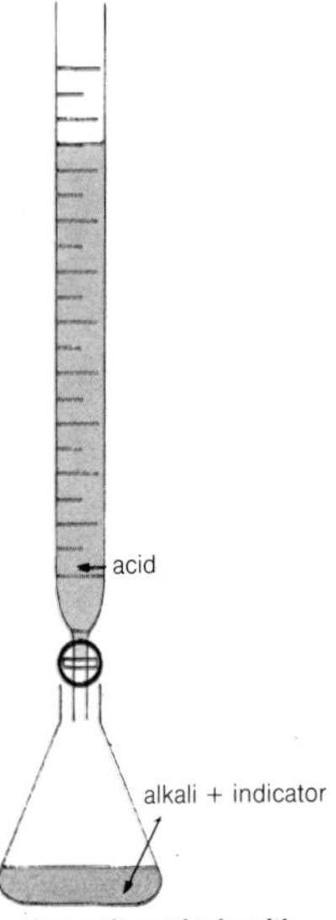

volumetric analysis with acid/alkali titration

composition (*n*) the chemical composition of a substance gives the proportion of the elements united in the substance. The proportions can be given as a percentage composition of elements by mass, or as the proportion of atoms of each element in a covalent (p.136) molecule (p.77). The composition of a mixture gives the proportion of substances in it. **composed of** (*v*).

stoichiometric (*adj*) describes compounds which obey the law of constant proportions (p.76), or processes of analysis (↑) which measure mass or volume and assume (p.222) the law. **stoichiometry** (*n*).

structure[1] (*n*) the arrangement of connected parts of a whole with the parts depending on each other to form the structure, e.g. the structure of an atom (p.110) shows the arrangement of electrons, protons and neutrons; the structure of a crystal shows the arrangement of ions (p.123) and their dependence on each other. **structural** (*adj*).

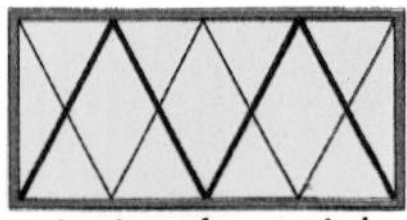
a structure of connected and dependent parts making a whole

constitution (*n*) the constitution of a compound (p.8) gives the number and position (p.211) in relation (p.232) to each of the other atoms forming a molecule (p.77), a radical (p.45) or an ion (p.123). The relative positions of the atoms can be shown by structural or graphic formulae (p.181). **constitute** (*v*), **constituent** (*n*).

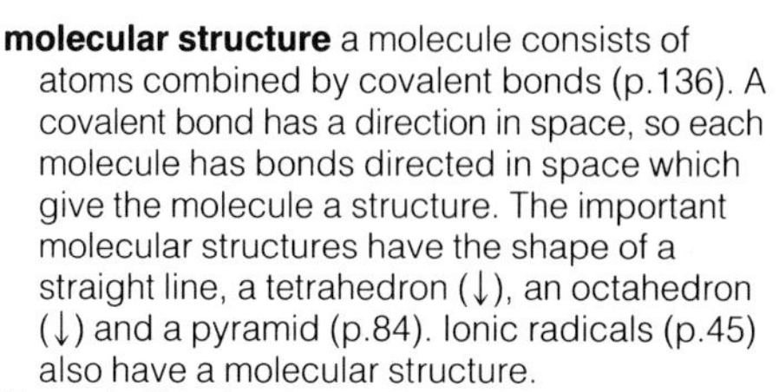

molecular structure a molecule consists of atoms combined by covalent bonds (p.136). A covalent bond has a direction in space, so each molecule has bonds directed in space which give the molecule a structure. The important molecular structures have the shape of a straight line, a tetrahedron (↓), an octahedron (↓) and a pyramid (p.84). Ionic radicals (p.45) also have a molecular structure.

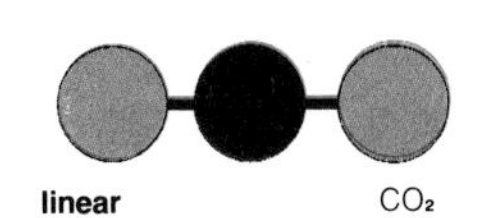

linear CO_2

linear (*adj*) in the shape of a straight line. A linear molecule has three atoms in a straight line, e.g. carbon dioxide, nitrogen monoxide. The central atom has no lone pairs (p.133) of electrons.

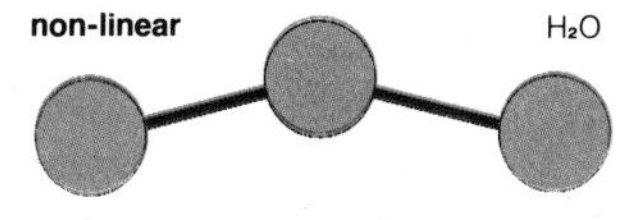

non-linear H_2O

trigonal planar CO_3^{2-}

non-linear (*adj*) in the shape of a bent line, *see diagram*. A non-linear molecule has three atoms joined by two bonds which are at an angle to each other, e.g. water has two hydrogen atoms combined with one oxygen atom and the angle between the covalent bonds is 104.5°. The central, oxygen atom has two lone pairs (p.133) of electrons.

trigonal planar (*adj*) describes a molecule with one central atom and three other atoms joined to it. All four atoms are in the same plane, e.g. in the carbonate ion (CO_3^{2-}), *see p.49*, the angle between the bonds is 120°. The central, carbon atom has no lone pairs (p.133) of electrons.

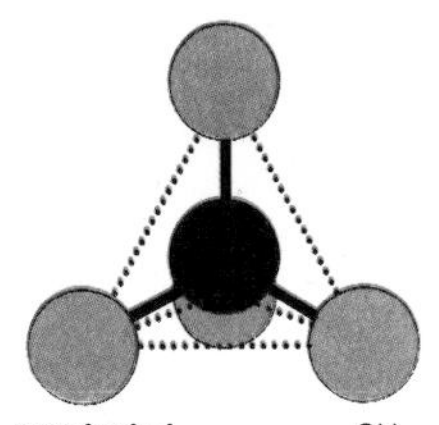

tetrahedral CH_4

tetrahedral (*adj*) in the shape of a tetrahedron, *see diagram*. A central atom is joined to four other atoms, and all the covalent bonds (p.136) have equal angles of 109.5° between each pair of bonds, e.g. methane (CH_4) has a tetrahedral structure. The central, carbon atom has no lone pairs (p.133) of electrons. **tetrahedron** (*n*).

octahedral (*adj*) in the shape of an octahedron, *see diagram*. A central atom is joined to six other atoms, e.g. the hexacyanoferrate ions (p.53) have an octahedral structure. The central atom has no lone pairs (p.133) of electrons in an octahedral structure.

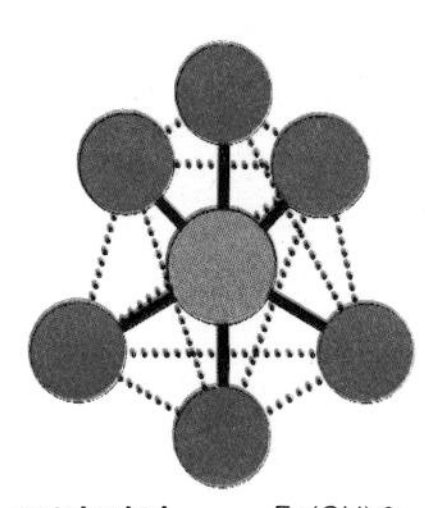

octahedral $Fe(CH)_6^{3-}$

pyramidal (*adj*) in the shape of a pyramid. A pyramid in molecular structure can have a triangular base or a square base. **pyramid** (*n*).

trigonal pyramidal in the shape of a pyramid with a triangular base. A central atom is joined to three other atoms, e.g. ammonia (NH_3) has a trigonal pyramidal structure. The central, nitrogen atom has one lone pair (p.133) of electrons, which completes a tetrahedral structure.

square pyramidal in the shape of a pyramid with a square base. A central atom is joined to five other atoms. This structure is not common. The central atom has one lone pair (p.133) of electrons.

square planar in the shape of a square. A central atom is joined to four other atoms and all five atoms are in the same plane, e.g. the tetrachloroplatinate (VI) ion ($PtCl_4^{2-}$). The central, platinum atom has two lone pairs (p.133) of electrons, which complete an octahedral (p.83) structure.

trigonal bipyramidal in the shape of two triangular pyramids joined at their bases. A central atom is joined to five other atoms, e.g. phosphorous pentachloride (PCl_5) has a trigonal bipyramidal structure. The central, phosphorous atom has no lone pairs (p.133) of electrons.

SHAPE	NUMBER OF BONDS	NUMBER OF LONE PAIRS	EXAMPLE
linear	2	0	CO^2
trigonal planar	3	0	NO_3
tetrahedral	4	0	CH_4
trigonal pyramidal	3	1	NH_3
non-linear	2	2	H_2O
trigonal bipyramidal	5	0	PCl_5
octahedral	6	0	$Fe(CN)_6^{3-}$
square pyramidal	5	1	—
square planar	4	2	$PtCl_4^{2-}$

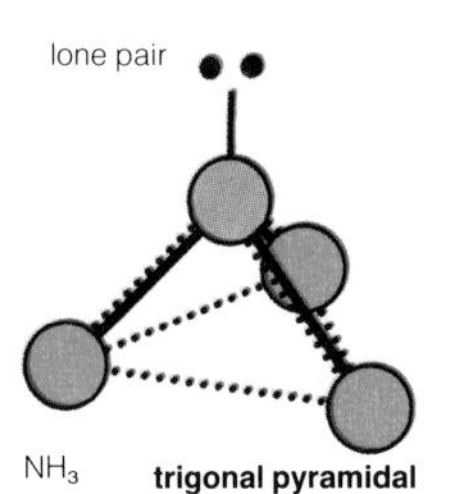

trigonal pyramidal

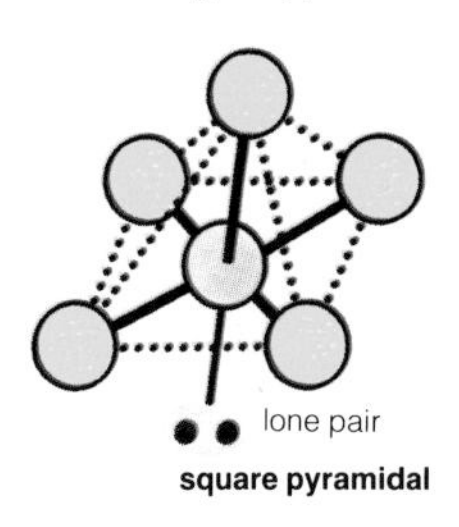

square pyramidal

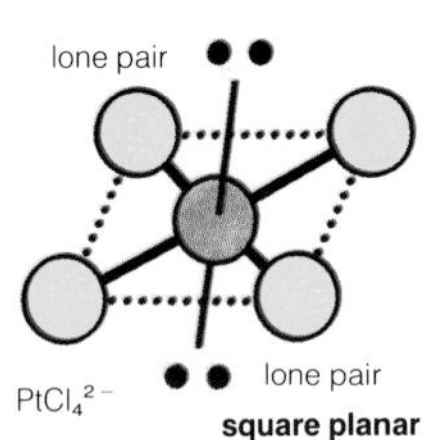

square planar

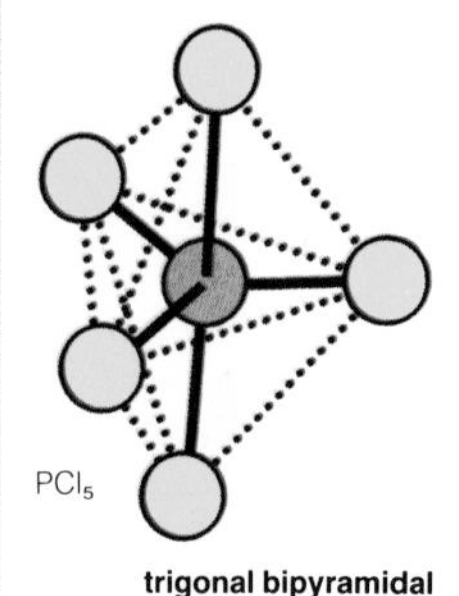

trigonal bipyramidal

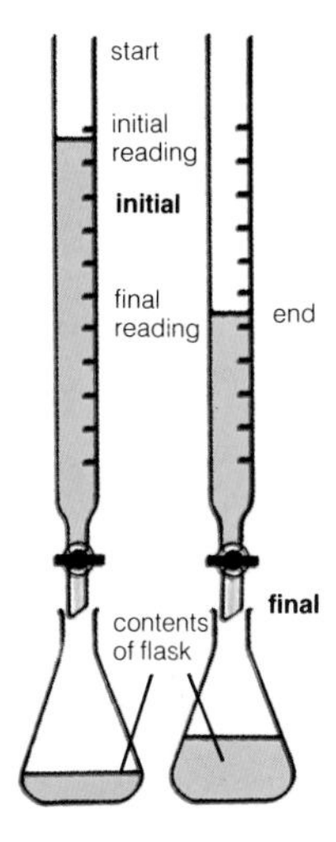

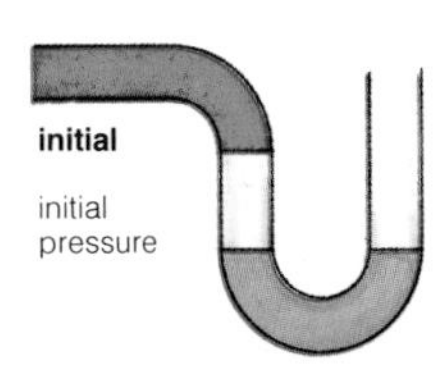

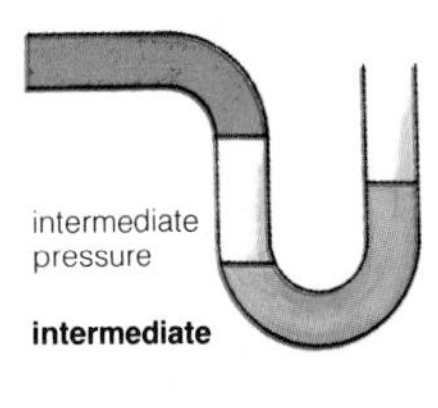

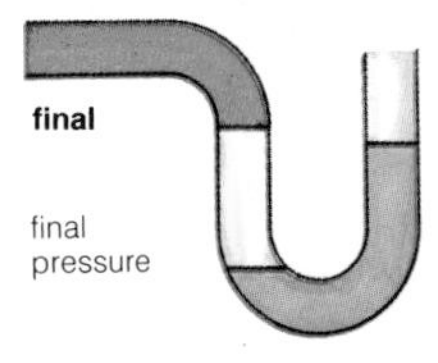

content (*n*) (1) the amount of a substance in a mixture, particularly the amount of a metal in a mineral, e.g. the silver content of an ore (p.154). (2) the amount a vessel contains. Usually given as the contents of the vessel.

available (*adj*) describes anything that can be obtained if it is needed, e.g. a gas is available from a generator (p.27); carbon dioxide in the air is available to plants; hydrogen in an acid is available from a chemical reaction. Anything available does not necessarily exist (p.213) uncombined.

initial (*adj*) the first part of a process, the first event in time, e.g. the first reading on a burette is the initial reading, taken before the contents (↑) are run into a flask; the mass of a substance before it undergoes (p.213) a chemical change is its initial mass; the first temperature recorded in an experiment is the initial temperature.

intermediate (*adj*) describes any part, or event, between the initial (↑) and final (↓) part or event.

final (*adj*) the last part of a process, the last event in time, e.g. the final reading on a burette after titration is finished; the final mass left after a substance has undergone (p.213) chemical change; the final temperature recorded at the end of an experiment.

qualitative (*adj*) describes an observation (p.42) in which no actual measurement is made, but quantities (p.81) are compared (p.224), e.g. warm, hot, red-hot, are qualitative observations on temperature; a rapid evolution (p.40) of hydrogen is a qualitative observation (by comparison with a slow evolution). **quality** (*n*).

quantitative (*adj*) describes an observation with a measurement, e.g. a quantitative measurement of the heat given out in a chemical reaction, measured in joules. **quantity** (*n*).

predict (*v*) to say what is going to happen in the future and to be sure it is correct, e.g. to predict the volume of a gas after a change in pressure, using Boyle's law (p.105). Predictions are made by using chemical laws. **prediction** (*n*), **predictable** (*adj*).

solute (*n*) a solid or a gas which dissolves (p.30) in a liquid.

solvent (*n*) the liquid in which a solute (↑) is dissolved (p.30).

solution (*n*) the result of dissolving a solute (↑) in a solvent (↑). If it is a homogenous (p.54) mixture of two or more substances it is called a *true solution*. Liquids other than water can be used as solvents (↑) to make a solution, but water is the usual solvent. For example, iodine can be dissolved in trichloromethane (chloroform). If a solvent other than water is used, it must be stated (p.222). A solution can also be made by dissolving one liquid in another; the liquid in greater quantity is the solvent, e.g. a solution of ethanol (ethyl alcohol) in water.

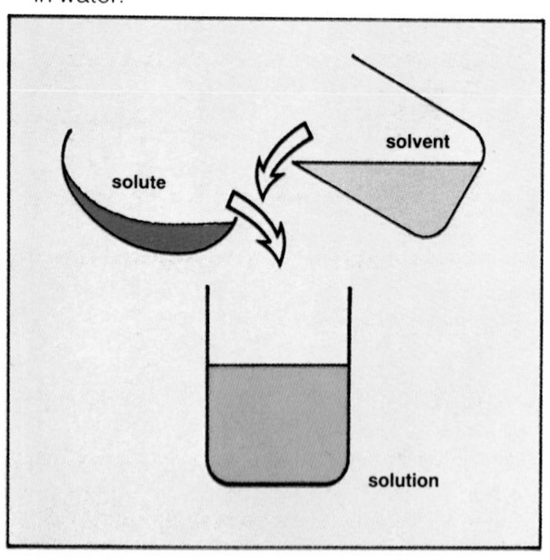

suspension (*n*) finely divided (p.13) particles of an insoluble (p.17) substance suspended (p.31) in a liquid forming an homogenous (p.54) mixture. For example, clay shaken up with water forms a suspension; lime added to water forms a solution (↑) and the excess solid forms a suspension. When a suspension is filtered, the solid substance is collected as a residue (p.31).

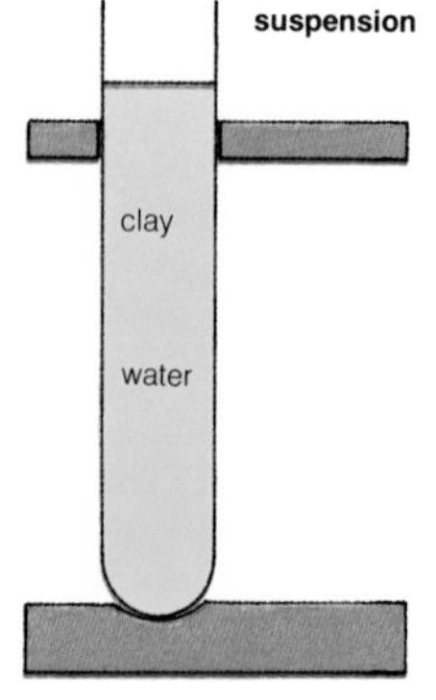

solubility (*n*) (1) the property of being soluble (p.17); or a qualitative (p.85) observation on how soluble a substance is. (2) the mass in grams of a solid which will dissolve in 100 grams of solvent (↑) at a stated (p.222) temperature in the presence of excess solute (↑). (3) the volume of gas, measured in cm^3, which will saturate (↓) 100 g of solvent at a stated temperature. Water is considered to be the usual solvent, any other solvent must be named. **soluble** (*adj*).

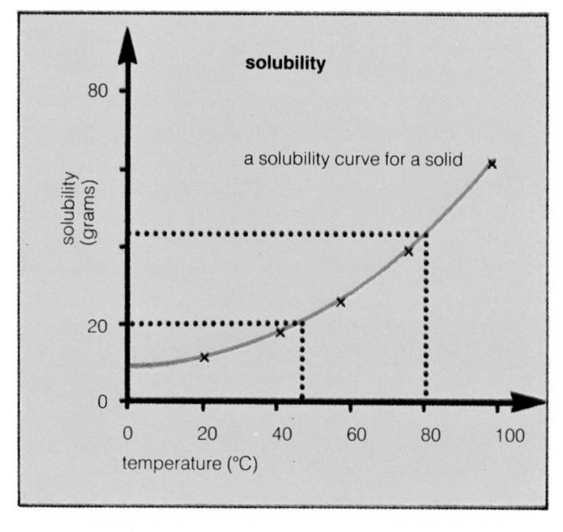

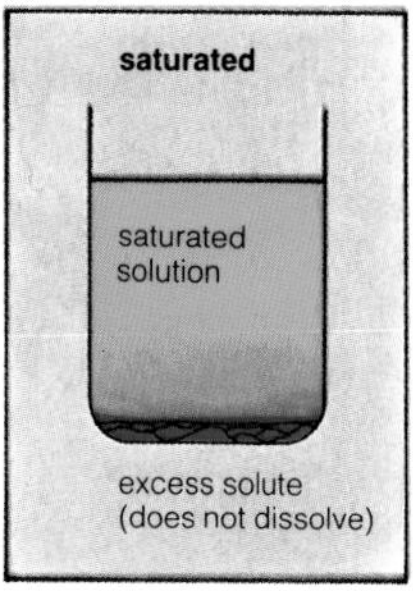

saturated

saturated solution

excess solute (does not dissolve)

saturated[1] (*adj*) (1) describes a solution (↑) which will not dissolve any more of a solute (↑). (2) describes air containing water vapour when the air will not hold any more water vapour at a particular temperature. **saturation** (*n*), **saturate** (*v*).

unsaturated[1] (*adj*) describes any solution (↑), or air, which is not saturated (↑) as more solute (↑) can be dissolved (p.30).

supersaturated (*adj*) an unstable (p.75) state of a solution (↑) which contains more solute (↑) at a given temperature than it should contain. This can happen when a warm, saturated (↑) solution is suddenly cooled. **supersaturate** (*v*), **supersaturation** (*n*).

aqueous (*adj*) describes a solution (p.86) with water as the solvent (p.86), e.g. an aqueous solution of potassium hydroxide is made by dissolving potassium hydroxide in water.

non-aqueous (*adj*) describes a solution (p.86) with a solvent (p.86) other than water; the solvent is usually named, e.g. an alcoholic (ethanolic) solution of potassium hydroxide.

concentrated (*adj*) describes a solution (p.86) containing a high proportion of solute (p.86). **concentration** (*n*), **concentration** (*v*).

dilute (*adj*) describes a solution (p.86) containing a low proportion of solute (p.86). **dilution** (*n*), **dilute** (*v*).

molar concentration a term used incorrectly, but often used to mean the concentration in moles (p.80) in 1 dm^3 of solution (p.86). It is better to use the word *concentration* and say mol dm^{-3}

molarity (*n*) the number of moles (p.80) of solute (p.86) in one cubic decimetre of solution (p.86), e.g. the mass of 1 mole of sodium hydroxide is 40 g. A solution of sodium hydroxide containing 4 g dm^{-3} contains 0.1 mol dm^{-3}, so the molarity is 0.1.

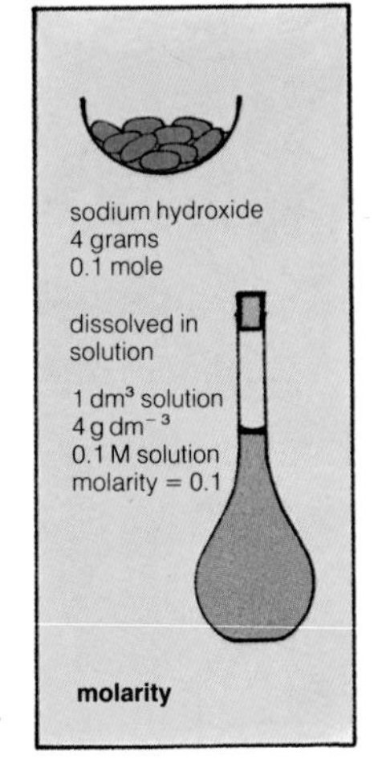

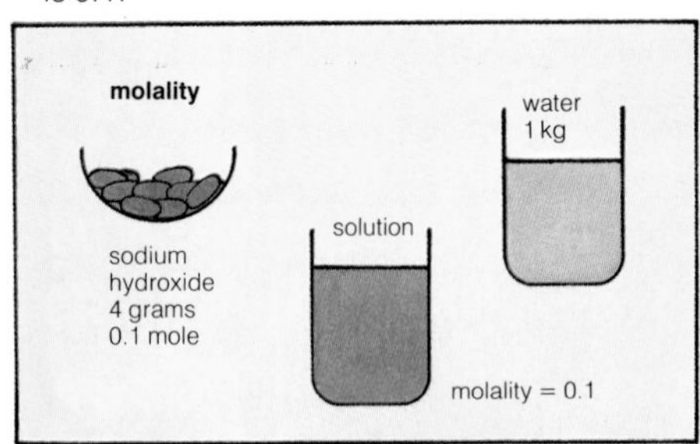

molality (*n*) a method of measuring concentration, giving the number of moles (p.80) of solute (p.86) in one kilogram of solvent (p.86).

M-value (*n*) a method of describing concentration of a solution (p.86) giving the number of moles of solute (p.86) in one cubic decimetre of solution, i.e. the value of the molarity (↑). If a solution has a molarity of 0.1 it is described as 0.1 M.

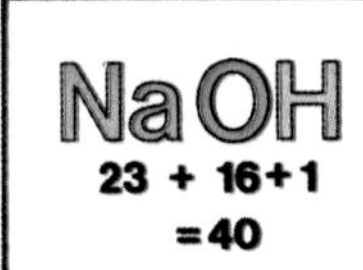

gram molecular weight

gram molecule the molecular weight of a compound stated (p.222) in grams. It is found by adding the atomic weights (p.114) of the elements in the compound. This measurement is no longer used; instead the amount of a compound is given in moles (p.80).

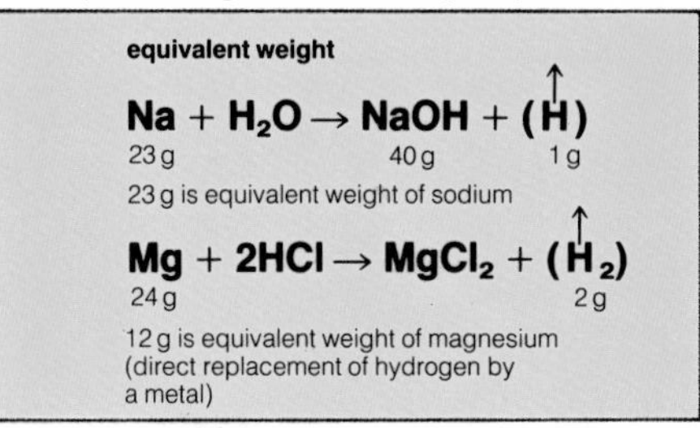

equivalent weight the mass of an element which will combine (p.64) with or displace (p.68), directly or indirectly, 1 gram of hydrogen. The equivalent weight of an acid is the mass, in grams, of an acid that contains 1 g of replaceable (p.68) hydrogen. The equivalent weight of an alkali, or a base, is that mass that neutralizes (p.67) the equivalent weight of an acid. The equivalent weight of an oxidizing agent is the mass of the agent which provides 8 g oxygen. This measurement is no longer used.

normal solution

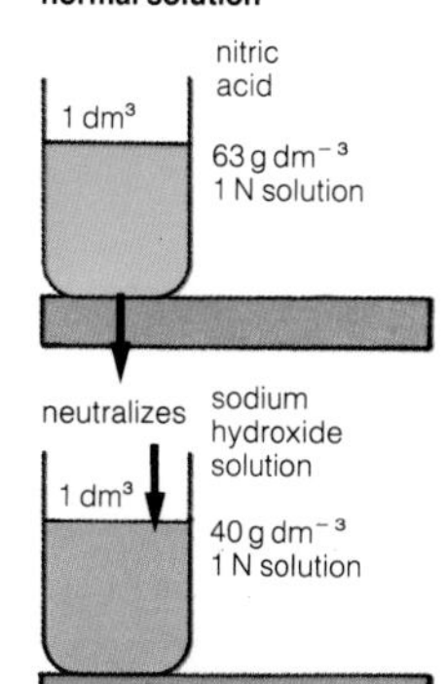

normal solution a solution (p.86) containing the equivalent weight (↑) of a substance dissolved in one cubic decimetre of solution, e.g. 98 g sulphuric acid contains 2 g replaceable hydrogen, hence the equivalent weight of sulphuric acid is 49. A normal solution of sulphuric acid contains 49 g dm^{-3}

normality (*n*) the fraction of the equivalent weight (↑) of a solute (p.86) dissolved in one cubic decimetre of solution (p.86), e.g. if 1 dm^3 of solution contains 98 g sulphuric acid, then it contains 2 equivalent weights and its normality is 2. This is usually written as 2N.

standard solution a solution (p.86) whose concentration is known accurately. The concentration can be given in grams per cubic decimetre, as a molarity (↑) or as a normality (↑).

solvation (*n*) a process in which molecules of a solvent (p.86) become attached to ions (p.123), or molecules (p.77), of solutes (p.86). **solvate** (*v*).

hydration (*n*) solvation (↑) when water is the solvent (p.86). *See aqua-ion (p.132).* **hydrate** (*v*), **dehydrate** (*v*).

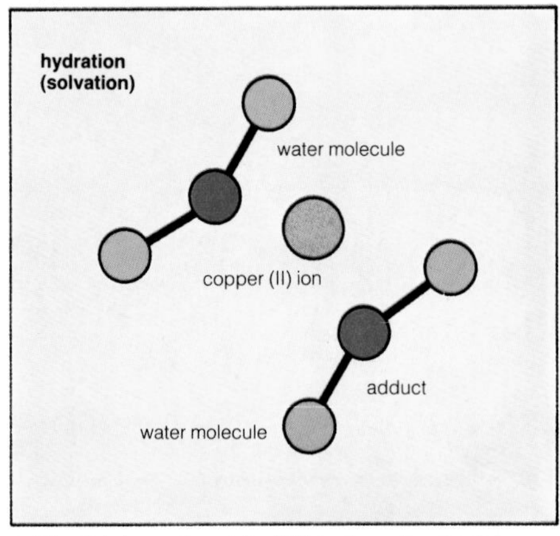

adduct (*n*) a solvent (p.86) molecule which is combined (p.64) with either an ion or a molecule in solution.

hydrate (*n*) a crystal (↓) with molecules of water combined with ions (p.123) or radicals (p.45) in its crystal structure, e.g. zinc sulphate has seven water molecules combined with its ions: $ZnSO_4.7H_2O$. **hydrated** (*adj*).

water of crystallization water molecules in hydrates (↑). A particular crystalline substance always has the same number of molecules of water of crystallization combined with the ions (p.123) of a molecule of that substance, e.g. $CuSO_4.5H_2O$; $CuCl_2.2H_2O$; $Na_2SO_4.10H_2O$; $Na_2CO_3.10H_2O$; $Na_2CO_3.H_2O$.

mother liquor the solution left after crystals (↓) have formed.

supernatant (*adj*) describes the liquid above a precipitate (p.30) or a sediment (p.31). A supernatant liquid is separated from a sediment by decantation (p.31).

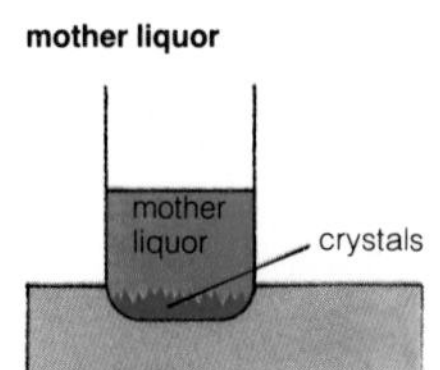

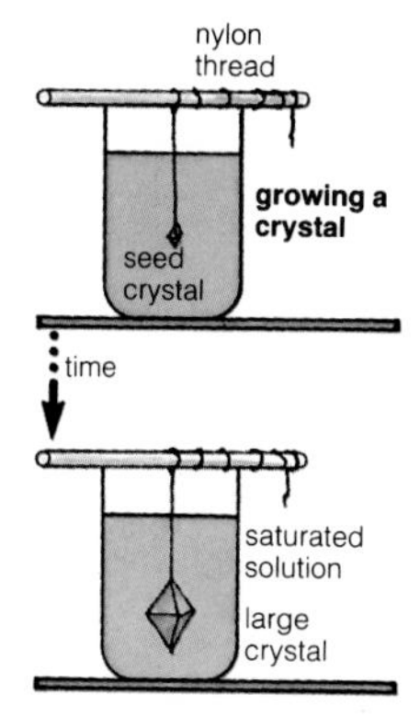

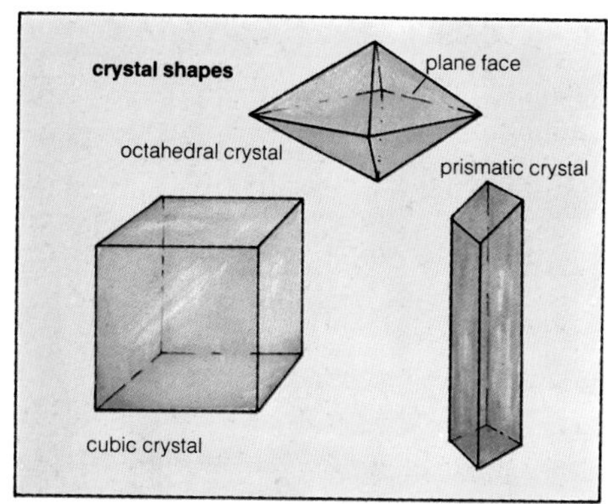

crystal (*n*) a solid substance (p.8) with a regular shape. It has plane (p.92) faces, which are always at the same angle for similar sides in all crystals of the substance. The crystal shape is a property (p.9) of a crystalline substance. **crystalline** (*adj*), **crystalloid** (*n*), **crystallize** (*v*).

crystalloid (*n*) a substance which forms crystals (↑) and forms a true solution (p.86) in water. A crystalloid passes through a permeable membrane (p.99). *See colloid (p.98).*

crystallization (*n*) the process of crystals (↑) forming in a solution of a crystalline (↑) substance; also the production of crystals in an experiment (p.42).

recrystallization (*n*) to form crystals (↑) of a substance, then to dissolve the crystals in a solvent and to crystallize (↑) again. This makes sure the crystals consist only of the pure substance.

fractional crystallization a process for separating two crystalline (↑) solids which have solubilities of nearly the same value. The substances undergo recrystallization (↑) many times. After each crystallization the crystals will be richer in one of the substances and the mother liquor (p.90) richer in the other. At the end of separation (p.34) pure specimens (p.43) of both substances can be obtained.

crystallization

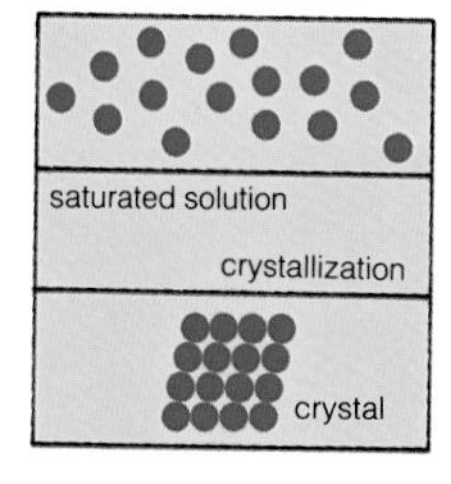

crystallization

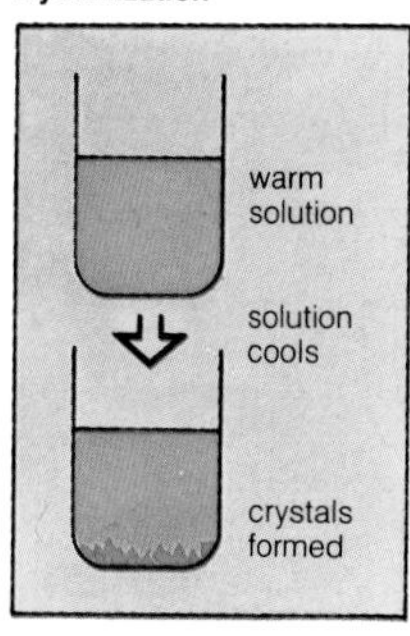

polymorphism (*n*) the state of a solid substance (p.8) existing (p.213) in two, or more, crystalline (p.15) forms. Elements and compounds can be polymorphic. *See allotropy (p.118).* Examples of polymorphism are: mercury (II) oxide which has a red and a yellow crystalline form; sodium carbonate which has several hydrates, the decahydrate ($10H_2O$) forms monoclinic crystals, the heptahydrate ($7H_2O$) forms rhombic crystals and the transition point (↓) between these two crystalline forms is 32°C. **polymorphic** (*adj*).

enantiotropy (*n*) polymorphism (↑) in which two stable (p.74) crystalline forms exist, one below a transition point (↓) and one above; the change between the two forms is reversible (p.216), e.g. sulphur has two crystalline forms, rhombic S_α and monoclinic S_β. S_α is stable below 96°C while S_β is stable above 96°C. 96°C is the transition point (↓). **enantiotropic** (*adj*).

monotropy (*n*) polymorphism (↑) in which there is only one stable (p.74) crystalline form and other forms are unstable, e.g. red phosphorus is the stable form of phosphorus and white phosphorus is unstable. There is no transition point (↓) between the two forms. **monotropic** (*adj*).

transition point (*n*) the temperature at which one crystalline (p.15) form of a substance changes to another crystalline form in a reversible (p.216) change. At the transition temperature, both crystalline forms can exist (p.213) together.

lattice (*n*) a regular arrangement of points in space with a pattern (↓) that can be recognized.

crystal lattice (*n*) a lattice (↑) with atoms, molecules, or ions at the points of the lattice. A lattice has a pattern in three dimensions and reaches to the faces of the crystal (p.91).

isomorphism (*n*) the state of crystals of two different substances having the same crystal lattice (↑), hence the same crystal shape. **isomorphic** (*adj*).

plane (*n*) a flat surface. Two planes meet in a straight line. Three planes can meet in the same straight line, or at a point. **planar** (*adj*).

axis (*n*) (*n.pl. axes*) (1) a line drawn through an object which divides it into two equal halves. (2) a line about which an object turns. **axial** (*adj*).

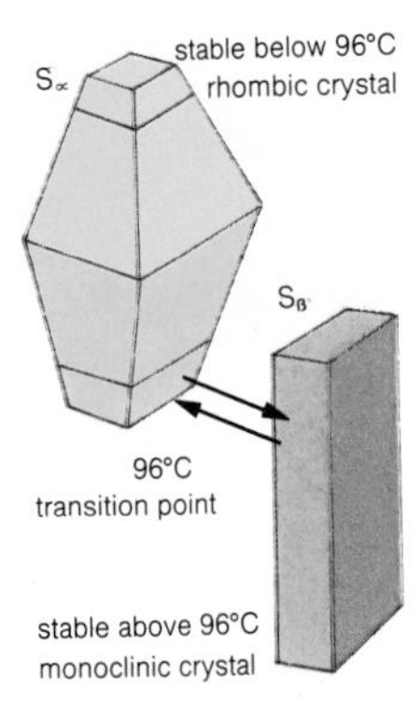

enantiotropy of sulphur

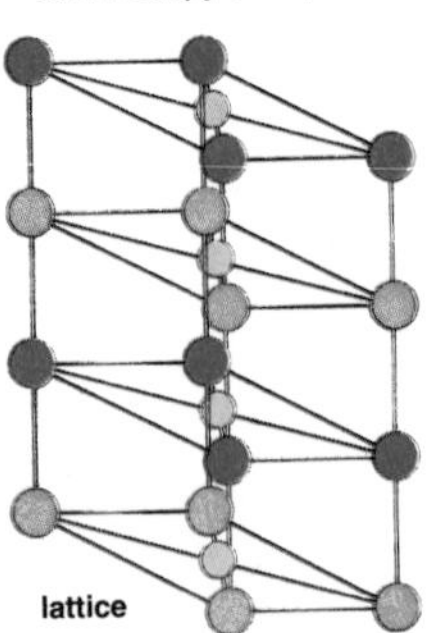
lattice

two planes meet in a line

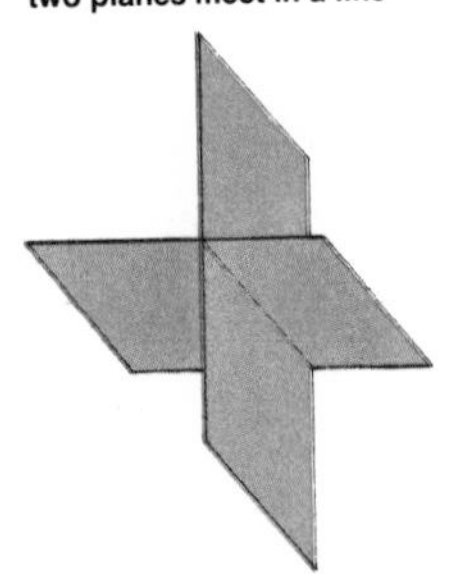

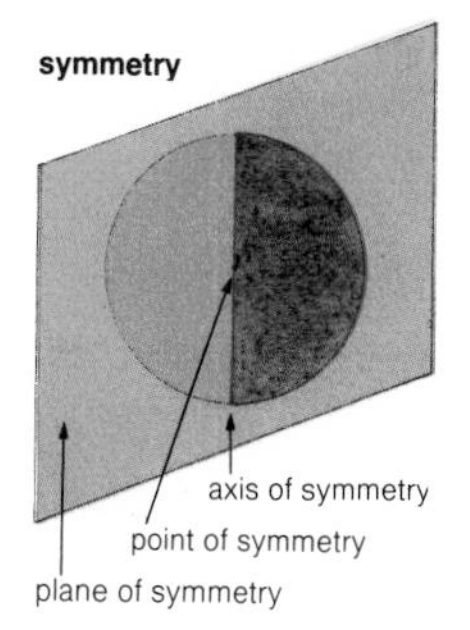

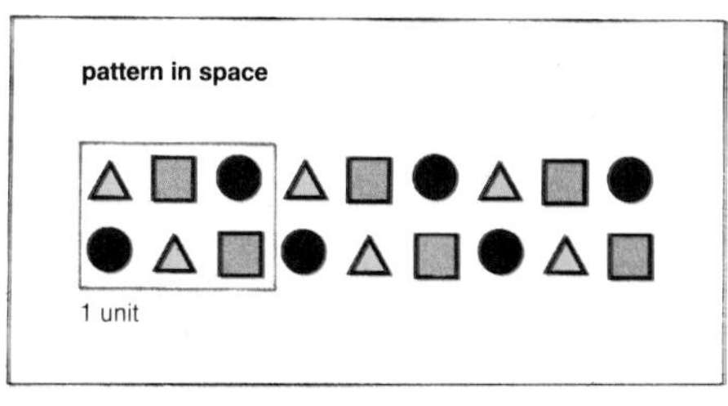

pattern (*n*) a regular arrangement of points, objects, or shapes which recur (p.217) in space; each arrangement is called a unit of the pattern. Processes or events which occur regularly in time form a dynamic pattern.

symmetry (*n*) the state of having a regular shape such that a line or plane can be drawn which divides the shape into two equal halves, e.g. a circle has symmetry as any line through the centre divides it into two equal halves. **symmetrical** (*adj*), **asymmetrical** (*adj*).

crystal symmetry symmetry (↑) of the lattice (↑) or shape of a crystal (p.91). Crystals can have planes, lines or centres of symmetry.

crystal face (*n*) a plane surface of a crystal.

plane of symmetry a plane which divides the lattice (↑) or the shape of a crystal into two equal halves.

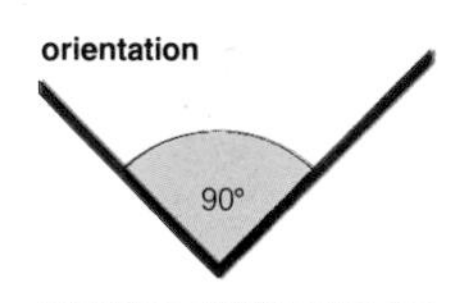

orientation of 90° to other line

orientation (*n*) the direction in which an object points in relation to its surroundings (p.103), e.g. the orientation of one line in relation to another line which it meets at a point.

perfect crystal (*n*) a crystal (p.91) which has no breaks in the lattice (↑), i.e. no atoms, ions or molecules missing from the lattice.

irregular (*adj*) describes any arrangement, in space or in time, which is not regular. The opposite of regular.

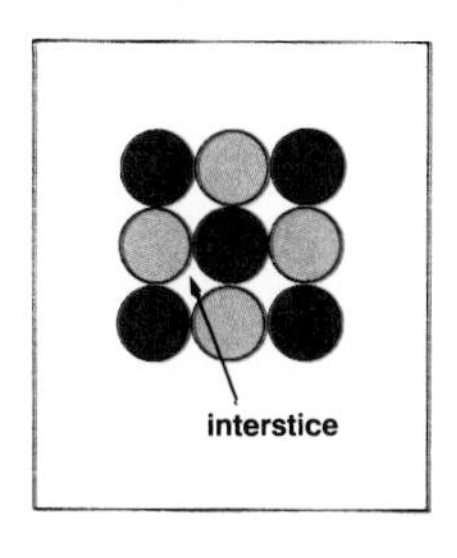

interstice (*n*) a small space between the ions, atoms or molecules in a crystal lattice (↑), e.g. in steel, there are atoms of carbon in the interstices of the iron atoms forming the lattice of the metal crystals. *See metal crystals (p.95).* **interstitial** (*adj*).

cleavage plane a plane in a crystal (p.91) along which a blow will separate the crystal into two parts with a clean cut. The crystal is *cleaved* by the blow. Along any other plane the crystal will be shattered (↓). **cleave** (*v*), **cleavage** (*n*).

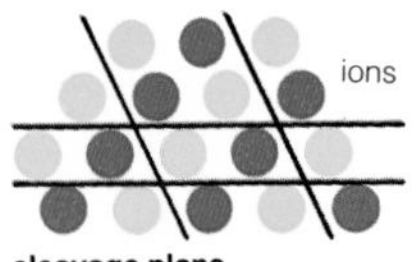

cleavage plane

shatter (*v*) to break a hard, or brittle (p.14) object into very many small pieces by a hard blow, e.g. a cup dropped on the ground is shattered.

slip plane a plane in a crystal (p.91), particularly a metal crystal (↓), along which one part of the crystal will move relative to the other part, when a force acts on the crystal. One part of the crystal lattice *slips* over the other part. Slip planes in metal crystals cause metals to be *ductile* (p.14).

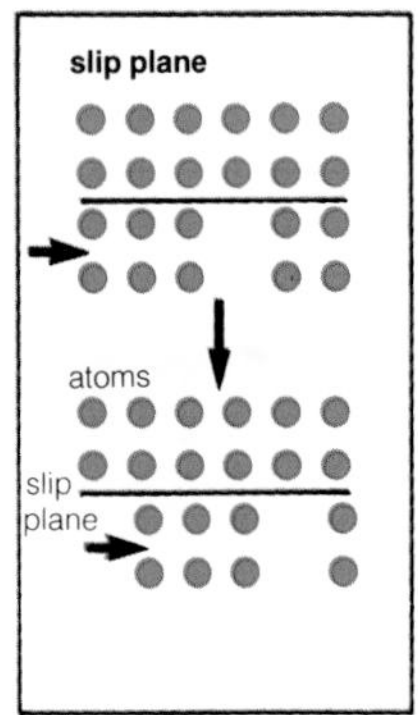

giant structure a crystal lattice (p.92) with no separate molecules (p.77). The ions (p.123) or atoms (p.110) forming the lattice are all joined, one to another, by bonds (p.133), so that the crystal appears to be one very large molecule. Many giant structures are held together by ionic bonds.

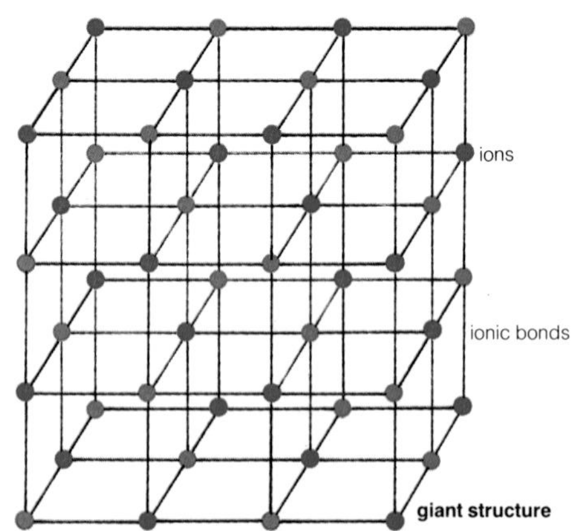

giant structure

molecular crystal a crystal consisting (p.55) of separate molecules held together by weak van der Waals' (p.137) forces. The crystal is very soft and has a low melting point (p.12). Many organic compounds form molecular crystals.

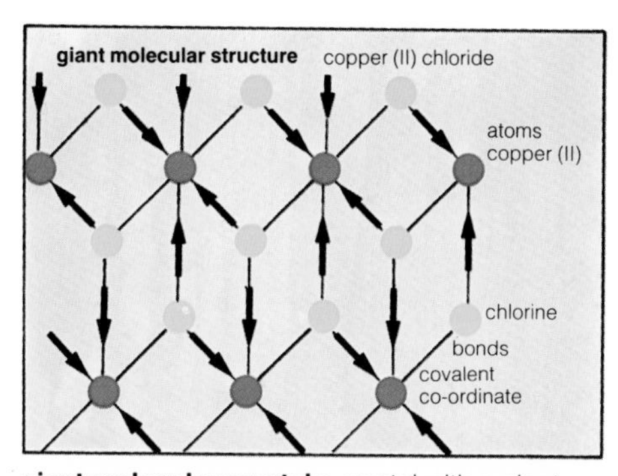

giant molecular crystal a crystal with a giant structure (↑) composed of atoms joined by covalent bonds (p.136). A giant molecular crystal is very hard and has a very high melting point (p.12). Aluminium oxide and anhydrous copper (II) chloride form giant molecular crystals.

metal crystal a giant structure (↑) formed by metals. The crystal lattice consists (p.55) of positive ions of metals. A cloud of electrons can move throughout the lattice and holds the positive ions together. The cloud of freely moving electrons accounts for the ability of metals to conduct electric current (p.122) and heat.

close packing an arrangement of atoms put as close to each other as possible. This forms a close packed lattice of one layer (p.18). Two arrangements of close packing are found in metal crystals (↑), hexagonal (↓) and cubic (↓).

hexagonal close packing a second layer of atoms lies in the hollows formed by groups of three atoms in the first layer of close packing (↑). A third layer of atoms similarly lies in the hollows formed by groups of three atoms in the second layer; each atom in the third layer is directly below an atom in the first layer.

cubic close packing the atoms in the first two layers are arranged in the same way as the atoms in hexagonal close packing (↑). The third layer of atoms lies in the hollows formed by groups of three atoms, but these atoms are not below the atoms in the first layer.

close packing

close packing of atoms

hexagonal close packing

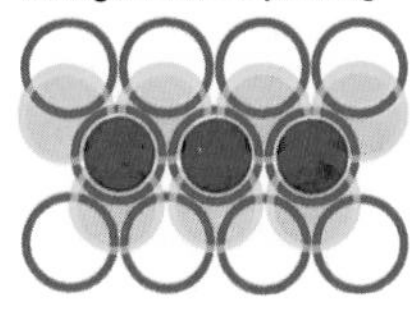

● top layer
● middle layer
○ bottom layer

cubic close packing

● top layer
● middle layer
○ bottom layer

crystal systems crystals are grouped into systems according to their symmetry (p.93).

cubic system a system with three equal axes of symmetry (a = b = c). The angle between any two axes is 90°.

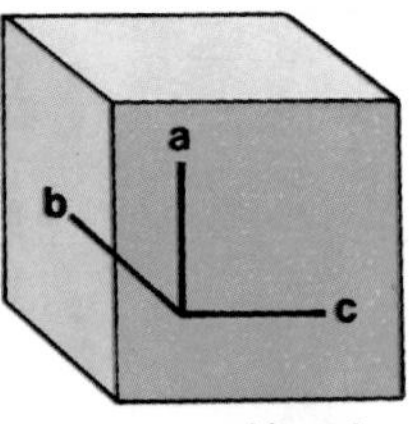

a = b = c **cubic system**

tetragonal system a system with two equal axes of symmetry and a third axis which is longer or shorter (a = b ≠ c). The angle between any two axes is 90°.

tetragonal system

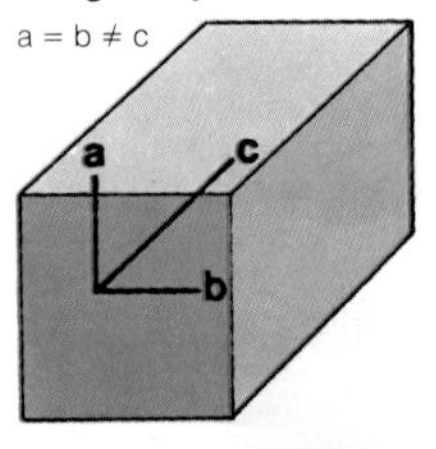

orthorhombic system a system with three unequal axes of symmetry (a ≠ b ≠ c). The angle between any two axes is 90°.

orthorhombic system

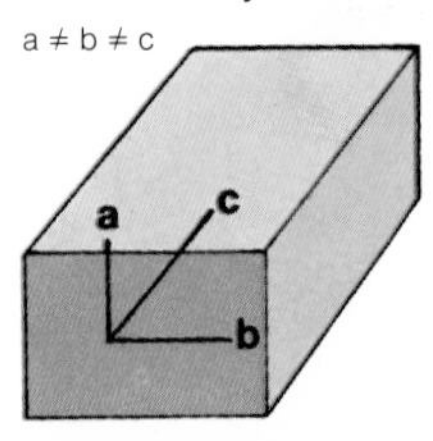

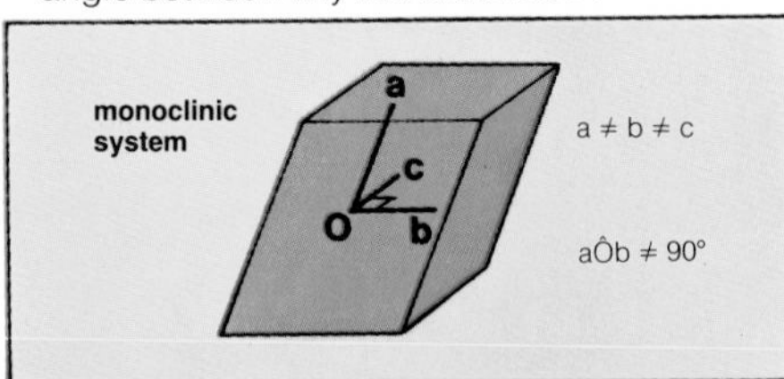

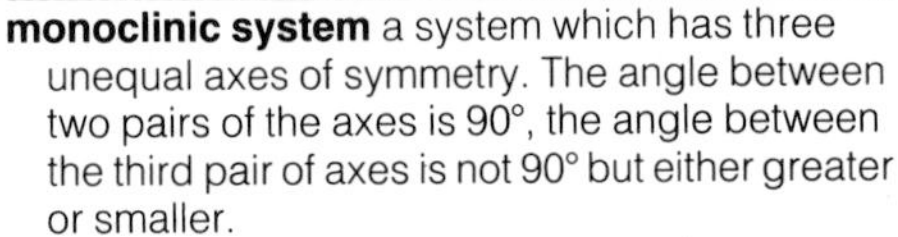

monoclinic system a system which has three unequal axes of symmetry. The angle between two pairs of the axes is 90°, the angle between the third pair of axes is not 90° but either greater or smaller.

triclinic system a system which has three unequal axes of symmetry. The angle between any two axes is not 90°.

hexagonal system a system which has three equal axes of symmetry at an angle of 120° to each other with a fourth axis of symmetry at an angle of 90°, and of a different length.

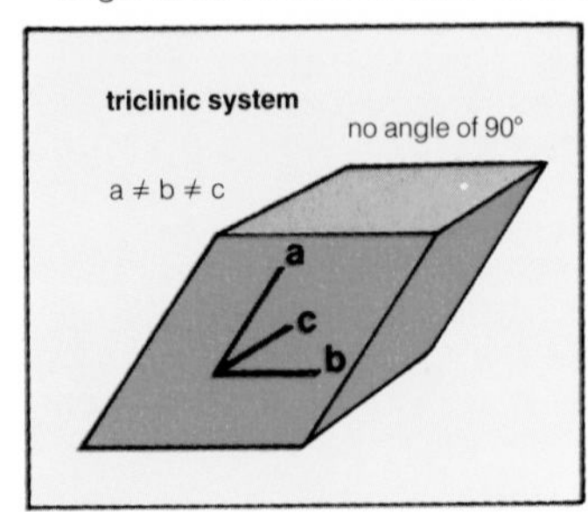

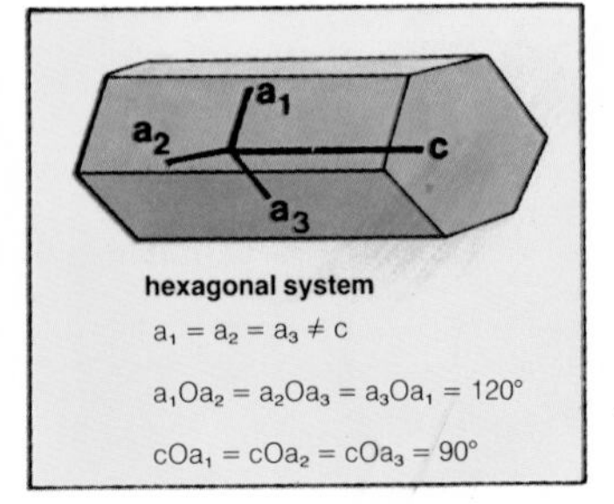

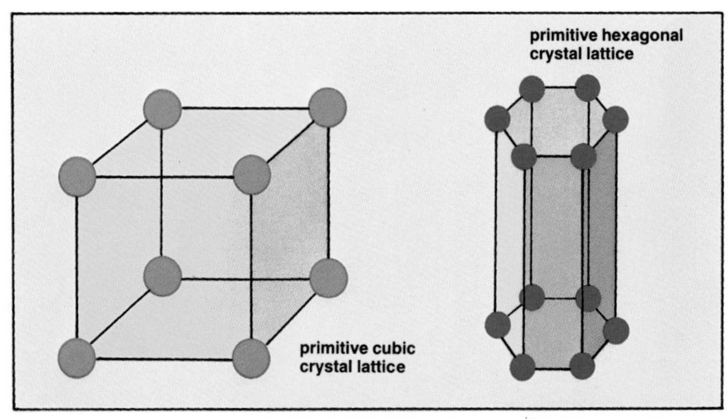

primitive structure this is the simplest crystal structure with an atom (p.110) or ion (p.123) at each corner of the figure for each of the crystal structures. *See diagram*, which shows a primitive cubic lattice and a primitive hexagonal lattice.

body-centred lattice this structure is a primitive structure (↑) with an additional atom or ion at the centre of the figure.

face-centred lattice this structure is a primitive structure (↑) with additional atoms or ions at the centre of each face of the figure.

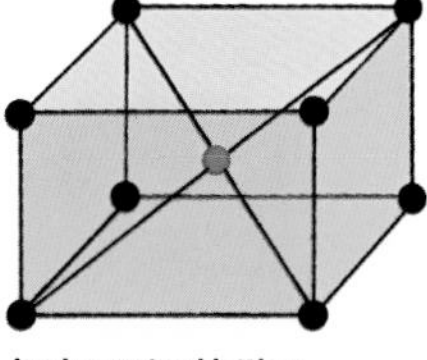
body-centred lattice

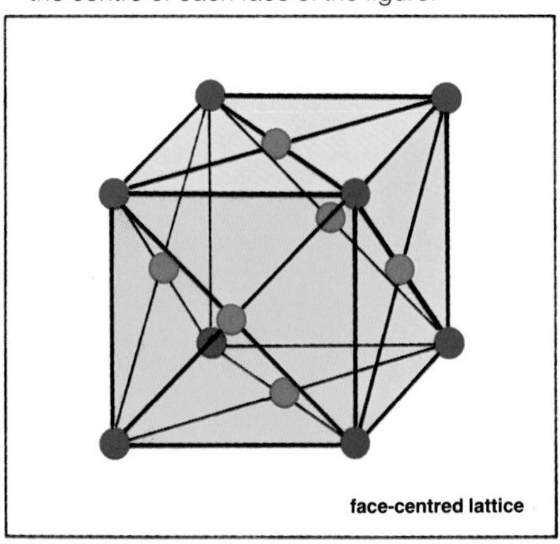
face-centred lattice

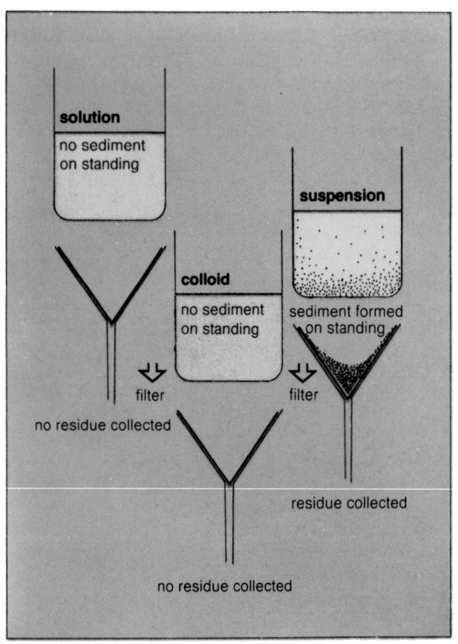

colloid (*n*) a constituent (p.54) of a disperse (↓) system (p.212) in which that constituent is dispersed throughout another constituent, e.g. a constituent (clay) is dispersed in water (the other constituent). The colloidal constituent is present (p.217) in particles which are between 1 nm and 100 nm in size. If the colloid is a solid and it is dispersed in a liquid, the size of the particles (p.13) is greater than those in a true solution (p.86), but smaller than those in a suspension (p.86). Solid colloidal particles dispersed in a liquid pass through a filter (p.30) but not through a permeable membrane (↓). Examples of colloids include: milk (a dispersion of colloidal particles of fat in water); foam (p.100); aerosol (p.100); sol (p.100); froth (p.100); smoke (p.100). **colloidal** (*adj*).

colloids
effect of a light beam on true solutions and colloids

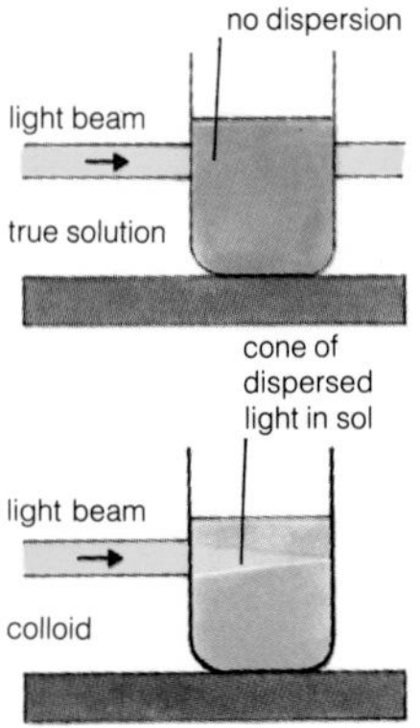

disperse (*v*) to spread particles (p.13), or similar small objects, over a large area or volume. The action points to the agent causing it, e.g. a liquid disperses a solid colloid (↑) over all its volume. **dispersion** (*n*), **disperse** (*adj*), **dispersed** (*adj*).

dispersion medium the liquid, or gas, that disperses (↑) a colloid (↑) over all its volume. It has a continuous phase whereas the disperse phase is discontinuous.

disperse phase the colloidal (↑) substance that is dispersed (↑) over all the dispersion medium (↑).

membrane (*n*) a thin piece of material such as parchment paper, cellophane, a bladder. A liquid can pass slowly through a membrane if the membrane is permeable (↓). A crystalloid (p.91) solid can also pass through a permeable membrane, but a colloidal (↑) solid and a suspension cannot pass through it.

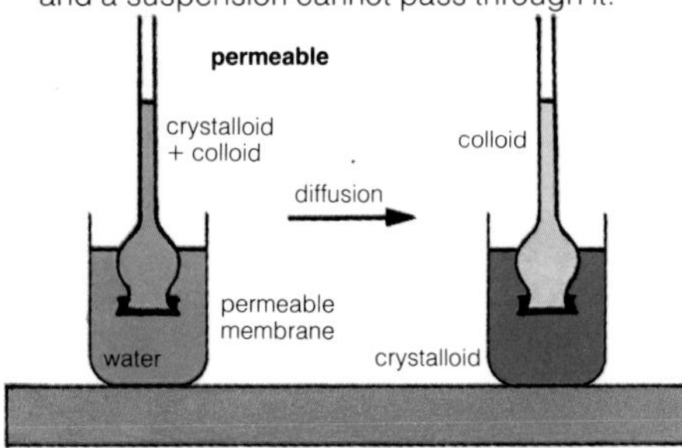

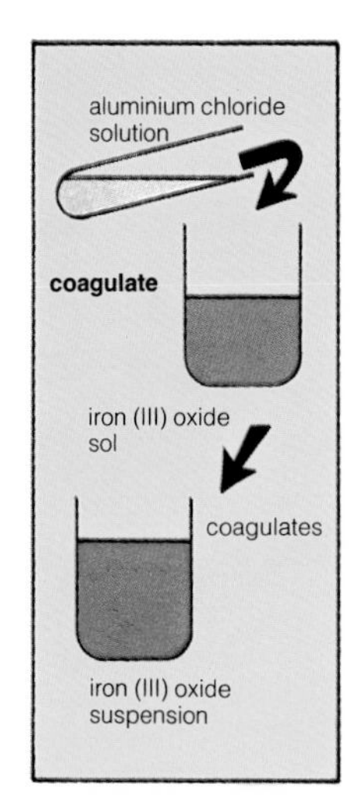

permeable (*adj*) describes any object which allows fluids (p.11) to pass through by diffusion (p.35). Permeable membranes (↑) allow molecules (p.77) of fluids to diffuse. Other substances which are finely divided, e.g. sand, also allow particles (p.13) to diffuse. **permeability** (*n*).

coagulate (*v*) to cause small particles to join together to form bigger particles or a mass of particles, e.g. a colloidal sol (p.100) can be coagulated to form a suspension (p.86); an iron (III) oxide sol in water is coagulated to a suspension by a solution of aluminium chloride. **coagulation** (*n*), **coagulated** (*adj*).

sol (*n*) a sol is formed by a colloidal (p.98) solid dispersed (p.99) in a liquid. The colloidal solid passes through a filter paper (p.30) but does not pass through a permeable membrane (p.99). Crystalloids (p.91) are separated from colloids by dialysis (p.34). Sols are described as lyophilic or lyophobic (↓).

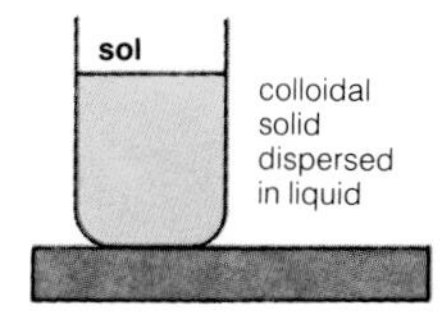

hydrosol (*n*) a sol (↑) in which the dispersion medium (p.99) is water.

emulsion (*n*) an emulsion is formed from two liquids, one of which is dispersed (p.99) in the other liquid in the colloidal state. If the emulsion is not stabilized (↓) the liquids separate out, e.g. coconut oil and water form an emulsion when shaken together; the two liquids separate on standing. Soap acts as an emulsifying agent (↓) to stabilize the emulsion. Milk is an emulsion with butter fat dispersed in a dilute sugar solution. **emulsify** (*v*), **emulsified** (*adj*).

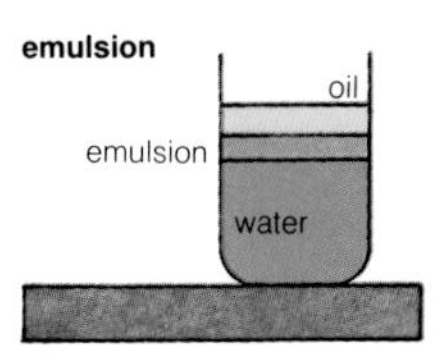

foam (*n*) a foam is formed from a gas dispersed (p.99) in a liquid. It consists of small bubbles of gas with a thin film (p.18) of liquid round the bubbles. A foam is usually stable. Foams are formed when a gas and a liquid are forced through a jet under pressure. **foam** (*v*), **foaming** (*adj*).

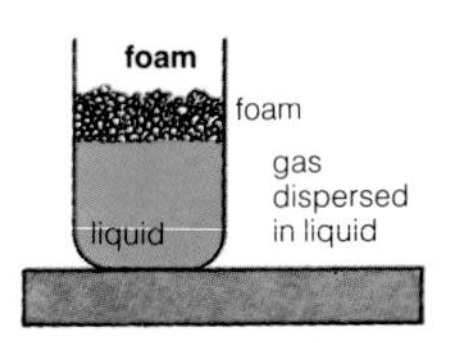

froth (*n*) a foam (↑) which has large bubbles and is less stable. Mixtures of ethanol and water form a froth, with carbon dioxide as the gas. **froth** (*v*).

aerosol (*n*) in an aerosol, a liquid is dispersed in a gas, usually air. It is formed by atomization (↓). A mist is an example of an aerosol. Many insecticides are used as aerosols.

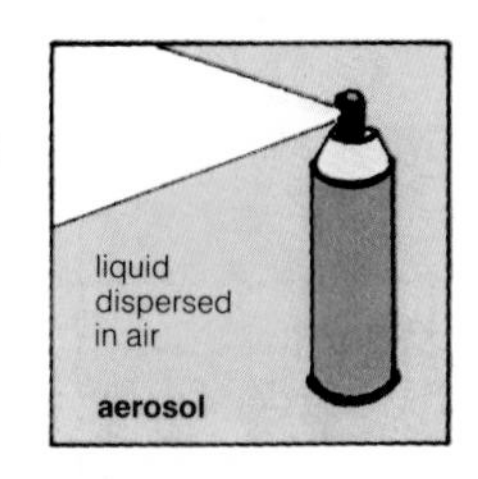

smoke (*n*) smoke is formed by solids dispersed in a gas, usually air, e.g. small particles (p.13) of carbon dispersed in air in wood smoke.

gel (*n*) an intermediate (p.85) stage (p.159) between a sol (↑) and a suspension. The colloidal particles (p.13) form long thin threads round the liquid dispersion medium (p.99). The result appears to be solid but it is easily deformed. A jelly is a gel. **gel** (*v*), **gelation** (*n*).

lyophilic (*adj*) describes a sol (↑) in which the disperse phase (p.99), a solid, has an attraction (p.124) for the dispersion medium (p.99), a liquid. The colloid (p.98) goes readily into solution; after coagulation (p.99), it is readily dispersed (p.99) again by adding more liquid, e.g. a starch sol is lyophilic.

lyophobic (*adj*) describes a sol (↑) in which the disperse phase (p.99), a solid, has a repulsion (p.124) for the dispersion medium (p.99), a liquid. The colloid (p.98) readily comes out of solution; after coagulation (p.99) it does not become dispersed (p.99) again, e.g. a gold sol.

hydrophilic (*adj*) describes a lyophilic (↑) sol with water as the dispersion medium (p.99).

hydrophobic (*adj*) describes a lyophobic (↑) sol with water as the dispersion medium (p.99).

stabilize (*v*) to make a sol, emulsion, or gel, stable, by preventing them from changing to a suspension, or from coagulating or from separating. **stable** (*adj*), **stabilizer** (*n*), **stabilization** (*n*).

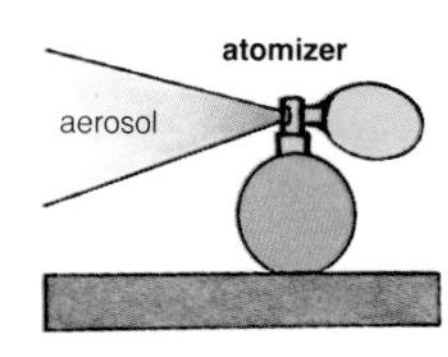

emulsifying agent an agent (p.63) which stabilizes (↑) an emulsion, e.g. soap stabilizes a coconut oil and water emulsion.

atomize (*v*) to blow a liquid through a fine (p.13) jet (p.29). This forms an aerosol (↑) with the liquid dispersed (p.99) in air. **atomizer** (*n*).

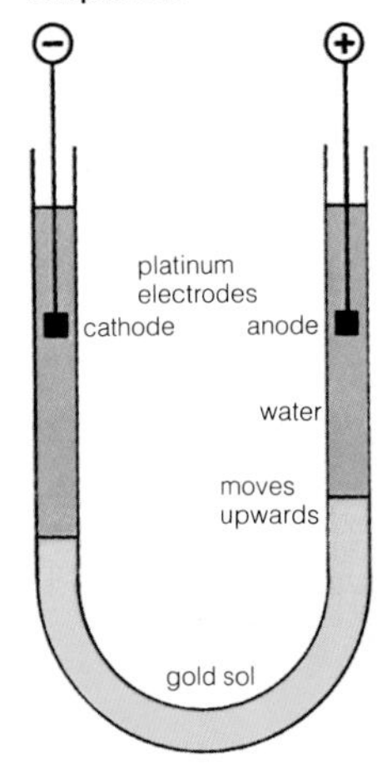

cataphoresis (*n*) the movement of colloidal (p.98) particles (p.13) towards electrodes in a lyophobic (↑) sol, or in a smoke (↑). Lyophilic (↑) sols may undergo (p.213) cataphoresis if the colloid has an electric charge. The colloidal particles may move either to the anode (p.123) or the cathode (p.123) depending on their charge, e.g. clay and metallic particles are negatively charged; iron (III) chloride and aluminium hydroxide are positively charged in aqueous (p.88) sols. Platinum electrodes are usually used in cataphoresis.

electrophoresis (*n*) another name for cataphoresis (↑).

thixotropy (*n*) a property of some colloidal liquids by which the liquid has a decreased viscosity at an increased rate of flow. **thixotropic** (*adj*).

pressure (*n*) the force per unit area acting on a surface. It is the force in newtons acting on an area measured in square metres. A force of 200 newtons acting on an area of 5 sq metres gives a pressure of $200\,N \div 5\,m^2 = 40\,N\,m^{-2}$. The unit of pressure is the pascal (Pa); a pressure of $1\,N\,m^{-2} = 1\,Pa$. Pressure in gases is also measured by the height of a column of mercury supported by the pressure; the height is given in millimetres (mm) of mercury. 1 mm of mercury = $133.322\,N\,m^{-2}$. **press** (*v*).

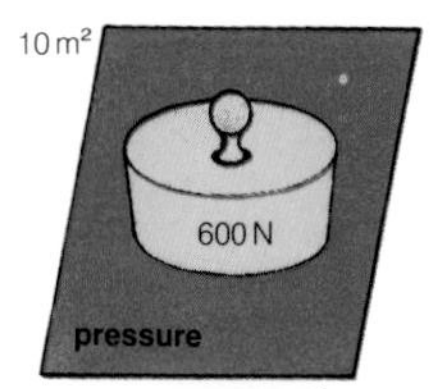

atmospheric pressure the pressure (↑) exerted (p.106) by the air in the atmosphere on the Earth's surface. It is measured by a barometer.

standard atmospheric pressure an atmospheric pressure (↑) of $101\,325\,N\,m^{-2}$ or 760 mm of mercury. The everyday variations of atmospheric pressure are above and below this value.

s.t.p. (*abbr*) an abbreviation for standard temperature (↓) and pressure (↑), i.e. 0°C or 273 K and $101\,325\,Nm^{-2}$.

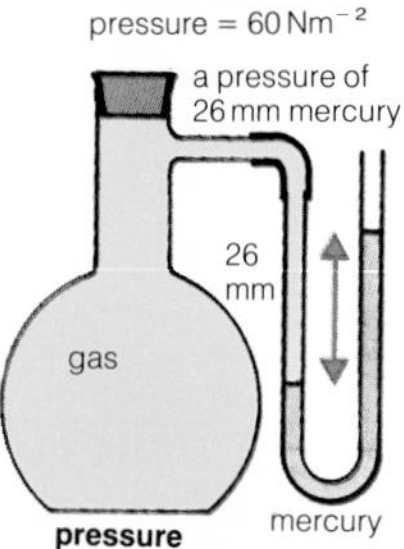

temperature (*n*) a physical property which determines the direction of the flow of heat between materials (p.8) in thermal (p.65) contact (p.217). Heat flows from a higher to a lower temperature.

Celsius scale a temperature scale (p.26) with 0° as the temperature of melting ice and 100° as the temperature of steam at standard atmospheric pressure (↑). One Celsius degree (1°C) is 1/100 of the temperature interval (p.220) between the ice point and the steam point. The symbol for Celsius temperature is θ.

absolute scale a temperature scale (p.26) with 0 as the absolute zero of temperature and a temperature interval of 1 kelvin. On this scale the temperature of melting ice is 273 K and of steam is 373 K. Absolute temperatures are changed to Celsius temperatures by: 273 + Celsius temperature = absolute temperature. The symbol for absolute temperature is T.

kelvin (*n*) the S.I. unit of temperature, measured from the triple point of water at which ice, water and water vapour all exist (p.213) together. One kelvin is 1/273.16 of the value of the triple point. A temperature interval of 1 K = 1°C (symbol = K).

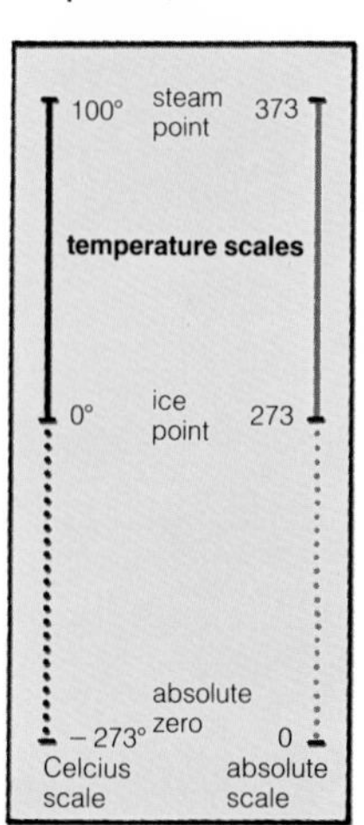

surroundings (*n.pl.*) all the objects and materials near to and around an object, or the objects and materials for an experiment, are the surroundings of that object or experiment. The surroundings may, or may not, have an effect on the object or experiment. **surround** (*v*).

factor (*n*) a possible cause of an effect or change. Some of the surroundings (↑) are factors, i.e. those that have an effect on an object or experiment. Factors for chemical reactions are temperature (↑), pressure (↑), concentration (p.81), catalysts (p.72), surface nature (p.19).

mercury barometer

using a mercury barometer to measure vapour pressure of a liquid

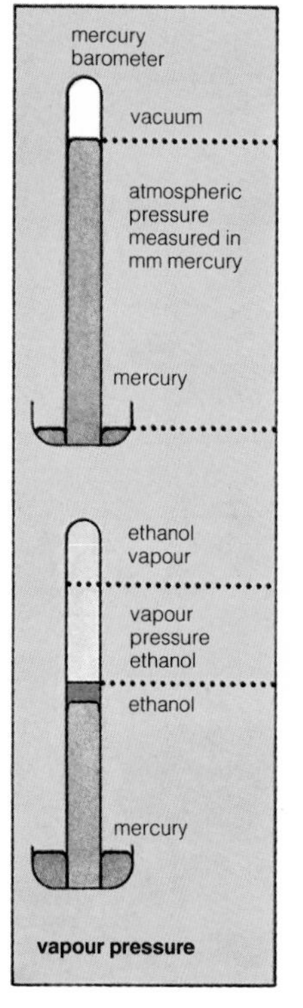

conditions (*n.pl.*) the magnitude of factors (↑) of a particular object or experiment. To contrast *factors* and *conditions*: (1) temperature is a *factor* for the rate of reaction (p.149) between substances; a temperature of 60°C is an actual *condition* in an experiment. (2) water is a *factor* in the rusting (p.61) of iron; the presence of water is a necessary *condition*. Conditions can be *necessary*, i.e. without them, there is no reaction or effect, or conditions may be *adverse*, under which the reaction or effect does not take place or takes place very slowly, or conditions may be *suitable*, under which the reaction or effect is normal (p.229).

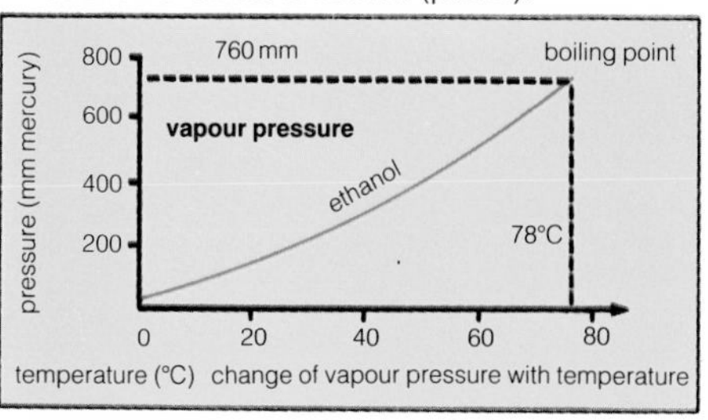

change of vapour pressure with temperature

vapour pressure the pressure (↑) exerted (p.106) by a saturated (p.87) vapour (p.11). Vapour pressure increases with temperature. A liquid boils when its vapour pressure is equal to the atmospheric pressure (↑).

ambient (*adj*) describes conditions of the surroundings (↑), e.g. the ambient temperature is the temperature of the surrounding air.

condensation[2] (*n*) a change of state (p.9) from vapour, or gas, to liquid caused by a decrease in temperature. **condense** (*v*).

liquefaction (*n*) a change of state (p.9) from vapour to liquid caused by an increase in pressure. **liquefy** (*v*).

permanent gas a gas which is not easily liquefied because it has to be first cooled to a very low temperature, e.g. hydrogen, nitrogen and oxygen are permanent gases.

noble gas one of the following gases: helium, neon, argon, krypton, xenon, and radon. They do not take part in chemical reactions and are usually considered to be inert (p.19). They are permanent gases (↑) and also monatomic (↓).

critical temperature a particular temperature for each gas, above which the gas cannot be liquefied (↑) by pressure (p.102) alone. The critical temperature for oxygen is −119°C (154 K), so oxygen has to be cooled below −119°C before it can be liquefied by pressure; for this reason, oxygen is a permanent gas (↑). Critical temperature has the symbol T_c.

critical pressure (*n*) the pressure sufficient to liquefy a gas at its critical temperature (↑).

monatomic (*adj*) describes a gas with molecules (p.77) consisting of one atom (p.110) only, e.g. neon and helium are monatomic gases.

diatomic (*adj*) describes a gas with molecules (p.77) consisting of two atoms (p.110).

polyatomic (*adj*) describes a gas with molecules (p.77) consisting of several atoms (p.110). Such gases are usually compounds, e.g. carbon dioxide.

atomicity (*n*) the number of atoms in a molecule of an element or a compound, when the element or compound consists of separate molecules. In most cases, the substances will be gaseous, e.g. helium has an atomicity of 1; hydrogen of 2; sulphur dioxide of 3.

gram-molecular volume the volume of one gram-molecule (the molecular weight in grams, calculated from the sum of the atomic weights in grams) of a substance. Molar volume is now used instead of gram-molecular volume.

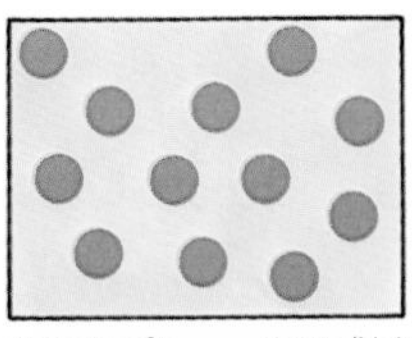

monatomic neon (Ne)

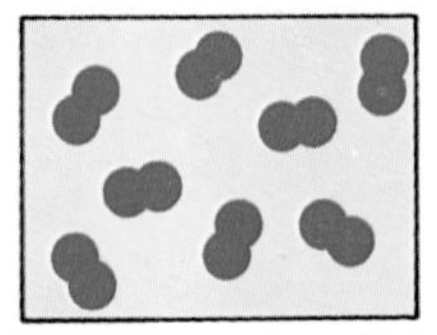

diatomic hydrogen (H_2)

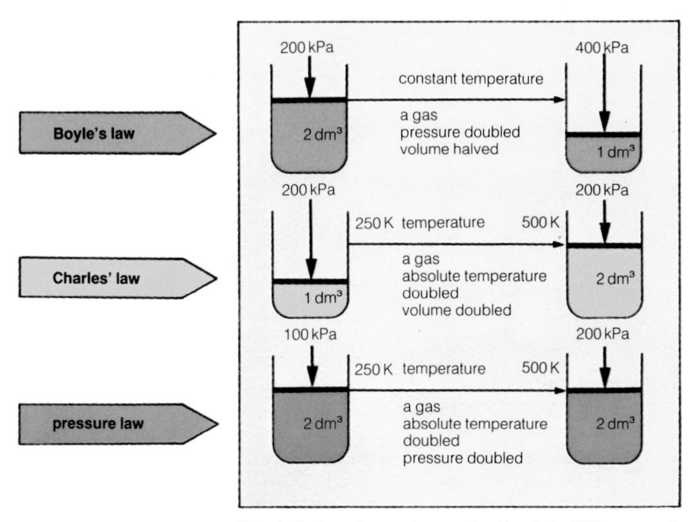

Boyle's law the volume of a fixed (p.79) mass of gas is inversely proportional to the pressure (p.102) at a constant (p.106) temperature; the relation can also be written as pressure times volume equals a constant. In symbols $pV = k$, where k is a constant. The law is obeyed (p.107) at low pressures, but real (p.107) gases do not obey the law at high pressures.

Charles' law the volume of a given mass of gas increases by 1/273 of its volume at 0°C for each degree Celsius rise in temperature, if the pressure (p.102) remains constant. In symbols

$$V_\theta = V_0 (1 + \theta/273)$$

The volume of the gas is proportional to the absolute temperature (p.102), so $V \propto T$ and hence $V_1/V_2 = T_1/T_2$.

pressure law the pressure (p.102) of a given mass of gas increases by 1/273 of its pressure at 0°C for each degree Celsius rise in temperature, if the volume remains constant. In symbols $p_\theta = p_0 (1 + \theta/273)$.
The pressure of the gas is proportional to the absolute temperature (p.102) so $p \propto T$ and hence $p_1/p_2 = T_1/T_2$.

gas equation the three laws, Boyle's, Charles' and pressure, can be combined into one equation which is $pV = kT$, where k is a constant (↓) depending on the mass of gas. The equation is usually written as $p_1V_1/p_2V_2 = T_1/T_2$. If the amount of gas is given as a mole fraction (p.80) with symbol n, then the equation becomes $pV = nRT$, where R is the same constant for all gases.

constant¹ (*n*) (1) the value of a physical quantity (p.81) which cannot be changed, e.g. the speed of light is a constant; the Avogadro constant (p.80). (2) the unchanging value of a physical quantity under experimental conditions (p.103), e.g. the boiling point of a liquid at s.t.p. (p.102).

constant² (*adj*) describes an unchanging value of a physical quantity which is controlled (p.221) by an observer.

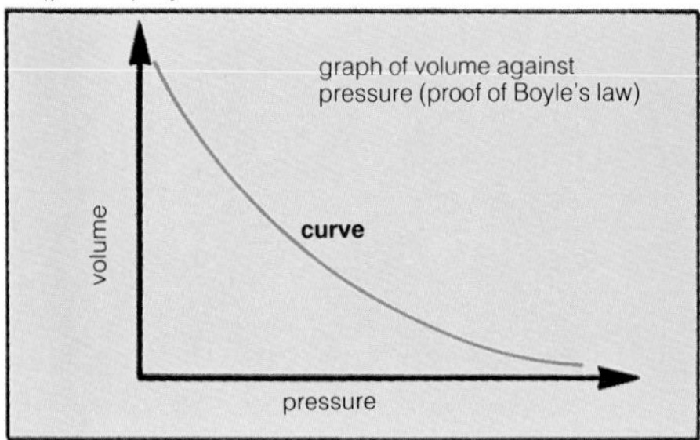

curve (*n*) the line drawn in a graph showing the relation (p.232) between two quantities (p.81).

derive (*v*) to obtain a value or a statement (p.222) by several steps of experimental work or deduction (p.222). The work must start from something initially (p.85) known. For example, Boyle's law is derived from experiments on gases, measuring volumes and pressures, plotting (p.39) graphs and then deducing the relation between pressure and volume. The statement of the law is then derived. **derivation** (*n*), **derivative** (*n*), **derived** (*adj*).

exert (*v*) to bring into effect, e.g. a gas exerts a pressure on the walls of a vessel (p.25) so the pressure has an effect on the walls.

ideal (*adj*) (1) describes a gas which obeys (↓) Boyle's law (p.105) at all pressures. Such gases do not exist (p.213). Real (↓) gases are ideal gases at low pressures. (2) describes an object, arrangement or theory (p.76) that is perfect; such an object, arrangement or theory is usually part of a hypothesis (p.108).

non-ideal (*adj*) (1) describes a gas which does not obey Boyle's law at all pressures. (2) describes any object or arrangement which is not ideal (↑).

real (*adj*) (1) describes any object that exists (p.213), in particular any gas that exists, e.g. gases such as hydrogen, oxygen and nitrogen are all real gases. (2) describes any object that is what it is said to be, e.g. a bag made of real leather not *imitation* leather.

conform (*v*) to follow a pattern (p.93) of behaviour, e.g. gases conform with Boyle's law, but the degree (p.227) of conformity becomes less at high pressures. A material or substance (p.8) can conform with a law (p.109), or any other statement (p.222) of its properties, but the agreement between experimental (p.42) results (p.39) and the prediction (p.85) is not necessarily exact. **conformity** (*n*).

obey (*v*) to be in exact agreement with a law, so that experimental (p.42) results (p.39) are in exact agreement with predictions (p.85), e.g. an ideal (↑) gas obeys Boyle's law; chemical compounds obey the law of constant proportions (p.76). To contrast *conform* and *obey*: if experimental results are approximately (p.79) in agreement with predictions, then a substance *conforms* to the law, if the results are in exact agreement, a substance *obeys* the law.

Graham's law the rate of diffusion (p.35), or effusion (p.35), of a gas at a constant temperature is inversely (p.233) proportional (p.76) to the square root of its density (p.12). The greater the relative molecular mass (p.114) of a gas, the slower is its rate of diffusion, or effusion.

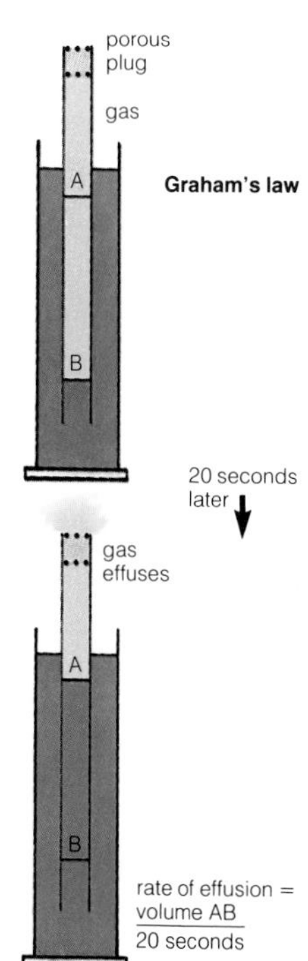

Graham's law

Avogadro's hypothesis equal volumes of all gases at the same temperature and pressure contain the same number of molecules. *See Avogadro constant (p.80).* The molar volume (p.80) of a gas at s.t.p. (p.102) contains the Avogadro number of molecules.

Avogadro's principle another name of Avogadro's hypothesis.

hypothesis (*n*) a statement which cannot be proved experimentally (p.42). From it, however, laws can be deduced (p.222) and proved.

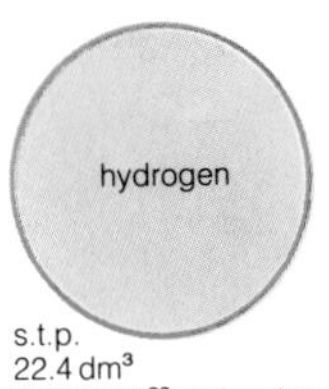

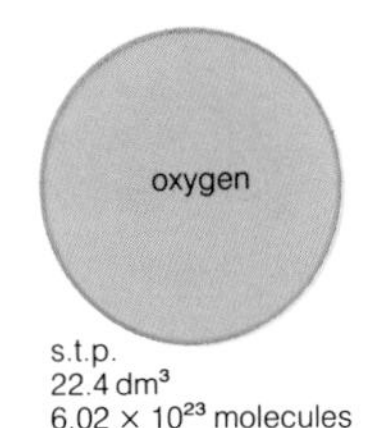

Avogadro's hypothesis

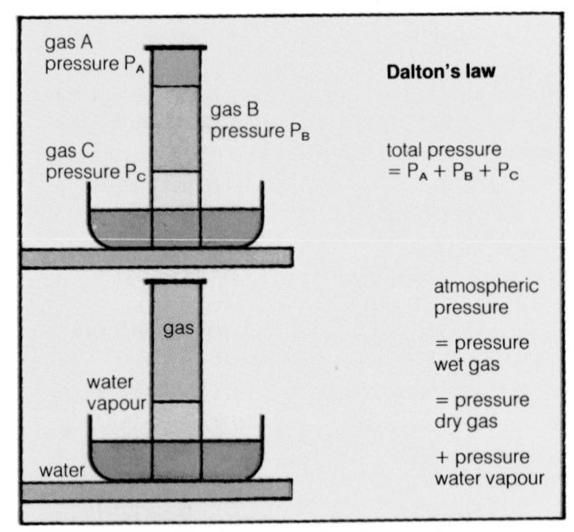

Dalton's law in a mixture of gases which do not react (p.62), the total pressure exerted (p.106) by the mixture of gases is equal to the sum of the partial pressures (↓) of each gas. Also called **Dalton's law of partial pressures**.

partial pressure the partial pressure of a gas in a mixture of gases is equal to the pressure which each gas would exert (p.106) if it alone filled the whole volume of the mixture.

kinetic theory a theory (p.76) which explains the behaviour of liquids and gases by considering them to consist (p.55) of molecules (p.77) in random (p.223) motion; the molecules hit each other and the walls of the containing vessel (p.25). The molecules are considered to be completely elastic (p.14).

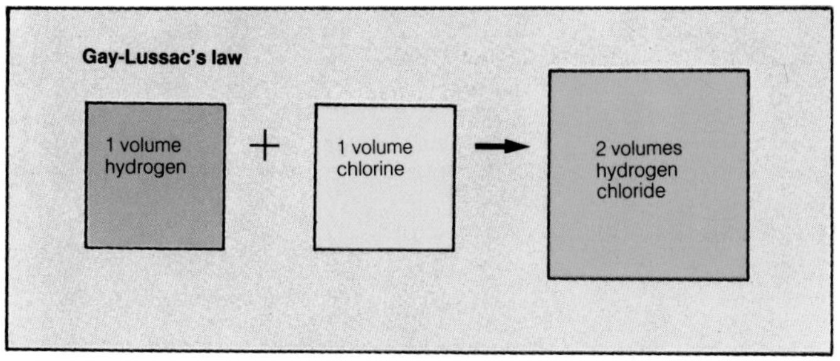

Gay-Lussac's law the volumes of gases which react, and the volumes of the products if gaseous, are in the ratio (p.79) of small whole numbers if pressure and temperature are kept constant. For example, one volume of hydrogen combines (p.64) with one volume of chlorine to form two volumes of hydrogen chloride. One and two are simple whole numbers.

law (*n*) a statement about the properties of materials and substances (p.8) that is accepted as true. From it, predictions (p.85) about the behaviour of substances can be made. A law is universal (p.212) or limited (↓), e.g. the law of constant proportions (p.76) is a universal law as no substances are excepted from it; Boyle's law is a limited law as it depends on the condition of temperature being constant.

limited[1] (*adj*) describes a law (↑), or theory (p.76), which is true only under particular conditions. **limitations** (*n*), **limit** (*v*).

gas laws statements about the behaviour of gases under particular conditions. The important laws are those of Boyle (p.105), Charles (p.105) and Dalton (↑).

Brownian motion the random (p.223) motion of molecules in liquids (p.10) and gases. The motion is seen when pollen grains are suspended (p.31) in water, or when smoke particles are seen in air. In both cases, the motion of molecules causes the motion of the pollen grains or smoke particles.

atom (*n*) the smallest particle (↓) of an element (p.8) that has the properties of that element. Atoms combine chemically to form molecules (p.77). An atom consists of a nucleus (↓) with electrons (↓) around it. **atomic** (*adj*).

subatomic (*adj*) describes particles (↓) smaller than an atom (↑).

electron (*n*) a subatomic (↑) particle with a negative electric charge (p.138). It is the smallest subatomic particle with a mass of 9.109×10^{-31} kg.

proton (*n*) a subatomic (↑) particle with a positive charge (p.138). The charge is equal and opposite to that of an electron (↑). The mass of a proton is 1840 times that of an electron.

neutron (*n*) a subatomic (↑) particle with a mass almost equal to the mass of a proton (↑). It has no electric charge (p.138).

nucleus (*n*) the nucleus of an atom (↑) consists (p.55) of protons (↑) and neutrons (↑), except for hydrogen, which has a nucleus consisting of one proton. The nucleus has a positive electric charge, the size of which depends upon the number of protons it contains. The nucleus provides the mass of an atom. **nuclear** (*adj*).

nuclear (*adj*) describes anything to do with the nucleus (↑) of an atom.

particle[2] (*n*) a piece of substance so small that it is considered to have mass but not size or volume, that is, it is like a point. Atoms (↑) and molecules (p.77) are particles in this sense.

positron (*n*) a subatomic (↑) particle with a mass the same as the mass of an electron (↑), but with a positive charge equal and opposite to the charge on an electron.

meson (*n*) subatomic (↑) particles with a mass about half that of a proton (↑) and with zero spin; they appear to bind nuclear particles together.

orbit (*n*) in the first models (p.223) of atoms, the electrons (↑) were considered to be in orbits round the nucleus (↑). An orbit is a circular path round a nucleus. This idea is no longer considered to be true.

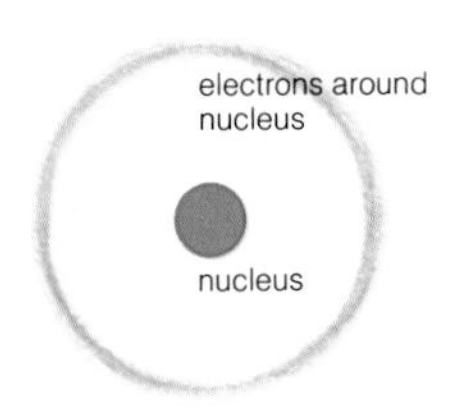

structure of an atom

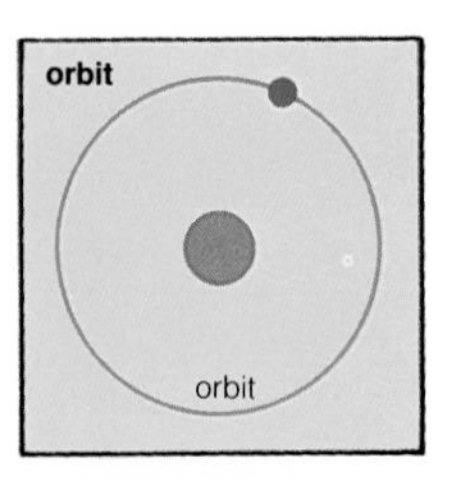

shell (*n*) a spherical (↓) space around a nucleus (↑); it contains extranuclear (p.113) electrons with energy levels (p.152) of that shell. An atom possesses different shells of increasing radii (↓) around its nucleus (↑). The radius of a shell is 100 000 times greater than the radius of the nucleus.

sphere (*n*) a solid in the shape of a ball. **spherical** (*adj*).

radius (*n*) (*radii n.pl.*) (1) the line joining the centre of a circle to the line of the circle; the line drawn from the centre of a sphere to its surface. (2) the length of the line described above, e.g. a circle with a radius of 5 cm.

orbital (*n*) a space in which there can be one or two electrons (↑) but not more. The space is where there is a probability (p.223) of finding one or both electrons. The probability varies (p.218) over the space, and can be shown in various ways by a diagram (p.29). The position and motion of an electron in an orbital cannot be described, only the probability can be described. One way of showing probability is by shading, *see diagram*. Another way is by drawing the limits (p.211) of the orbital. Two electrons in an orbital do not interfere (p.216) with each other; two electrons form a stable (p.74) orbital, while one electron in an orbital is active in forming chemical bonds (p.133). The orbitals commonly taking part in bonds are s-, p- and d-orbitals (p.112).

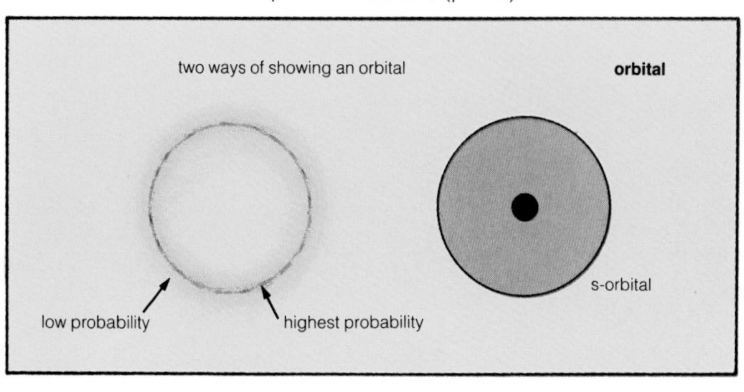

s-orbital (*n*) an orbital (p.111) in an atomic shell (p.111) containing one or two electrons (p.110). There is only one s-orbital in a shell. An s-orbital is spherically (p.111) symmetrical (p.93) about a nucleus. It has no directional properties.

p-orbital (*n*) an orbital (p.111) containing one or two electrons (p.110). There are three p-orbitals in an atomic shell (p.111). The three orbitals are at right angles to each other along three axes (p.92) and have directional properties.

d-orbital (*n*) an orbital (p.111) containing one or two electrons (p.110). There are five d-orbitals in an atomic shell (p.111). Four of the orbitals have the same shape, and one has a different shape. All five orbitals have directional properties.

s-electron (*n*) an electron (p.110) in an s-orbital (↑). s-electrons have the least energy in a shell (p.111). Two s-electrons fill the s-orbital before an electron goes into a p-orbital.

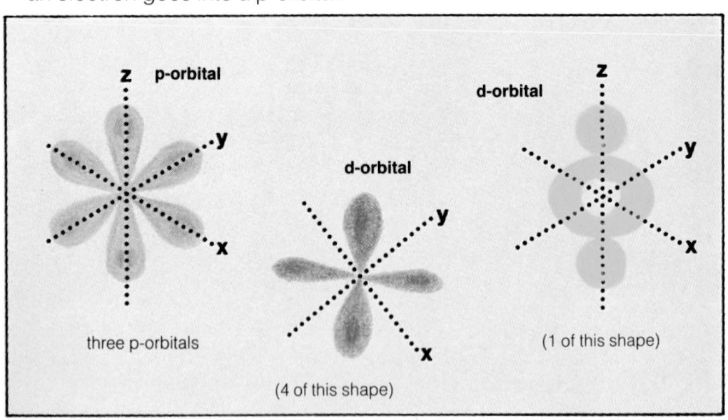

p-electron (*n*) an electron (p.110) in a p-orbital (↑). One p-electron goes into each of the three p-orbitals before the fourth electron goes into one p-orbital to fill it. A total of six p-electrons fill the three p-orbitals before an electron goes into a d-orbital (↑).

d-electron (*n*) an electron (p.110) in a d-orbital (↑). One d-electron goes into each of the five d-orbitals before the sixth electron goes into one d-orbital to fill it.

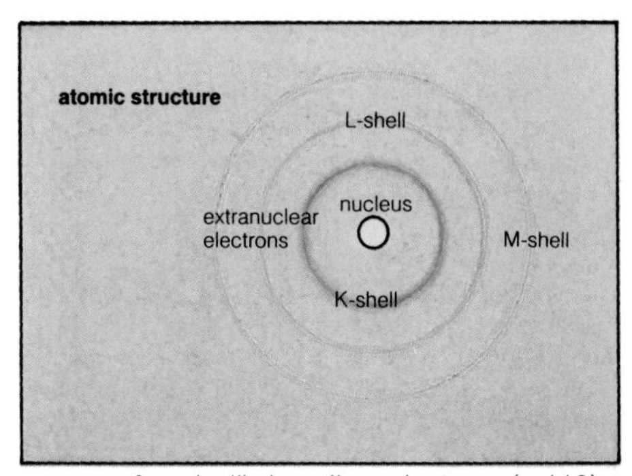

extranuclear (*adj*) describes electrons (p.110) outside the nucleus (p.110) of an atom.

atomic structure the structure of the nucleus (p.110) and of the extranuclear (↑) electrons. The extranuclear electrons are arranged in atomic shells (p.111).

K-shell (*n*) the innermost atomic shell (p.111). It contains only one s-orbital. This arrangement completes the shell.

L-shell (*n*) the next atomic shell to the K-shell (↑). It contains one s-orbital and three p-orbitals. This arrangement completes the shell.

M-shell (*n*) the next atomic shell to the L-shell (↑). It contains one s-orbital, three p-orbitals and five d-orbitals. This completes the shell. There are further shells beyond the M-shell.

atomic number a number equal to the number of protons (p.110) in the nucleus (p.110) of an atom of an element. It has the symbol Z.

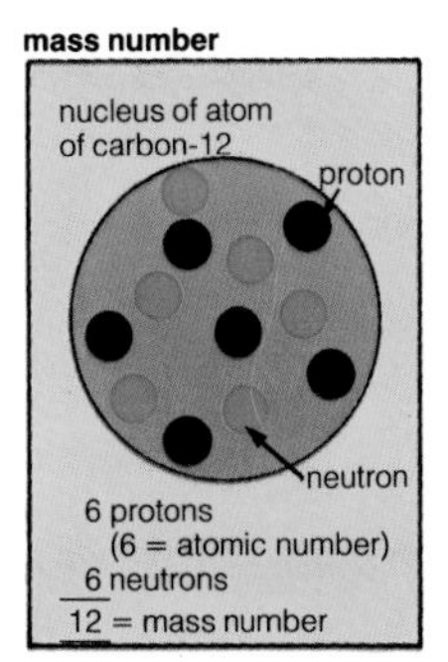

mass number a number equal to the sum of the number of protons (p.110) and number of neutrons (p.110) in the nucleus (p.110) of an atom of an element. It is the mass number of an isotope (p.114). The symbol for mass number is A.

relative atomic mass the ratio (p.79) of the mass of one atom of an element to one-twelfth of the mass of one atom of carbon-12. Most elements consist of isotopes (p.114) of its atoms, hence the relative atomic mass measures the average mass per atom of the normal isotopic composition of an element.

isotope (*n*) atoms of the same element (p.116) which have different mass numbers (p.113) are isotopes of the element. As all atoms of an element have the same number of protons (p.110), isotopes differ in having different numbers of neutrons (p.110) in the nucleus (p.110) of the atoms, e.g. there are two isotopes of carbon, one with a mass number of 12 and one with a mass number of 13. Both isotopes have six protons in their nuclei, so one isotope has six neutrons and the other seven neutrons. Isotopes are shown by writing the mass number after the element, e.g. carbon-12 is the isotope of carbon with a mass number of 12. **isotopic** (*adj*).

isotopic ratio the ratio (p.79) of the different isotopes (↑) in a specimen (p.43) of an element. For elements obtained from natural sources (p.138) the isotopic ratio is always the same, e.g. the isotopic ratio for carbon is 98.9% C-12 and 1.1% C-13. The relative atomic mass of carbon is thus slightly greater than 12.

relative isotopic mass (*n*) (of an isotope) the ratio of the mass of one atom of the isotope to 1/12 of the mass of one atom of carbon-12.

isotopic weight (*n*) another name for relative isotopic mass (↑), no longer used.

atomic mass unit a mass equal to one-twelfth of the mass of an atom of the isotope (↑) carbon-12. Its value is 1.66043×10^{-27} kg. The symbol for atomic mass unit is *amu*.

atomic weight (*n*) the ratio of the mass of one atom of an element to one-sixteenth of the mass of an atom of oxygen. Relative atomic mass is now used instead of atomic weight, and the method of measuring is different with a standard of carbon-12 instead of oxygen.

relative molecular mass (*n*) the ratio (p.79) of the mass of one molecule of a substance (p.8) to the mass of one atom of carbon-12. The relative molecular mass of a compound is calculated by adding the relative atomic masses of all the atoms in a molecule of the compound. Relative molecular mass is used with covalent compounds, but not with ionic compounds. *See relative formula mass (p.78).*

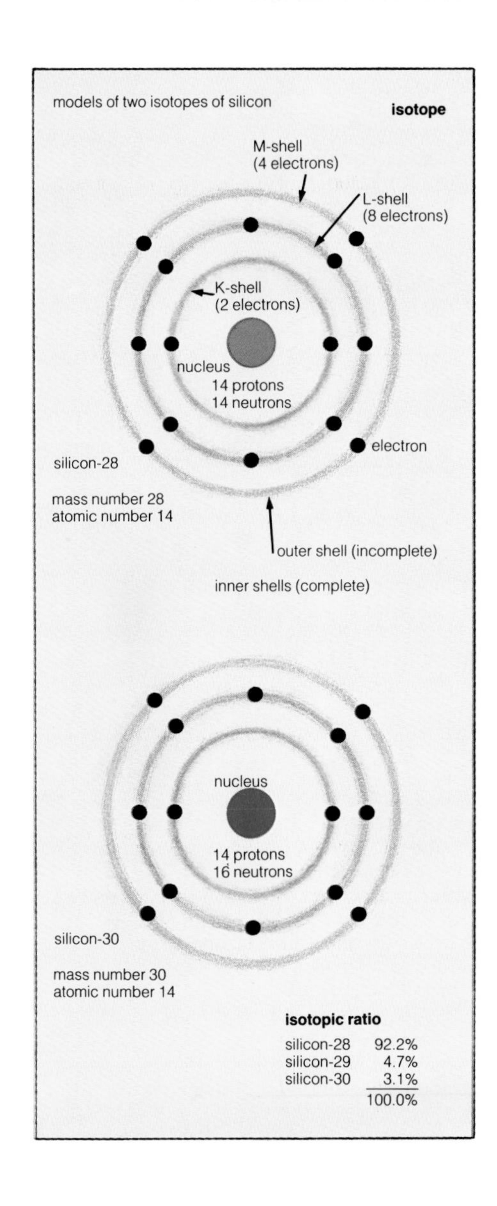
models of two isotopes of silicon
isotope
M-shell
(4 electrons)
L-shell
(8 electrons)
K-shell
(2 electrons)
nucleus
14 protons
14 neutrons
electron
silicon-28
mass number 28
atomic number 14
outer shell (incomplete)
inner shells (complete)
nucleus
14 protons
16 neutrons
silicon-30
mass number 30
atomic number 14
isotopic ratio
silicon-28 92.2%
silicon-29 4.7%
silicon-30 3.1%
100.0%

element[2] (*n*) a substance (p.8) with all its atoms (p.110) having the same positive charge on the nucleus (p.110), i.e. the nucleus of all atoms has the same number of protons (p.110) and thus the same atomic number, and this determines the chemical nature of the element. **elementary** (*adj*).

metal (*n*) an element (↑) which forms positive ions (p.123) in chemical reactions. The general physical properties of metals are: (a) they conduct (p.122) electric current and heat; (b) they are lustrous (p.16), ductile (p.14) and malleable (p.14). The general chemical properties of metals are: (a) they form basic (p.46) oxides; (b) they form compounds with non-metals which are salts. Metals possess these properties in varying (p.218) degrees (p.227). All metals, except mercury, are solids. **metallic** (*adj*).

non-metal (*n*) an element (↑) which is not a metal (↑). Non-metals are solids or gases, except bromine which is a liquid. Their physical properties depend upon their structure (p.82). Solid non-metals are neither ductile (p.14) nor malleable (p.14), but are usually brittle (p.14). They do not conduct electric current, except for graphite (p.118), nor heat. They generally form negative ions (p.123) in chemical reactions, except for hydrogen which forms positive ions. Their oxides are generally acidic, but some are neutral, e.g. carbon monoxide.

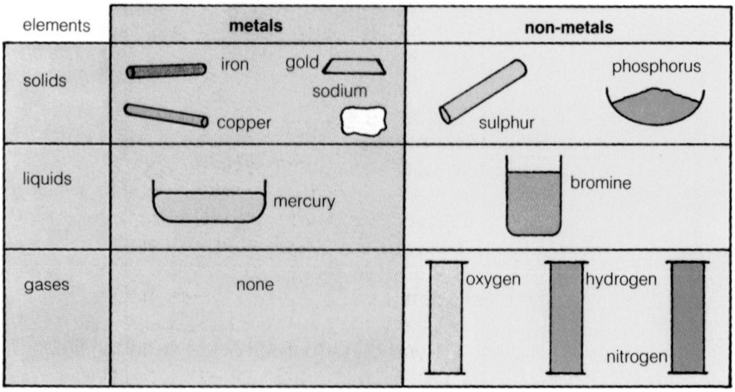

elements	**metals**	**non-metals**
solids	iron, gold, sodium, copper	sulphur, phosphorus
liquids	mercury	bromine
gases	none	oxygen, hydrogen, nitrogen

metalloid (*n*) an element with some of the properties of metals and some of the properties of non-metals, e.g. antimony and arsenic are metalloids.

alkali metal a metal (↑) which forms an ion (p.123) with an oxidation state (p.135) of +1 only. The oxide is very soluble in water forming a hydroxide (p.48). Alkali metals are chemically very reactive (p.62); they are the elements of group I of the periodic table (p.119) and have one s-electron (p.112) in their outer shells (p.111).

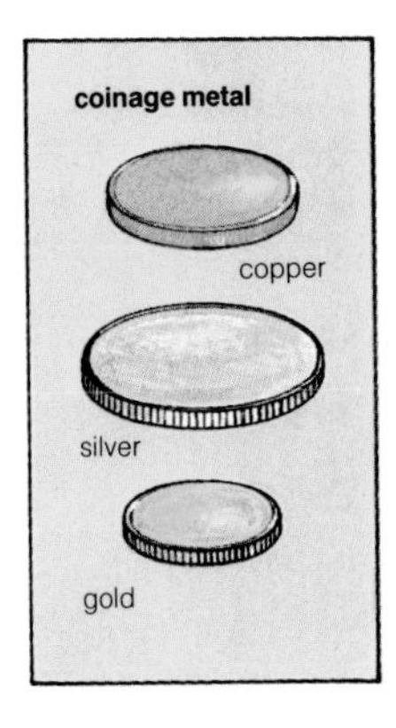

alkaline earth metal a metal (↑) which forms an ion (p.123) with an oxidation state (p.135) of +2 only. The oxide is sparingly soluble in water forming a hydroxide (p.48). Alkaline earth metals are chemically reactive (p.62); they are the elements of group II of the periodic table (p.119) and have two s-electrons (p.112) in their outer shell (p.111).

coinage metal a metal (↑) which which does not oxidize (p.70) readily in the air, e.g. copper, silver, gold. The coinage metals are not chemically reactive (p.62); they form positive ions with variable (p.218) oxidation states (p.135); they are all transitional elements (p.121).

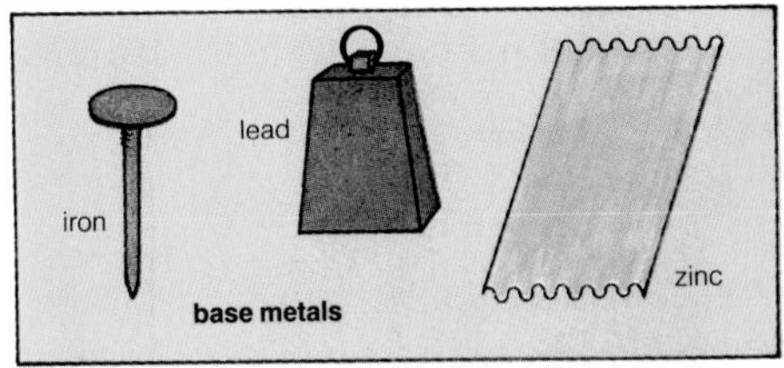

base metal a metal (↑) which is oxidized (p.70) by heating in air and is acted upon by mineral acids (p.55). Base metals are used for making articles for everyday use. For example, iron, lead, tin, zinc; some are transitional metals, e.g. iron, tin, lead.

halogen (*n*) a non-metal (↑), which forms a negative ion (p.123) with an electrovalency of −1 and is in group VII of the periodic table (p.119). The halogens are fluorine, chlorine, bromine and iodine.

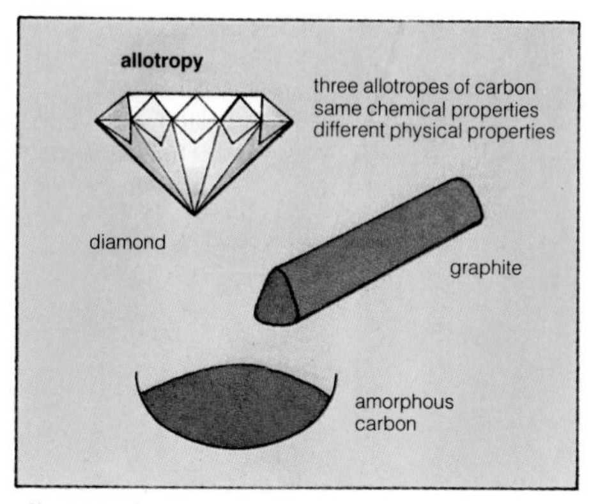

allotropy (*n*) the existence of two or more different forms of an element without a change of state (p.9). If the forms are crystalline (p.15) they are polymorphic (p.92) as well as allotropic, e.g. sulphur exists in five allotropes (↓); carbon has two crystalline allotropes. **allotropic** (*adj*), **allotrope** (*n*).

dynamic allotropy a kind of allotropy in which the allotropes are in a dynamic equilibrium (p.150) with each other, e.g. liquid sulphur has three allotropes (↓) which show dynamic allotropy.

allotrope (*n*) one of the forms of an element showing allotropy (↑).

carbon (*n*) a non-metal (p.116) with atomic number (p.113) 6, relative atomic mass (p.113) of 12.01 and in group IV of the periodic table (↓). *See isotope (p.114)*. Carbon is one of the most important elements as it is a constituent (p.54) of all living things. It occurs (p.63) in two crystalline forms, diamond (↓) and graphite (↓). **carbonic** (*adj*), **carbonaceous** (*adj*).

diamond (*n*) a crystalline form of carbon with a tetrahedral (p.83) lattice (p.92). It is the hardest natural substance; it does not conduct electric current.

graphite (*n*) a crystalline form of carbon with a hexagonal (p.96) lattice (p.92). It is a soft substance that marks paper. It is the only non-metal that conducts electric current.

periodic system if the elements are arranged in order of increasing atomic number (p.113), a periodicity (p.120) is seen in their properties. For example, the elements with atomic numbers 3, 11, 19 are all chemically active metals; atomic numbers 9, 17, 35, 53 are all chemically active non-metals with an electrovalency (p.134) of − 1; atomic numbers 4, 12, 20, 38, 56 are all metals with an oxidation state (p.135) of II. The arithmetical difference between any pair of these numbers is 8 or 18. This suggests a relation (p.232) between the extranuclear (p.113) structure of the electrons and the periodicity of properties. The periodic system is shown by a periodic table (↓).

periodic table the periodic table is shown on p.120 and endpapers. The atomic number (p.113) of each element is shown and the table is arranged in periods (p.120) and groups (↓).

group (*n*) a vertical column of the periodic table (↑). Group I contains the alkali metals (p.117), group II the alkaline earth metals (p.117) and group VII the halogens (p.117). The elements in a group have very similar properties. For metals, chemical reactivity increases from elements with low atomic numbers (p.113) to those with high atomic numbers. For non-metals, chemical reactivity decreases (p.219) from elements with low atomic numbers to those with high atomic numbers. The elements in group I have one s-electron (p.112) in their outer shell (p.111), and those in group II have two s-electrons; elements in group VII have two s-electrons and five p-electrons (p.112).

group (I) alkali metals			
atomic number	relative atomic mass	element	chemical activity
3	6.94	lithium	increases ↓
11	22.98	sodium	
19	39.10	potassium	
37	85.47	rubidium	
55	132.91	caesium	
group (VII) halogens			
9	18.99	fluorine	↑
17	35.45	chlorine	
35	79.91	bromine	
53	126.90	iodine	increases

period	group 1	group 2		group 3	group 4	group 5	group 6	group 7	group 8
1	1 H 1.01								2 He 4.00
2	3 Li 6.94	4 Be 9.01		5 B 10.81	6 C 12.01	7 N 14.01	8 O 16.00	9 F 19.00	10 Ne 20.18
3	11 Na 22.99	12 Mg 24.31		13 Al 26.98	14 Si 28.09	15 P 30.97	16 S 32.06	17 Cl 35.45	18 Ar 39.45
4	19 Pk 39.10	20 Ca 40.08		31 Ga 69.72	32 Ge 72.59	33 As 74.92	34 Se 78.96	35 Br 79.90	36 Kr 83.80
5	37 Rb 85.47	38 Sr 87.62	transition elements	49 In 114.82	50 Sn 118.69	51 Sb 121.75	52 Te 127.90	53 I 126.90	54 Xe 131.30
6	55 Cs 132.91	56 Ba 137.33		81 Th 204.37	82 Pb 207.2	83 Bi 208.98	84 Po (209)	85 At (210)	86 Rn (222)
7	87 Fr (223)	88 Ra 226.03							

atomic number
symbol
relative atomic mass

periodic table

period (*n*) a horizontal row of the periodic table (p.119). It represents the gradual filling of s-orbitals (p.112) and p-orbitals (p.112), and for the transitional elements, the d-orbitals (p.112). The elements in a period change gradually from a characteristic (p.9) metal on the left of the period to a characteristic non-metal on the right. **periodic** (*adj*), **periodicity** (*n*).

periodicity (*n*) the regular occurrence (p.63) of similar chemical properties with increasing atomic number (p.113). This is chemical periodicity.

classify (*v*) to put materials, substances (p.8), objects, processes (p.157) into classes. In chemistry, to put elements into groups (p.119) of the periodic system; to put compounds into classes such as acids, alkalis, etc. **classification** (*n*).

period	transition elements									
4	21 Sc 44.96	22 Ti 47.90	23 V 50.94	24 Cr 52.00	25 Mn 54.94	26 Fe 55.85	27 Co 58.99	28 Ni 58.70	29 Cu 63.55	30 Zn 65.38
5	39 Y 88.91	40 Zr 91.22	41 Nb 92.91	42 Mo 95.94	43 Tc 98.91	44 Ru 101.07	45 Rh 102.91	46 Pd 106.4	47 Ag 107.87	48 Cd 112.41
6	57• La 138.91	72 Hf 178.49	73 Ta 180.95	74 W 183.85	75 Re 186.21	76 Os 190.2	77 Ir 192.22	78 Pt 195.09	79 Au 196.97	80 Hg 200.59
7	89•• Ac 227.03	104 Rf (261)	105 Hn (260)	106 (263)						

atomic number
symbol
relative atomic mass

•Lanthanide elements 57–71
••Actinide elements 89–92
transition elements

transition element an element in one of three periods (↑), each period containing 10 transitional elements. The first three periods of the periodic table (p.119) contain no transition elements, but periods 4, 5, and 6 have them. The transition elements are formed by the five d-orbitals (p.112) being filled in turn. The extranuclear (p.113) electrons cause the transitional elements to possess variable (p.218) electrovalencies (p.134) and to have the characteristics (p.9) of metals. These elements form complex ions (p.132) and many form coloured compounds.

s-block elements the elements of groups I and II. They have 1 or 2 s-electrons (p.112) in their outer shell (p.111).

p-block elements the elements of groups III to VIII inclusive. They have 1 – 6 p-electrons (p.112) in their outer shell (p.111).

d-block elements the transitional elements. They have s-electrons (p.112) in their outer shell (p.111) and 1 – 10 d-electrons (p.112) in the next inner shell. The number and arrangement of the d-electrons determine the nature of the element.

f-block elements an inner transition series. They have f-electrons in the outer shell and 1–14 electrons in an f-orbital in an inner shell.

transition element
extranuclear electrons of manganese

shell	s-electron	p-electrons	d-electrons
K	2		shell complete
L	2	2 2 2 / 6	
M	2	2 2 2 / 6	1 1 1 1 1 / 5
N	2		

current (*n*) electric charges (p.138) in motion form a current. A flow of electrons is the current of electric charge. Electric current is measured in amperes; it is said to flow from positive to negative.

conduct (*v*) to give direction to the flow of a fluid (p.11), electric current (↑) or heat, e.g. a pipe conducts water; a copper wire conducts an electric current. **conductor** (*n*), **conducting** (*adj*).

conductivity (*n*) a measure of the ability of a solution of given concentration to conduct electric current; its symbol is κ. The conductance of a solution is equal to 1/resistance; its symbol is G. $\kappa = \frac{G}{a}$ where a is a constant (p.106) for a particular electrolytic cell (↓).

electrolyte (*n*) a compound (p.8) which when dissolved in water will conduct (↑) electric current (↑). An electrolyte will also conduct electric current when molten (p.10). Acids, alkalis, inorganic salts (p.46) are generally electrolytes. The electrolyte is decomposed (p.65) by the current.

non-electrolyte (*n*) a compound (p.8) which when dissolved in a solvent (p.86) does not conduct electric current (↑). When molten, a non-electrolyte also does not conduct an electric current. Organic (p.55) compounds generally are non-electrolytes.

electrolytic cell a vessel in which electrolysis (↓) takes place. *See voltameter, p.129.*

electrolysis (*n*) the decomposition (p.65) of an electrolyte (↑) caused by passing an electric current through the solution. **electrolytic** (*adj*).

electrolytic (*adj*) describes any process (p.157) connected with electrolysis (↑).

electrode (*n*) a piece of material that conducts (↑) an electric current (↑) used in an electrolytic cell (↑) or voltameter (p.129) or simple cell (p.129). Electric current enters and leaves the solution of the electrolyte through the electrodes; the electrodes are connected (p.24) to a source (p.138) of electric current in an electrolytic cell or voltameter.

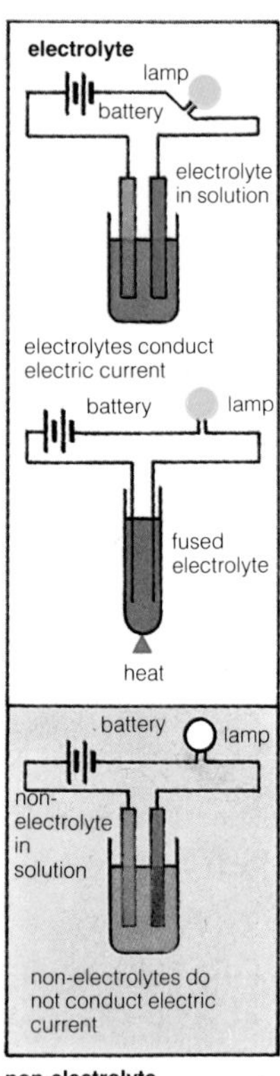

non-electrolyte

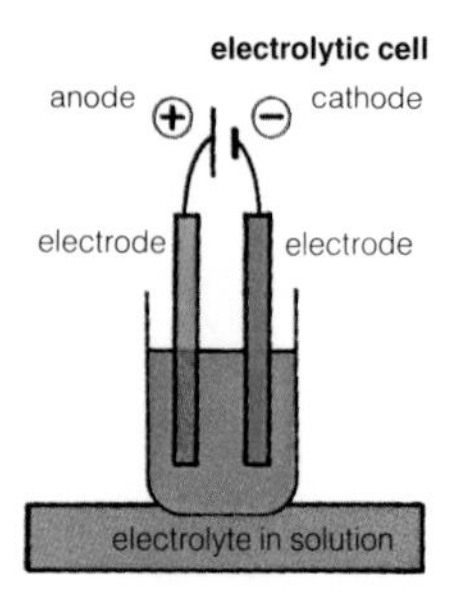

electrolytic cell

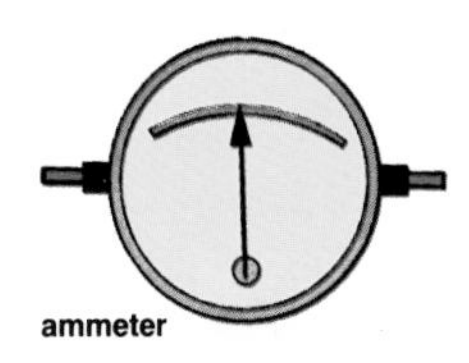
ammeter

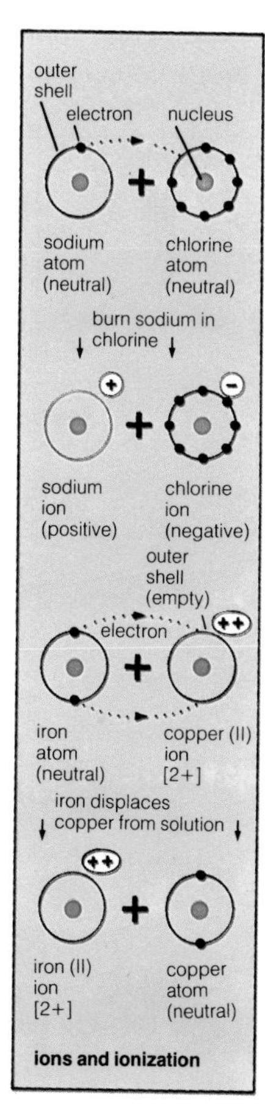

ions and ionization

ammeter (*n*) an instrument (p.23) for measuring the strength of an electric current (↑) in amperes.

anode (*n*) a positive electrode (↑); the electrode at which electric current is said to enter an electrolytic cell; the electrode from which electric current leaves a simple cell. **anodic** (*adj*).

cathode (*n*) a negative electrode (↑); the electrode from which current is said to leave an electrolytic cell; the electrode at which electric current flows into a simple cell after conduction by conductors.

electrolyze (*v*) to pass an electric current (↑) through a solution of an electrolyte (↑), or a molten (p.10) electrolyte, in order to decompose it, e.g. copper metal is obtained by electrolyzing a solution of copper (II) sulphate. **electrolysis** (*n*), **electrolytic** (*adj*).

ion (*n*) an ion is formed by an atom gaining or losing one or more electrons, or by an acid radical (p.45) gaining one or more electrons. An atom, or a group of atoms in an acid radical, is neutral; the loss or gain of electrons gives the ion an electric charge. Loss of electrons produces a positive charge; gain of electrons produces a negative charge. The magnitude of the charge depends upon the number of electrons concerned. The process of forming ions is ionization (↓). **ionize** (*v*).

ionization (*n*) the process of forming ions (↑). Ions are formed: (1) when energy is supplied to atoms; the energy can be heat or radiation; a high voltage (p.126) will also cause ions to be formed. (2) the attraction for electrons by atoms of some elements, e.g. a chlorine atom attracts one electron from a sodium atom. Crystals of electrolytes (↑) have a lattice consisting of ions; dissolving the crystals in water separates the ions, but this is not ionization. Hydrogen chloride is a covalent (p.136) compound, when it dissolves in water ionization takes place because the chlorine atom attracts an electron from the hydrogen atom. **ionize** (*v*), **ionized** (*adj*), **ionizing** (*adj*).

attract (*v*) to pull one object towards another object, e.g. a positive ion (p.123) attracts a negative ion towards itself; positive and negative electric charges attract each other. **attraction** (*n*), **attractive** (*adj*).

repel (*v*) to push one object away from another object, e.g. two positive ions (p.123) repel each other; two negative ions repel each other. **repulsion** (*n*), **repulsive** (*adj*).

repulsion (*n*) the force which causes objects to repel each other, e.g. a force of repulsion exists (p.213) between two positive electric charges and between two negative electric charges. **repulsive** (*adj*).

ionic theory the theory which explains the behaviour of electrolytes (p.122). It accounts for electrolysis (p.122) and the action of simple cells (p.129) by the existence (p.213) of ions (p.123) in all electrolytes.

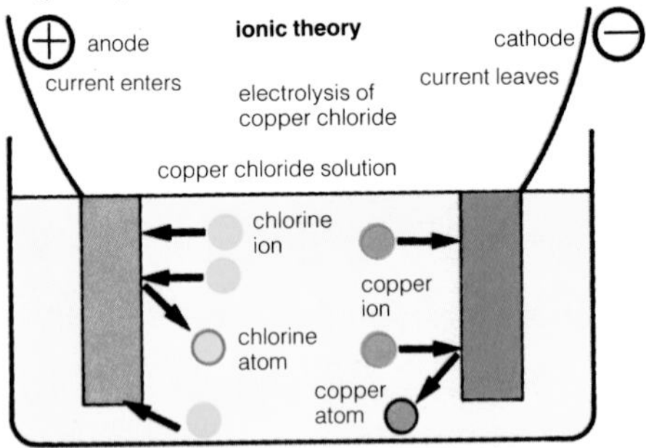

discharge (*v*) to give up an electric charge. When an ion (p.123) comes into contact (p.217) with an electrode (p.122), the ion either receives or gives electrons and becomes a neutral atom or acid radical (p.45); the ion has given its charge to the electrode and has become discharged. In the case of metals, it may be deposited on the electrode, in the case of non-metals, it is generally liberated (p.69) as a gas.

strength (*n*) a measure of the degree (p.227) of ionization (p.123) of an electrolyte (p.122) in a solution; also a measure of the characteristic reactivity (p.62) of acids (p.45) and alkalis (p.45).

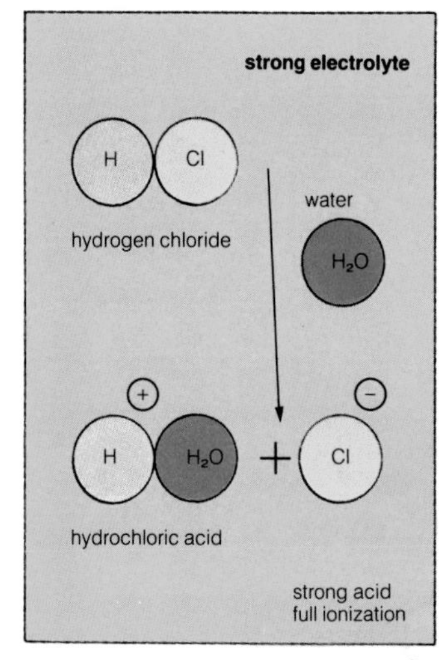

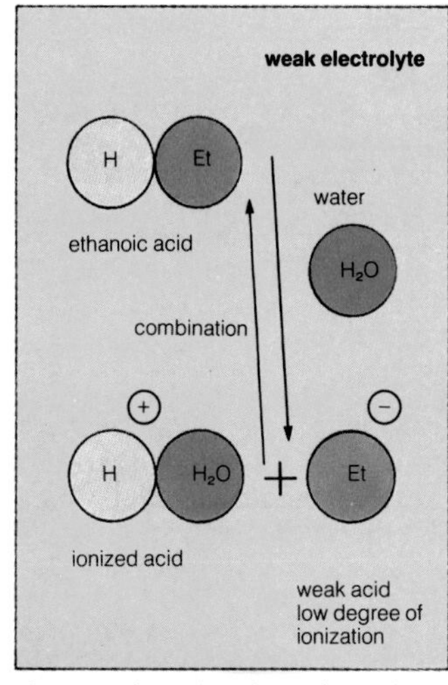

strong electrolyte an electrolyte (p.122) which is completely ionized (p.123) even in concentrated solutions. Strong acids (p.45) and strong alkalis (p.45) are strong electrolytes. Most inorganic salts are strong electrolytes, but *see hydrolysis*[1] *(p.66).* Strong electrolytes are good conductors (p.122) of electric current.

weak electrolyte an electrolyte which is only partly ionized (p.123) in water, or other ionizing solvents (p.86). At great dilutions the ionization becomes almost complete. Solutions of weak electrolytes are poor conductors (p.122) of electric current (p.122). Weak acids (p.45) and weak alkalis (p.45) are weak electrolytes. Examples of weak electrolytes are: ethanoic acid (and other organic acids); ammonia solution (a weak alkali).

anion (*n*) an ion which carries a negative charge; it is attracted to the anode (p.123) in electrolysis (p.122). Examples of anions are: chloride ion Cl^-; sulphate ion SO_4^{2-}; zincate (II) ion ZnO_2^{2-}.

cation (*n*) an ion which carries a positive charge; it is attracted to the cathode (p.123) in electrolysis (p.122). Examples of cations are: copper (II) Cu^{2+}; sodium Na^+; iron (III) Fe^{3+}.

voltage (*n*) a measurement in volts of the electrical force, or pressure, that drives an electric current (p.122) through an electrolyte (p.122) or conductor (p.122).

voltmeter (*n*) an instrument (p.23) that measures voltage (↑) in volts.

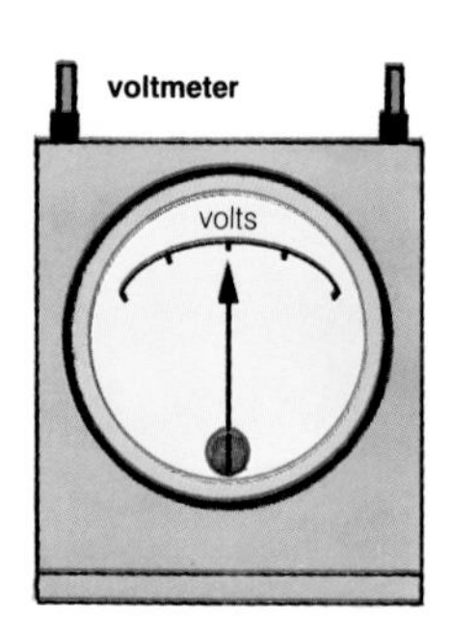

decomposition voltage the smallest voltage (↑) that will cause decomposition (p.65) of an electrolyte (p.122) by electrolysis (p.122) using platinum or other electrodes (↓). This voltage is necessary to overcome (p.213) the effects of polarization (↓) and overvoltage (↓). With some electrodes and electrolytes it is 0 volts.

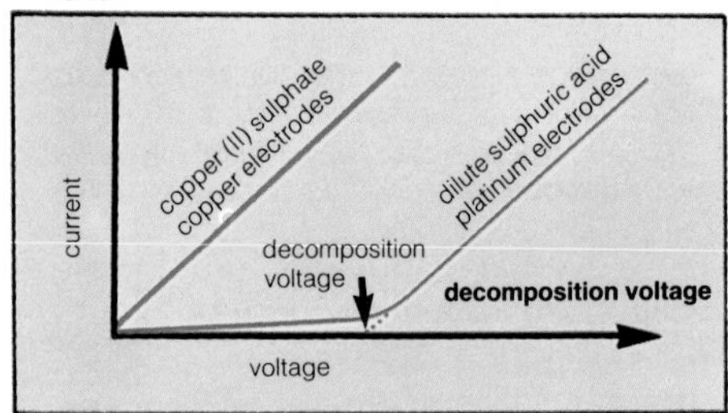

platinum electrode an electrode consisting of the metal platinum; such electrodes produce polarization (↓) and overvoltage (↓). *Platinized electrodes* consist of platinum electrodes with a layer of *platinum black*; such electrodes do not produce polarization.

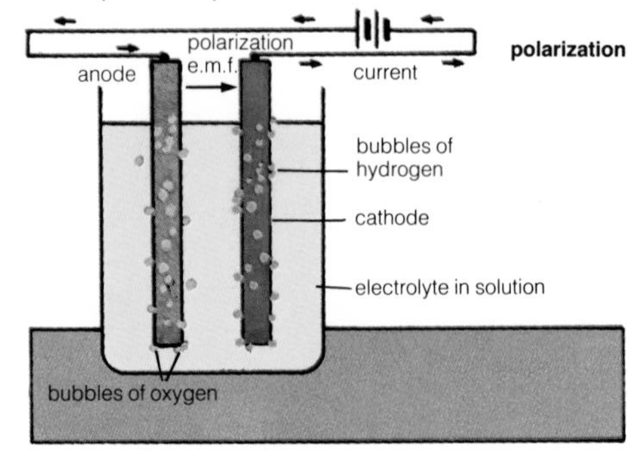

polarization (*n*) the production of hydrogen and oxygen gases which collect on an electrode, changing the nature of the electrode. In electrolysis (p.122) a primary cell (p.129) is formed with hydrogen and oxygen electrodes; this produces an *electromotive force* (p.129) acting against the electrolyzing current and causes a decomposition voltage (↑) to be necessary. In primary cells, the electromotive force produced by polarization acts against the electromotive force of the cell and reduces its voltage (↑).

overvoltage (*n*) the further voltage necessary to release (p.69) a gas at an electrode above its *electrode potential* (p.128). The magnitude of the overvoltage depends on the nature of the electrode and the gas being discharged (p.124). The cause of overvoltage has not been explained.

electrodeposit (*n*) a coat (↓) of metal on an electrode (p.122) produced during electrolysis by the discharge (p.124) of ions of the metal at the electrode. **electrodeposition** (*n*).

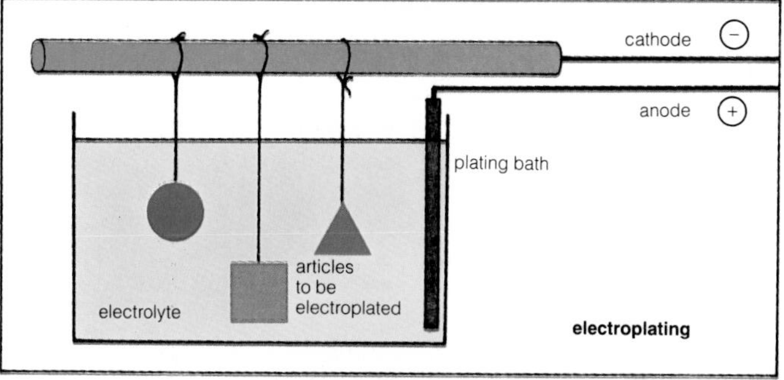

electroplating (*n*) the process (p.157) of putting an electrodeposit (↑) on an article. In electroplating with silver, a metal article is made the cathode (p.123) of a *plating bath* and a solution of a silver salt is used as the electrolyte. A thin coat of silver is formed on the article.

plate (*v*) to carry out electroplating (↑), especially with silver.

coat (*n*) a thin outer layer on an object. **coat** (*v*).

anodic (*adj*) describes anything happening at an anode (p.123) during electrolysis (p.122).

cathodic (*adj*) describes anything happening at a cathode (p.123) during electrolysis (p.122).

cathodic reduction a process (p.157) at a cathode (p.123) during electrolysis, in which a positive ion (p.123) gains one or more electrons from the cathode, e.g. hydrogen ions gain one electron to become hydrogen atoms, the ion is reduced (p.219) to the atom and discharged (p.124).

anodic oxidation a process (p.157) at an anode (p.123) during electrolysis (p.122), in which a negative ion (p.123) loses one or more electrons to the anode. A metal anode may go into solution, as its atoms can lose one or more electrons to become positive ions, e.g. negative ions:
$2Cl^- \rightarrow Cl_2 + 2e$ (chlorine discharged)
atoms:
$Cu \rightarrow Cu^{2+} + 2e$ (electrode goes into solution).

anodize (*v*) to form a layer of oxide (p.48) on a metal by electrolysis (p.122). The metal is made the anode (p.123) of an electrolytic cell (p.122), a suitable electrolyte is used and oxygen is discharged (p.124) at the anode forming a coat of the oxide of the metal. With aluminium, dyes (p.162) can be absorbed in the oxide, producing coloured surfaces. **anodizing** (*n*).

electrochemical (*adj*) describes any effect concerned with the electrical properties of solutions (p.86) and the ions (p.123) in solution.

electrochemical equivalent the mass in grams of an element liberated (p.69) during electrolysis (p.122) by 1 coulomb of electric charge. (1 coulomb of charge is the quantity of charge from a current of 1 ampere flowing for 1 second).

electrode potential when a metal is put in a solution containing ions (p.123) of the metal, a voltage (p.126) difference forms between the metal and the solution; this is an electrode potential. An electrode potential is also formed between a metal and any electrolyte. Two metals will have two electrode potentials and the difference between these potentials produces the electromotive force (↓) of a primary cell.

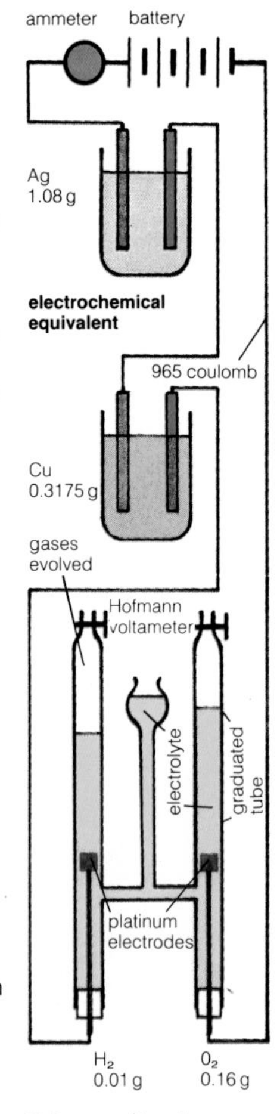

Hofmann voltameter

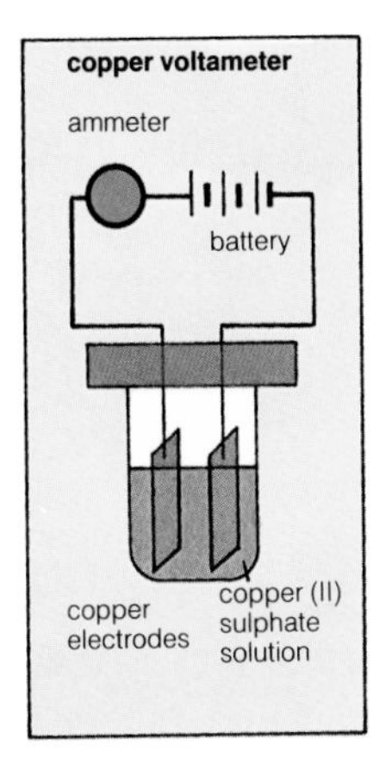

series[1] a way of connecting (p.24) parts of an electric circuit so that the current flows through each part in turn. The parts are said to be connected in series.

parallel (*n*) a way of connecting (p.24) parts of an electric circuit so that the current divides to go through each part at the same time. The parts are said to be connected in parallel.

voltameter (*n*) an apparatus for determining the electrochemical equivalent (↑) of an element. The electrodes are large metal plates, weighed before and after passing a measured electric current for a measured time. A Hoffman voltameter is used to determine the electrochemical equivalent of gases. *See electrolytic cell (p.122).*

faraday (*n*) the quantity of electric current that liberates (p.69) or forms 1 mole of monovalent (p.137) ions. Its value is 96 487 coulombs. *See electrochemical equivalent (↑).* One faraday of electric charge contains one mole of electrons.

inactive electrode an electrode which will not give ions (p.123) to the electrolyte (p.122) in an electrolytic cell (p.122), e.g. carbon and platinum are inactive electrodes.

active electrode an electrode which can give ions (p.123) to the electrolyte (p.122) in an electrolytic cell (p.122) if it is the anode. The cathode is not so important. For example, copper forms an active anode and may produce copper ions during electrolysis (p.122).

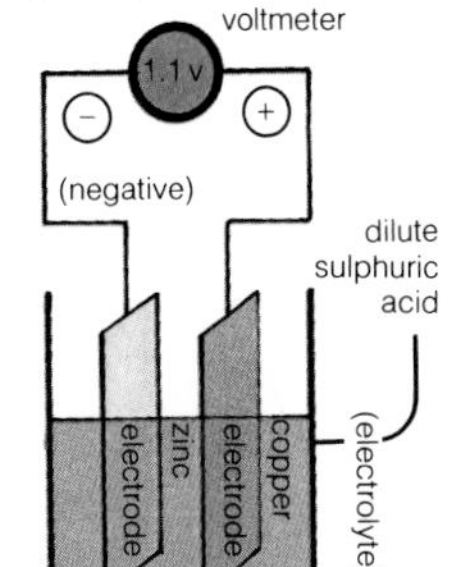

primary cell a source (p.138) of electric current. It is formed by putting two different metals into an electrolyte. The source of the electrical energy is the *electrode potentials* (↑) between each of the electrodes and the electrolyte in the cell. For example, a simple voltaic cell has an electrode of zinc and an electrode of copper in an electrolyte of dilute sulphuric acid; it has an electromotive force of 1.1 volt.

electromotive force the electrical force of a primary cell (↑) which drives current through a connected circuit. The symbol for electromotive force is *E*, and the unit for it is the volt. The electromotive force is produced by the two *electrode potentials* (↑) of the cell. (*abbr.*) **e.m.f.**

hydrogen electrode a platinum electrode (p.126) is platinized with platinum black and hydrogen gas is put in contact with the electrode at standard pressure (p.102). This electrode is placed in a 1 M (p.88) solution of hydrogen ions (p.123) at 25°C to form a hydrogen electrode. The *standard electrode potential* (↓) of the hydrogen electrode is fixed at 0 volts.

standard electrode potential the electrode potential (p.128) of a metal, or a gas, in contact with 1 M (p.88) solution of its ions at 25°C; the magnitude is determined (p.222) against the standard hydrogen electrode (↑) of 0 volts.

electrochemical series a list of elements in order of their standard electrode potentials (↑) using a scale with the hydrogen electrode (↑) as zero. In this series, a metal such as zinc has a negative electrode potential as in a primary cell with a zinc and a hydrogen electrode, the zinc would be the cathode. *See below for the electrochemical series*. Similarly, in a primary cell with a chlorine and a hydrogen electrode, the chlorine would be the anode and so chlorine has a positive electrode potential.

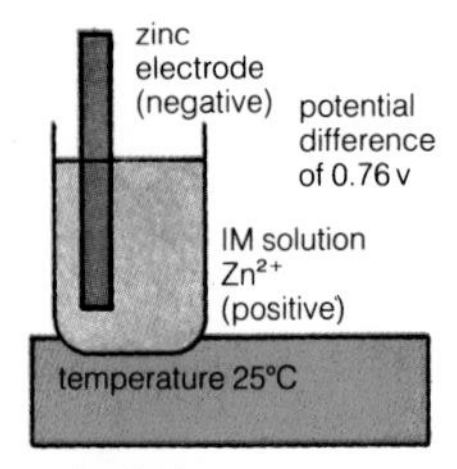

standard electrode potential

electrochemical series

element	electrode potential	ionic reaction
potassium	−2.92 v	$K^+ + e \rightarrow K$
calcium	−2.87 v	$Ca^{2+} + 2e \rightarrow Ca$
sodium	−2.71 v	$Na^+ + e \rightarrow Na$
magnesium	−2.38 v	$Mg^{2+} + 2e \rightarrow Mg$
aluminium	−1.66 v	$Al^3 + 3e \rightarrow Al$
zinc	−0.76 v	$Zn^{2+} + 2e \rightarrow Zn$
iron	−0.44 v	$Fe^{2+} + 2e \rightarrow Fe$
tin	−0.14 v	$Sn^{2+} + 2e \rightarrow Sn$
lead	−0.1 v	$Pb^{2+} + 2e \rightarrow Pb$
hydrogen	0 v	$H^+ + e \rightarrow H$
copper	+0.34 v	$Cu^{2+} + 2e \rightarrow Cu$
silver	+0.8 v	$Ag^+ + e \rightarrow Ag$

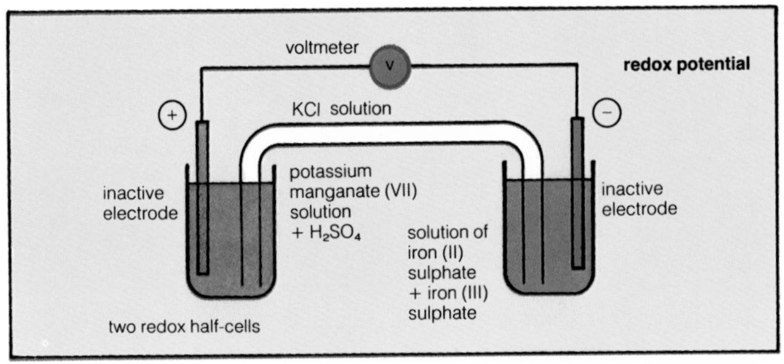

redox potential a metal acts as a reducing agent (p.71) and a non-metal, such as chlorine, acts as an oxidizing agent (p.71). The electrochemical series (↑) shows the relative (p.232) strengths of these agents. By using the apparatus of redox half cells, other ions can be used to determine (p.222) their electrode potential and these potentials are called redox potentials. Redox potentials show whether an agent is able, or not, to reduce or to oxidize an ion. For example, the redox potential of chlorine to chlorine ions is + 1.36 V; the redox potential of $(Cr_2O_7)^{2-}$ is 1.33 V and of $(MnO_4)^-$ is + 1.52 V. Hence potassium manganate (VII) will oxidize chlorine ions to chlorine, but potassium dichromate (VI) will not oxidize chlorine ions.

redox series a list of redox reagents, which include the electrochemical series, in order of their redox potentials, measured at 25°C, in 1 M (p.88) solutions, forms the redox series. Hydrogen is given a redox potential of zero.

redox series	
$Sn^{4+} + 2e \rightarrow Sn^{2+}$	+0.15 v
$Cu^{2+} + 2e \rightarrow Cu$	+0.34 v
$2H_2O + O_2 + 4e \rightarrow 4OH^-$	+0.44 v
$Fe^{3+} + e \rightarrow Fe^{2+}$	+0.77 v
$Cr_2O_7^{2-} + 14H_2^+ + 6e \rightarrow 2Cr^{3+} + 7H_2O$	+1.33 v
$Cl_2 + 2e \rightarrow 2Cl^-$	+1.36 v
$MnO_4^- + 8H^+ + 5e \rightarrow Mn^{2+} + 4H_2O$	+ 1.52 v

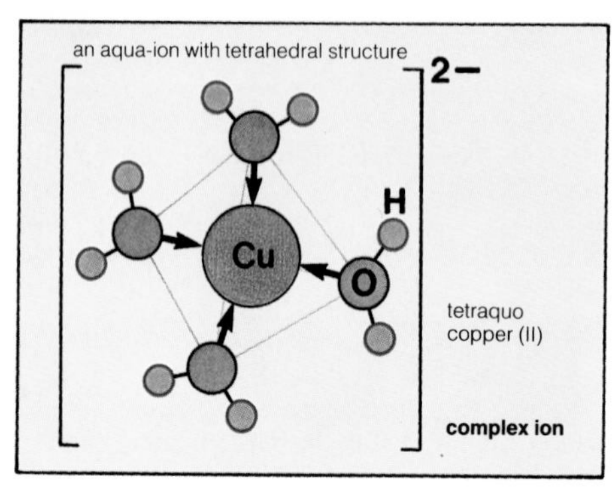

complex ion an ion (p.123) which has molecules (p.77) joined to it by coordinate bonds (p.136). A complex cation (p.125) has molecules or atoms such as water, a halogen, or cyanogen (CN) joined by coordinate bonds to the metal ion or to hydrogen and carries a positive charge, e.g. tetrammine copper (II) ion $(Cu(NH_4)_4)^{2+}$, tetrachlorodiaquo chromium (III) $(CrCl_4(H_2O)_2)^{3+}$. A complex anion (p.125) has molecules or atoms such as water, a halogen or cyanogen (CN) joined by coordinate bonds to a central metal atom and carries a negative charge, e.g. hexacyanoferrate (II) $(Fe(CN)_6)^{2-}$; trichlorocuprate (I) $(CuCl_3)^{2-}$. The central metal atom or ion in complex ions is usually a transitional element (p.121).

aqua-ion (*n*) a complex ion (↑) with water molecules joined to a metal ion by coordinate bonds, e.g. tetraquo copper (II) $Cu(H_2O)_4^{2+}$

hydroxyl ion the ion formed from the hydroxide radical (p.45). The ion produces the alkaline properties in a solution; it is discharged at an anode to produce oxygen:

$$4\,OH^- \rightarrow 2\,H_2O + O_2 + 4\,e$$

transfer (*v*) to move objects, materials, or energy from one place to another, without saying how they are moved, e.g. to transfer an electron from an ion to an electrode; to transfer a precipitate from a filter paper to a beaker. **transference** (*n*), **transferable** (*adj*).

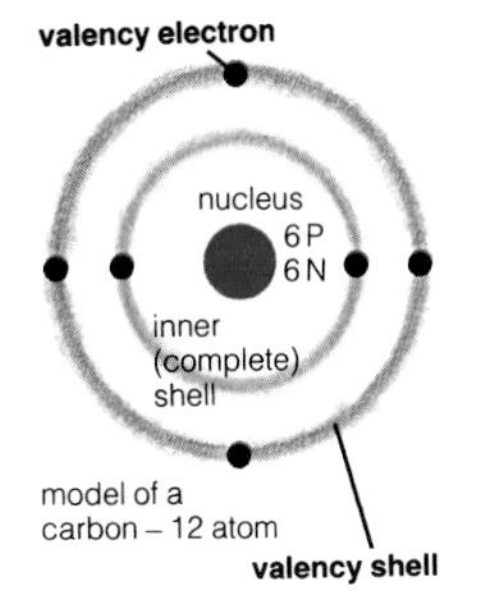

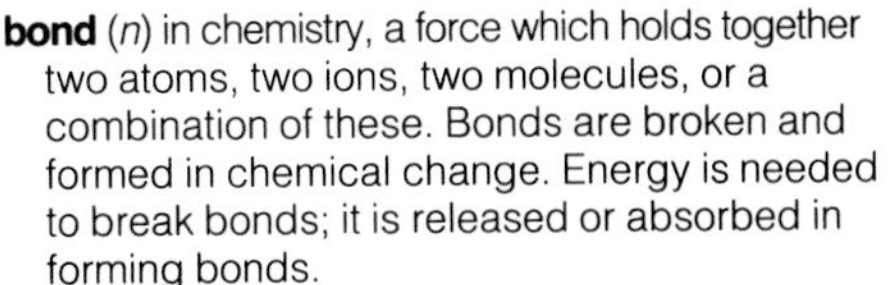

bond (*n*) in chemistry, a force which holds together two atoms, two ions, two molecules, or a combination of these. Bonds are broken and formed in chemical change. Energy is needed to break bonds; it is released or absorbed in forming bonds.

valency (*n*) the ability of an atom, and hence an element, to form bonds (↑), e.g. carbon, valency 4, can form 4 bonds. *Valency* has now been replaced (p.68) by *electrovalency* (p.134) and *covalency* (p.136).

valency electron an electron (p.110) which takes part in chemical bonds (↑).

valency shell the outside shell (p.111) of an atom which contains valency electrons (↑).

electron pair an orbital (p.111) can hold two electrons (p.110) and the orbital is then full. An orbital in a valency shell (↑) can lose electrons or gain electrons to form an ion (p.123). An orbital can gain an electron by sharing an electron from another orbital; this forms a covalent bond. Full orbitals are stable (p.74).

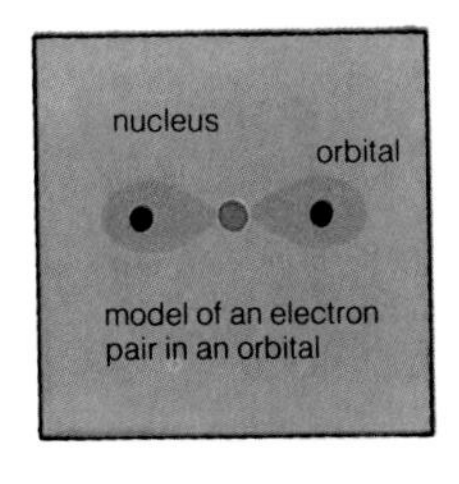

lone pair an orbital (p.111) containing two electrons (p.110) when the orbital is in a valency shell (↑). A lone pair can take part in a coordinate bond (p.136), e.g. nitrogen has 5 electrons in its valency shell, 2 electrons in one orbital forming a lone pair and 3 electrons, each in one orbital, available (p.85) to form bonds (↑).

octet (*n*) except for hydrogen and helium, the outer shell (p.111) of an atom has one s-orbital (p.112) and three p-orbitals (p.112). When these 4 orbitals each have 2 electrons, the orbitals are full and no electrons are available (p.85) to form bonds (↑). No chemical reactions can take place and the structure (p.82) is stable (p.74). An octet (8 electrons) structure is found in the noble gases (group VIII); this structure does not form bonds, the gases are inert.

shared electron

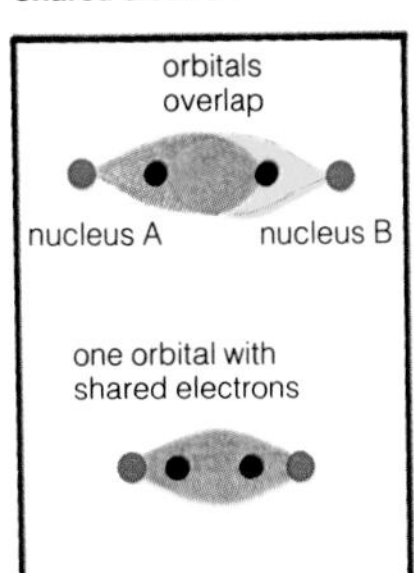

shared electron two orbitals (p.111), each with one electron (p.110), overlap (p.218). Each orbital shares its electron with the other orbital. In this way each orbital has its own electron and a shared electron, so each orbital can be considered to have two electrons.

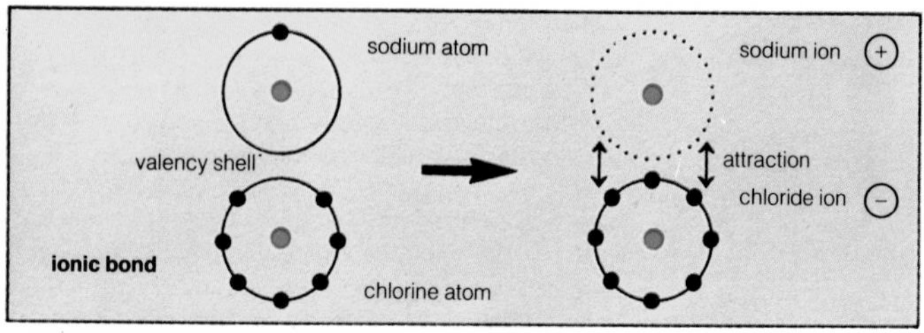

ionic bond a kind of chemical bond (p.133) formed between ions (p.123). It is usually formed between the ions of a metal and a non-metal (p.116). The metal atom loses one or more electrons (p.110) to form a positive ion and the non-metal atom gains one or more electrons to form a negative ion, e.g. sodium burns in chlorine to form sodium chloride; a sodium atom loses 1 electron to form a sodium ion and a chlorine atom gains 1 electron to form a chloride ion. In solution the ions are free; in a crystal (p.91) the ions are held together by the attraction (p.124) of a positive charge for a negative charge and the ions form a giant structure (p.94). Attraction between opposite charges forms an ionic bond.

electrovalent bond another name for an ionic bond (↑).

electrovalency (*n*) the number of ionic bonds (↑) an atom can form. This is equal to the number of electrons (p.110) which it can lose or gain in its valency shell (p.133) to form an ion. The electrovalency of an atom is also the electrovalency of the element. For example, a magnesium atom has two electrons in its valency shell, both can be removed (p.215) to form an ion with 2 positive charges, so magnesium has an electrovalency of $+2$ (this shows the ion is positive). A chlorine atom has 7 electrons in its valency shell and it can gain one electron to form an ion with 1 negative charge, so chlorine has an electrovalency of -1 (this shows the ion is negative). In both cases (magnesium and chlorine) the atoms form ions with a stable (p.74) octet (p.133) in the outside shell.

oxidation state this is determined by the charge on an ion (p.123) of an element (p.116). Copper, with an electrovalency (↑) of + 2, forms an ion Cu^{2+}; the oxidation state of copper is +2. Copper with an electrovalency of + 1 forms an ion Cu^{+}, the oxidation state is + 1. Similarly, the oxidation state of Cl^{-} is − 1 and of S^{-2} is − 2. The oxidation state indicates (p.38) the number of electrons that have been removed (p.215) from, or added to, an atom to form an ion, e.g. if the oxidation state is + 3, then three electrons have been removed from a neutral atom. *See oxidation number (p.78).*

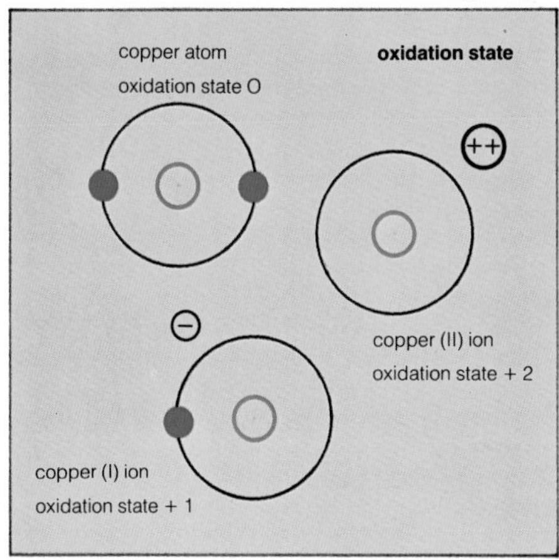

energy (*n*) energy can be used to do work. Chemical energy is available (p.85) in a material, or substance (p.8), from the energy stored in its chemical bonds (p.133). This energy is set free when the bonds are broken or new bonds made. Chemical energy is often transformed (p.144) into heat energy in a chemical reaction, e.g. in combustion (p.58); the heat energy can be used to do work as in a motor car engine. Chemical energy is transformed into electrical energy in a primary cell (p.129); the electrical energy can be used to do work. **energize** (*v*).

covalent bond a kind of chemical bond (p.133) formed between atoms (p.110) by shared electrons (p.133). Covalent bonds are usually formed by p-electrons (p.112) in p-orbitals (p.112). These orbitals are directed in space, so covalent bonds are directed in space and account for the shape of molecules. After an atom has formed covalent bonds, it has a complete octet (p.133) of electrons. Compounds with covalent bonds form molecular crystals (p.94) or very hard giant structures (p.94).

covalency (*n*) the number of covalent bonds (↑) an atom (p.110) can form. This is equal to the number of electrons (p.110) which are single electrons in orbitals (p.111); such electrons are available (p.85) to become shared electrons (p.133), e.g. oxygen has a covalency of 2 because it has two unpaired electrons (p.110) able to form two covalent bonds. **covalent** (*adj*).

bond energy the energy given out, or taken in, when a covalent bond is formed between two free atoms. It is also the energy needed to break the bond; and is thus a measure of the strength of the bond, e.g. the bond energy of the C–H bond is 415 kJ mol^{-1}

coordinate bond a kind of covalent bond (↑) formed between two atoms when one atom gives both electrons to form the shared electrons (p.133) of the bond. The electrons that are given are a lone pair (p.133). The ammonium radical is formed by a coordinate bond between the nitrogen atom of ammonia and a hydrogen ion, *see diagram*.

dative bond another name for coordinate bond (↑), not used now.

dative covalent bond another name for coordinate bond (↑), not used now.

semipolar bond another name for coordinate bond (↑), not used now.

donor (*n*) the atom which gives a lone pair in a coordinate bond (↑).

acceptor (*n*) the atom which receives a lone pair in a coordinate bond (↑).

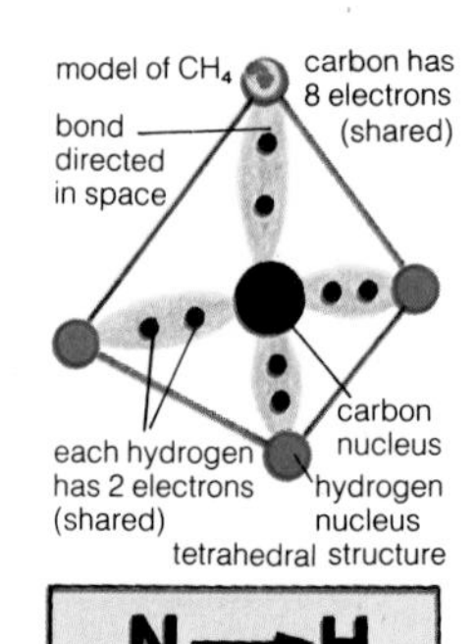

co-ordinate bond

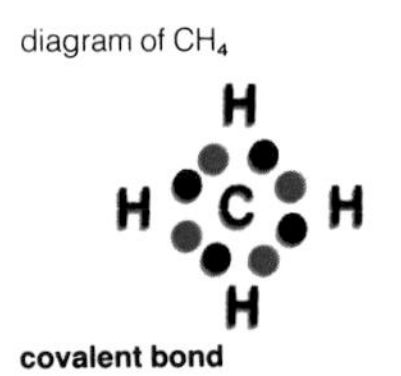

covalent bond

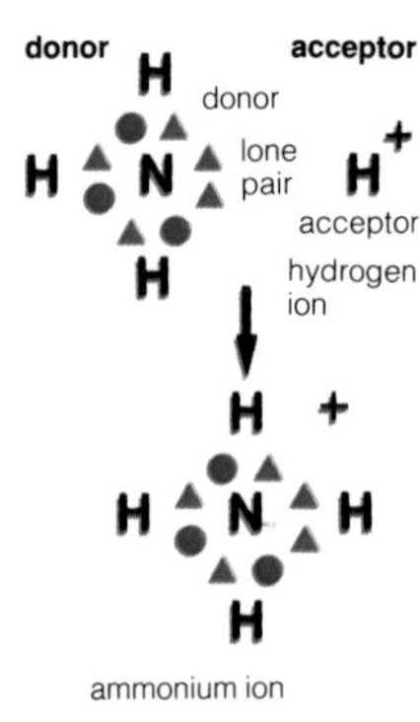

ammonium ion

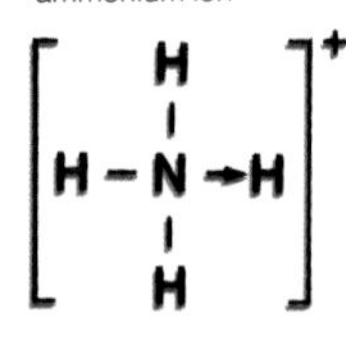

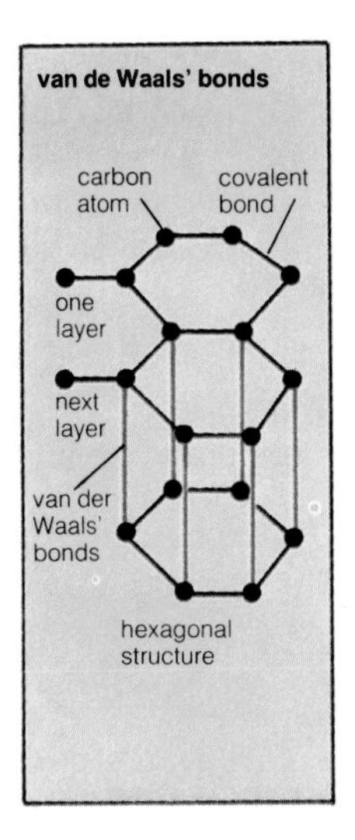

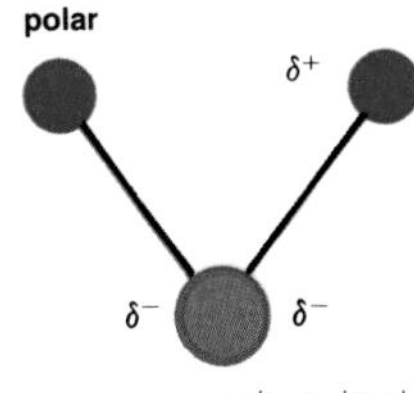

polar molecule of water (H_2O)

metallic bond a kind of chemical bond (p.133) found in metal crystals (p.95). The crystal consists (p.55) of positive ions of the metal in a lattice (p.92) with the valency electrons (p.133) free to move between the metal ions. The moving electrons form the metallic bond.

van der Waals' bond a weak chemical bond (p.133) which holds molecules together. It is an attractive (p.124) force arising from the movement of electrons in atoms and is ten to twenty times weaker than the attractive force between ions. *See molecular crystals (p.94).* Graphite is an example of molecules held by van der Waals' bonds. The carbon atoms form hexagons in layers, held by strong covalent bonds (↑). The layers are held together by van der Waals' bonds so that the layers slip over each other, explaining the softness of graphite.

polar (*adj*) describes a molecule (p.77) in which the covalent bonds (↑) have the pair of electrons forming the bond nearer to one atom than another. This makes one atom slightly negative and the other atom slightly positive, each with a fraction of the charge on an electron (shown by $\delta+$ or $\delta-$), e.g. the water molecule is polar, *see diagram*. If the substance is liquid, a polar solvent (p.86) is formed. Polar solvents dissolve ionic (p.123) compounds.

non-polar (*adj*) describes a molecule (p.77) which is not polar (↑). If the substance is liquid, a non-polar solvent (p.86) is formed. Non-polar solvents dissolve organic (p.55) compounds, which are generally non-polar themselves.

monovalent (*adj*) describes an element with a valency of 1; it is better described as monoelectrovalent or monocovalent.

divalent (*adj*) describes an element with a valency of 2; it is better described by the electrovalency or covalency.

tervalent (*adj*) describes an element with a valency of 3; it is better described by the electrovalency or covalency.

radioactivity (*n*) the property of spontaneous (p.75) nuclear (p.110) changes in which energy is released as radiation (↓) and a new nucleus is formed with a different number of protons and, in some cases, a different number of neutrons. **radioactive** (*adj*).

radiation (*n*) (1) a process in which energy is passed on in the form of a wave motion, with electromagnetic waves. Examples of radiation are light, ultra-violet light, X-rays, radio waves. (2) a form of energy, travelling in straight lines and causing ionization (p.123) of any material through which it passes, **radiant** (*adj*).

radioactive (*adj*) describes an element, or one of its compounds, which has the property of radioactivity (↑).

natural radioactivity the radioactivity (↑) of elements with atomic numbers (p.113) greater than 83. Such elements have compounds which occur (p.63) naturally (p.19).

artificial radioactivity the radioactivity of elements which have been made radioactive by bombardment (p.143) with particles (p.13) of high energy. Such elements have atomic numbers of 83 or less.

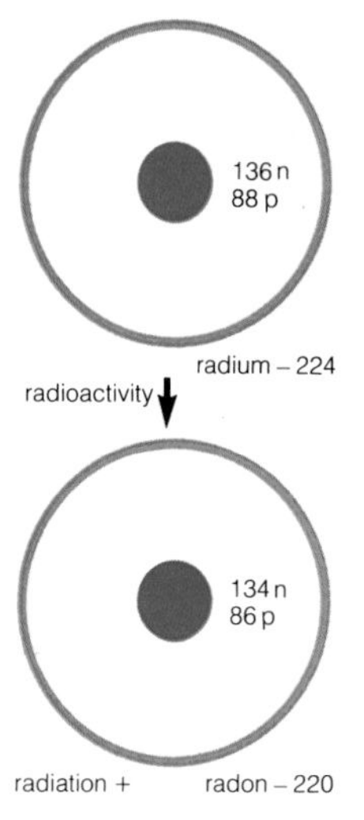

radioactivity 1

radioactive element
source emits rays
source
rays of radiation

radioactivity 2

charge (*n*) no description of charge can be made other than it exists (p.213) in two forms, *positive* and *negative*. It is a property of sub-atomic particles (p.13); an electric current is a flow of charge. **charged** (*adj*).

ray (*n*) (1) a line which shows the direction of radiation (↑). (2) a line of radiant (↑) energy. (3) a stream of charged (↑) particles (p.13).

source (*n*) a place from which something, such as radiation (↑), or an ore (p.154), has come, e.g. a radioactive element is a source of radiation; a primary cell is a source of electric current.

emit (*v*) to give out radiation, gas, odour, sound, from a source (↑), e.g. a bell emits sound when hit; a radioactive element emits radiation, **emission** (*n*), **emissive** (*adj*), **emitter** (*n*).

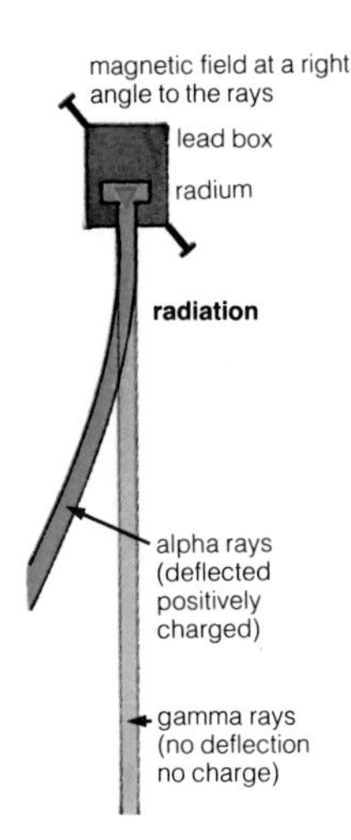

a radioactive source emits alpha rays or alpha rays and gamma rays

Becquerel rays the rays (↑) emitted by a radioactive source. The kind of ray is not described.

alpha ray a ray (↑) consisting (p.55) of alpha particles (↓).

beta ray a ray (↑) consisting (p.55) of beta particles (↓).

gamma ray a ray (↑) of gamma radiation (↓).

alpha particle the nucleus of a helium atom consisting (p.55) of two protons and two neutrons; it has a relative atomic mass of 4.0029 and a charge of + 2. An alpha particle has a range (p.140) in air of 7 cm and ionizes (p.123) gases through which it passes. Alpha particles cause scintillation (p.140) when they hit fluorescent (p.140) surfaces.

α-particle another way of writing alpha particle (↑).

beta particle an electron (p.110) emitted (↑) from the nucleus of a radioactive atom; it has a very high speed, up to 99% of the speed of light. Loss of an electron from a nucleus results in the nucleus increasing its number of protons by 1 and decreasing its number of neutrons by 1. A beta particle has a range (p.140) in air of 750 cm and can penetrate (p.144) thin pieces of metal foil. It ionizes (p.123) gases through which it passes but the effect is weaker than for that of alpha particles (↑).

ß-particle another way of writing beta particle.

gamma radiation radiation (↑) emitted by radioactive (↑) elements; it has wavelengths shorter than those of X-rays. Gamma radiation is emitted with either alpha or beta radiation. The radiation is very penetrating (p.144) with a range of 150 mm in lead; it carries no charge, so is not deflected by electric or magnetic fields. The wavelength of a radiation is a characteristic (p.9) of the nucleus emitting it. Gamma radiation has a very weak ionizing (p.123) effect.

ϒ-radiation a way of writing gamma radiation.

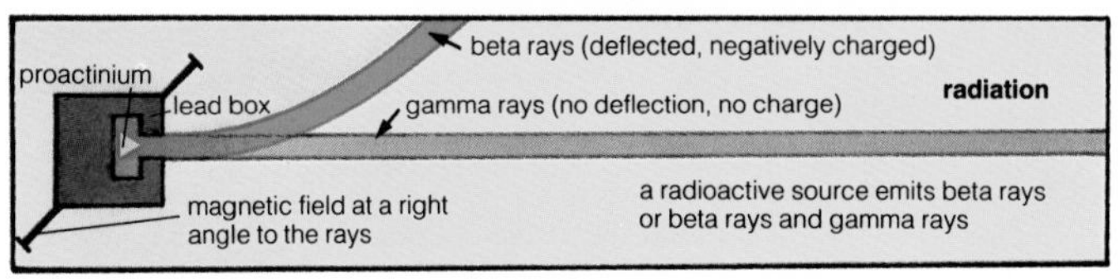

a radioactive source emits beta rays or beta rays and gamma rays

alpha emission the emission of an alpha particle (p.139) from the nucleus of a radioactive (p.138) atom. The resulting nucleus has an atomic number (p.113) which is 2 less and a mass number (p.113) which is 4 less, e.g.

$$^{220}_{86}Rn \rightarrow {}^{216}_{84}Po + {}^{4}_{2}He \ (\alpha\text{-particle})$$

Alpha emission may be accompanied (p.213) by gamma radiation; it is never accompanied by beta emission (↓).

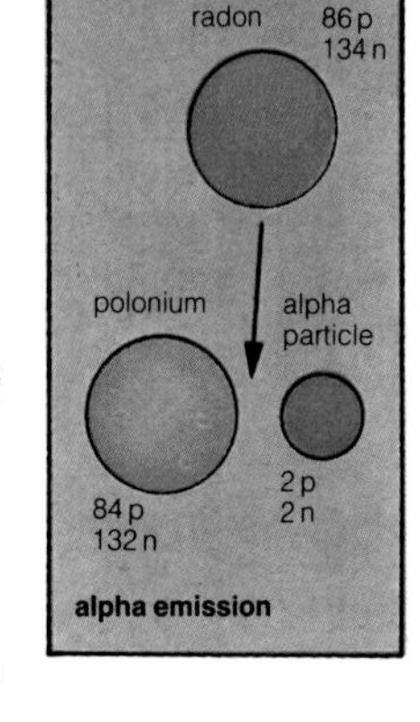

beta emission the emission of a beta particle (p.139) from the nucleus of a radioactive (p.138) atom. The resulting nucleus has an atomic number (p.113) which is 1 more, but the mass number (p.113) remains the same, e.g.

$$^{228}_{88}Ra \rightarrow {}^{228}_{89}Ac + {}^{0}_{-1}e$$

In the nucleus, a neutron change produces a proton and an electron; the proton remains and the electron is emitted. Beta emission may be accompanied (p.213) by gamma radiation; it is never accompanied by alpha emission (↑).

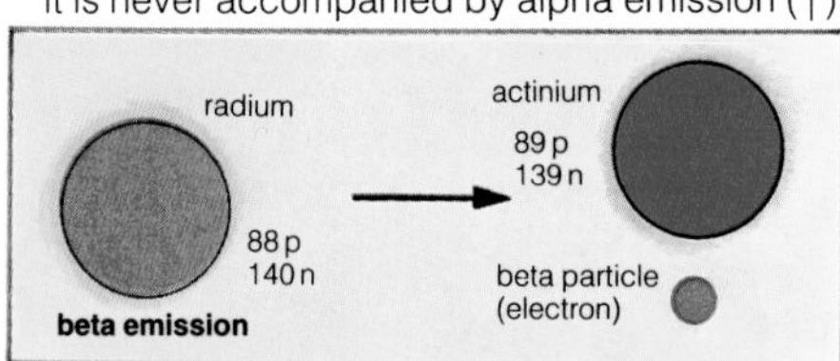

range (*n*) (1) the greatest distance an object can travel when it has been given energy, e.g. the range of an aeroplane on a tank full of fuel (p.160). (2) a set of values between an upper and lower limit, e.g. the temperature range between 0°C and 100°C. **range** (*v*).

scintillation (*n*) very short, bright, quick flashes of light. **scintillate** (*v*).

fluorescent (*adj*) a material which emits light when radiation, of a shorter wavelength than light, falls on the material. **fluorescence** (*n*).

spinthariscope (*n*) a device consisting of a surface coated (p.127) with zinc sulphide. When alpha particles hit the surface, it scintillates (↑). The number of scintillations can be counted, giving the number of alpha particles emitted by a radioactive source (p.138).

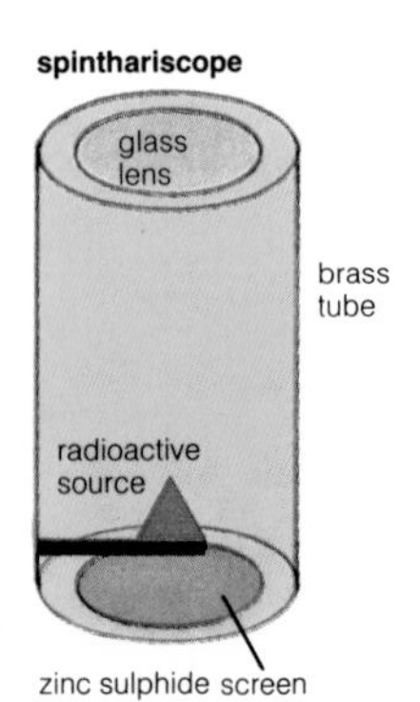

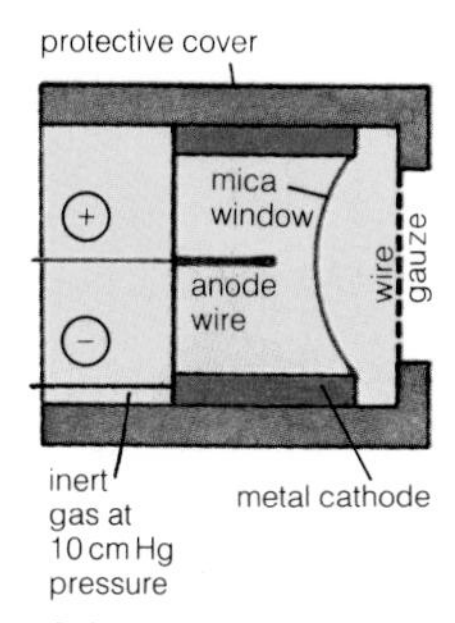

Geiger counter

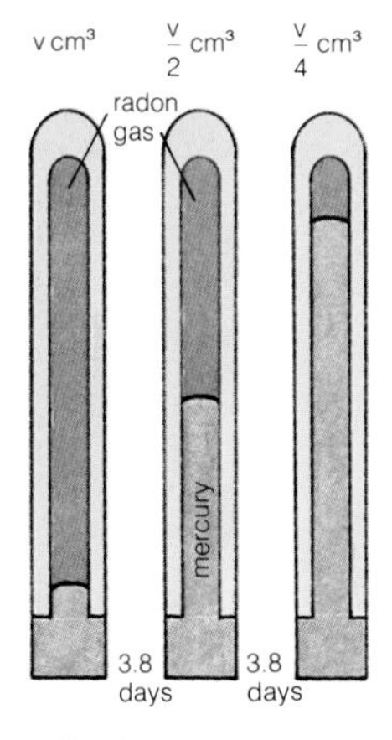

radioactive decay of radon
half-life period of 3.8 days

half-life

Geiger counter a device for the detection (p.225) of ionizing (p.123) radiations. It consists (p.55) of a metal tube, with a window, and a wire electrode (p.122) in the centre of the tube. The tube is filled with argon (an inert gas) at a low pressure. There is a high electric potential between the electrode and the tube, with the electrode made positive. When radiation enters through the window, the argon atoms are ionized and the electrode is discharged, giving a pulse of current, which can be detected.

radioactive disintegration the disintegration (p.65) of the nucleus of a radioactive element into two parts. The disintegration usually results in the formation of another nucleus and an alpha or beta particle.

radioactive decay the decrease, with time, of radioactivity (p.138) from a specimen of a radioactive element.

half-life (*n*) the time taken for one half of the atoms in a specimen of a radioactive element to disintegrate (↑). It is an important characteristic (p.9) of a radioactive element, and is a constant for that element, e.g. examples of half-lives are: uranium-238, 4.5×10^9 years; radium-226, 1620 years; radium-221, 30 seconds. The half-life period is independent of temperature, pressure, concentration and nature of the material.

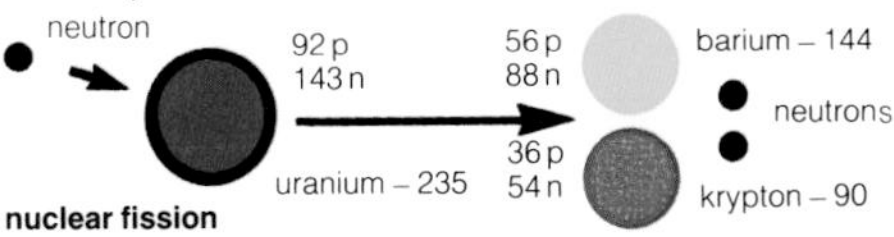

nuclear fission

nuclear fission the disintegration of a nucleus into smaller nuclei, brought about by bombardment (p.143) with neutrons, e.g. uranium-238 bombarded by slow neutrons splits into barium-144 and krypton-90, releasing two neutrons:

$$^{235}_{92}U + ^{1}_{0}n \rightarrow ^{144}_{56}Ba + ^{90}_{36}Kr + 2^{1}_{0}n$$

In this process, energy is released; 1 kg of uranium by nuclear fission produces approximately 10^{10}kJ.

fissile (*adj*) describes any element, or substance, which can undergo fission, particularly nuclear fission (↑).

nuclear fusion the combination of two nuclei of small mass number to produce one nucleus of a higher mass number; the nuclear reaction takes place only at very high temperatures (about 10^8 K). A large quantity of energy is released, e.g. $^2_1H + ^1_1H \rightarrow ^3_2He + 5 \times 10^5$ MJ. Deuterium bombarded with protons forms helium-3, releasing large quantities of energy.

curie (*n*) the unit for the rate of radioactive disintegration (p.141). 1 curie = 3.7×10^{10} atoms disintegrating per second. 2996 kg uranium-238 produce 3.7×10^{10} atomic disintegrations per second, hence this mass of uranium has an activity of 1 curie. The safe limit of radioactivity for human beings is estimated to be 10 microcuries. The symbol for curie is Ci.

radioactive series one of three groups of naturally occurring radioactive (p.138) elements, each series named after the element with which the series starts. The uranium series begins with uranium-238 and passes, by 15 nuclear changes, to lead-206, which is stable (p.74). The thorium series begins with thorium-232 and passes by 11 nuclear changes to lead-208, which is stable. The actinium series starts with protoactinium-231 and passes through 9 nuclear changes to lead-207, which is stable.

disintegration series alternative name for radioactive series (↑).

Fajans and Soddy law the emission of an alpha particle during radioactive change produces an element two places to the left in the periodic table (p.119). The emission of a beta particle, however, produces an element one place to the right in the periodic table.

nuclide (*n*) an atomic species which is defined by the number of protons and neutrons in the nucleus, and by its kind of radioactive decay and the half-life (p.141) of the rate of decay, e.g. radium-221 has 88 protons and 133 neutrons in its nucleus, it decays by alpha-emission with a half-life of 30 seconds. Radium-223 has 88 protons and 135 neutrons in its nucleus, it decays by alpha-emission accompanied (p.213) by gamma-radiation and has a half-life of 11.7 days.

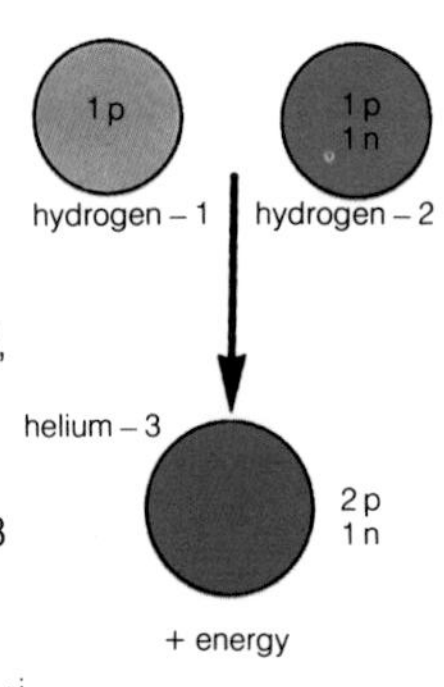

atomic fusion

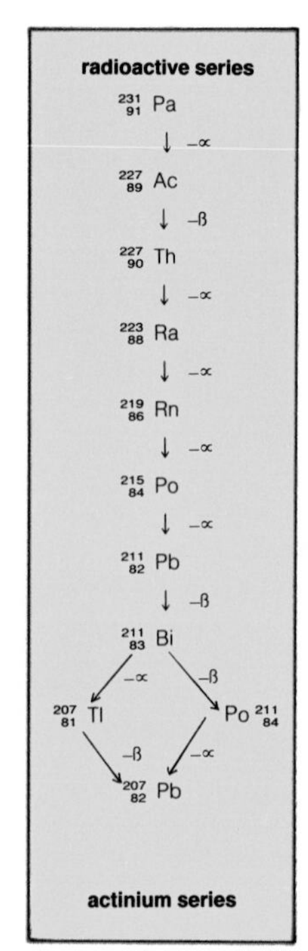

actinium series

chamber (*n*) a hollow space surrounded by walls, usually with a small entrance.
track (*n*) visible (p.42) signs that an object has travelled along a path. The signs are continuous.

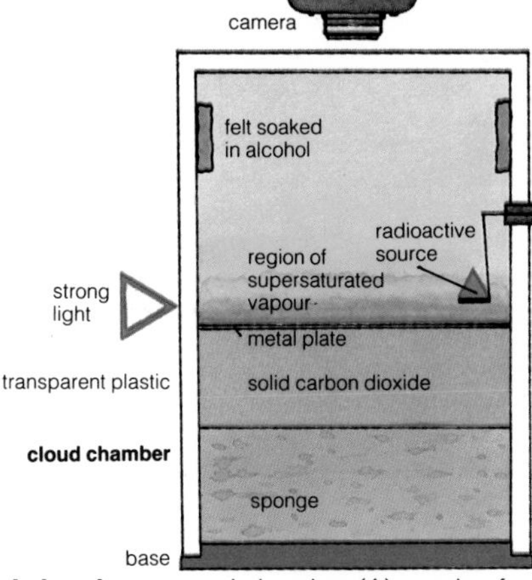

cloud chamber

cloud chamber a round chamber (↑), made of transparent (p.16) plastic which has a firm base. Above the base, solid carbon dioxide is supported by a sponge. A metal plate rests on the solid carbon dioxide. Round the top of the chamber is a strip of felt, soaked in alcohol (ethanol). The ethanol vaporizes (p.11) and saturates dust-free air in the chamber. The metal plate is very cold and forms a region of supersaturated vapour, just above the plate. A radioactive source produces radiation which ionizes (p.123) the molecules in the air. Drops of liquid condense on the ions, forming a track (↑) which is photographed by a camera, using strong light, as shown in the diagram. Condensation takes place only in the region of supersaturated vapour. Cloud chambers are used to study radiation from radioactive sources and nuclear reactions (p.144).
bombard (*v*) to hit a large object repeatedly with many small objects, each small object possessing considerable energy, **bombardment** (*n*).

tracks of alpha particles in a cloud chamber

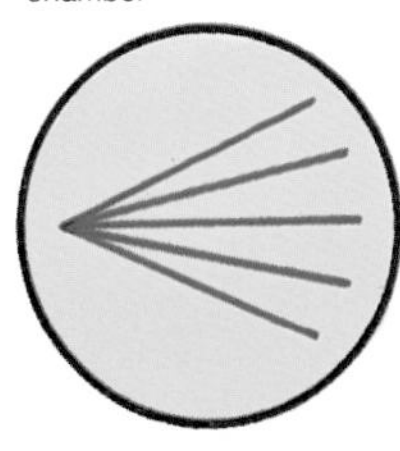

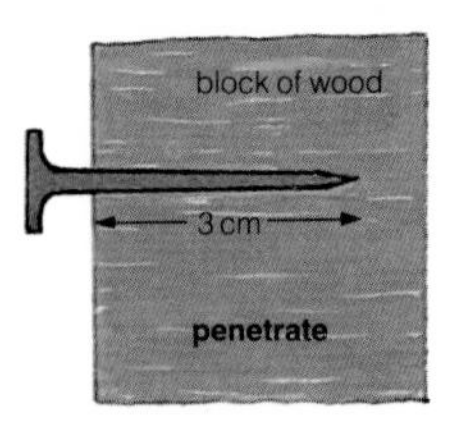

a nail penetrates a block of wood to a depth of 3 cm

penetrate (*v*) to go through an outer cover, using force or energy, or similarly, to reach inside a solid material, e.g. gamma rays penetrate steel; a neutron penetrates an atomic nucleus. **penetration** (*n*), **penetrating** (*adj*).

radio opaque (*adj*) describes a material which does not allow radiation (p.138) to pass through it because the radiation is absorbed. Radio opaqueness depends on the density of the material and on the wavelength of the radiation. The shorter the wavelength, the greater the penetration (↑). **radio opaqueness** (*n*).

transform (*v*) to change one form of energy into another form. **transformation** (*n*).

radiology (*n*) the study of X-rays and radioactivity (p.138). In particular, it is the use of radiations (p.138) in medical science for curing diseases. **radiologist** (*n*), **radiological** (*adj*).

nuclear reaction a reaction in which changes in the nucleus of an atom take place. Nuclear changes are caused by the bombardment (p.143) of a nucleus with subatomic (p.110) and other particles, e.g.

$$^{63}_{29}Cu + ^{1}_{1}H \rightarrow ^{63}_{30}Zn + ^{1}_{0}n$$

in which copper-63 is bombarded with protons and forms zinc-63 and neutrons. Aluminium-27 bombarded with alpha particles forms phosphorus-30 and neutrons:

$$^{27}_{13}Al + ^{4}_{2}He \rightarrow ^{30}_{15}P + ^{1}_{0}n$$

The products of nuclear reactions are generally radioactive isotopes (p.114) of the element.

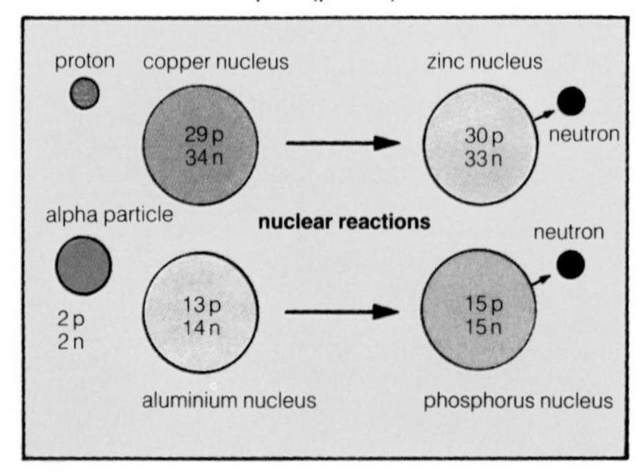

mass spectrograph

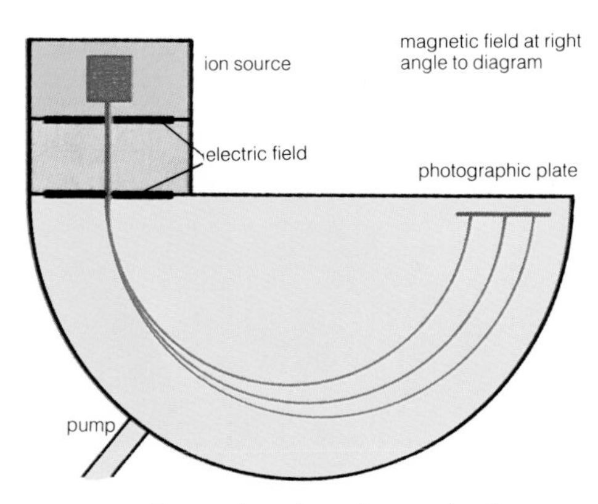

mass spectrograph

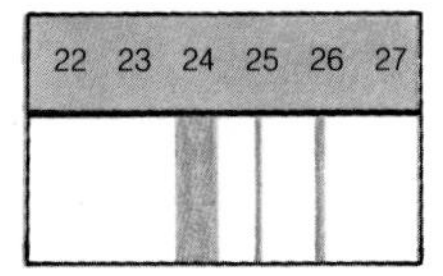

mass spectrograph of magnesium

mass spectrograph a piece of apparatus for determining (p.222) the mass number of isotopes (p.114). An ion (p.123) source, *see diagram*, produces positive ions of an element. The ions pass through two slits (p.211), S_1 and S_2, in electrodes. The electrodes are given a negative potential (voltage) to accelerate (p.219) the ions. The ions then enter a semi-circular chamber (p.143) which has a strong magnetic field acting at a right angle to the path of the ions. The ions follow a circular path, due to the magnetic field; the lower the mass number of the ion, the smaller is the radius of the path. The ions hit a photographic plate and produce a mark. The width of the mark corresponds to the abundance (p.231) of an isotope. The spectrograph is calibrated (p.26) using carbon-12, so mass numbers (p.113) can be read directly. The diagram shows a photograph of the isotopes of magnesium. The apparatus operates (p.157) with all the air pumped out, forming a high vacuum. **mass spectrography** (*n*).

mass spectrometer a piece of apparatus similar to a mass spectrograph (↑); it measures accurately the relative proportion (p.76) of each isotope in a naturally-occurring element.

heat of reaction the heat energy given out, or taken in, when a chemical reaction takes place between the masses of reactants (p.62) shown by the equation (p.78) for the reaction. The heat energy is measured in joules (p.153).

heat of combustion the heat energy given out when one mole of a substance is completely burned in oxygen, e.g. ethyne (acetylene) is burned completely to carbon dioxide and water, 1 mole of ethyne produces 1558 kJ of heat energy. This is written as:

$C_2H_2 + 3O_2 = 2CO_2 + H_2O$, $\Delta H = -1558$ kJ

The symbol (p.77) ΔH indicates (p.38) the heat change during the reaction. The negative sign shows heat is given out, i.e. lost by the reactants.

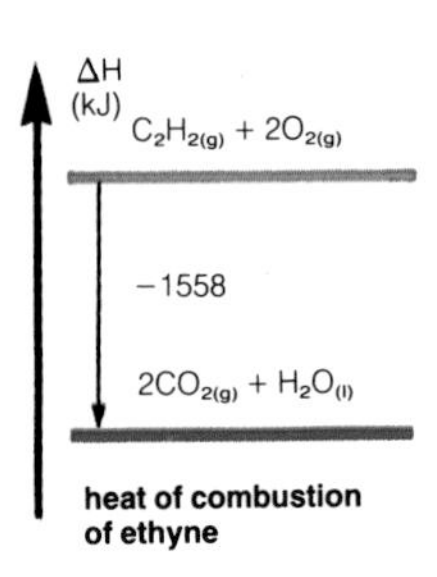

heat of combustion of ethyne

heat of neutralization the heat energy given out when one mole of hydrogen ions in an acid solution is neutralized by a base, with the reaction taking place in a dilute aqueous solution. In reactions between strong acids and strong bases, both reactants are completely ionized (p.123) and the reaction is between one mole of hydrogen ions and one mole of hydroxyl (p.132) ions. The heat of neutralization for strong acids with strong bases is 57.3 kJ; for weak acids or weak bases, the heat energy given out is less than 57.3 kJ, as energy is used to complete ionization of the acid or base.

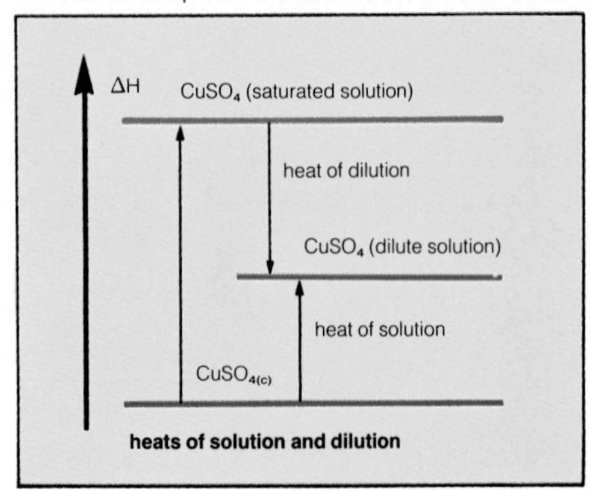

heats of solution and dilution

heat of dilution the change in heat energy when a solution is diluted (p.81). *See diagram.*

heat of solution the heat energy given out, or taken in, when one mole of a substance is dissolved in such a large volume of water that further dilution (p.81) produces no heat change.

heat of formation the heat change when one mole of a compound is formed from its elements, under stated conditions of temperature and pressure.

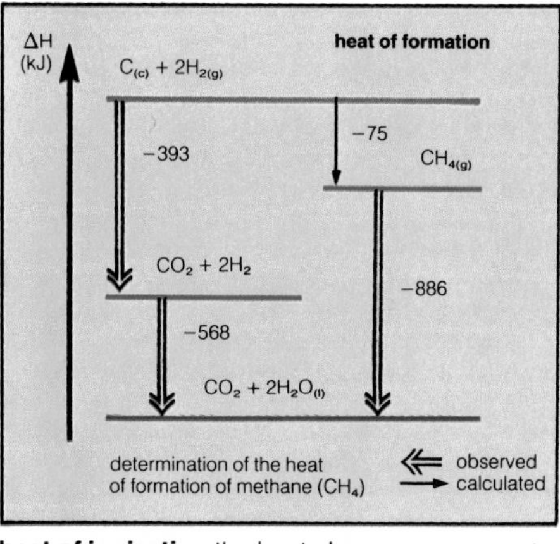

heat of ionization

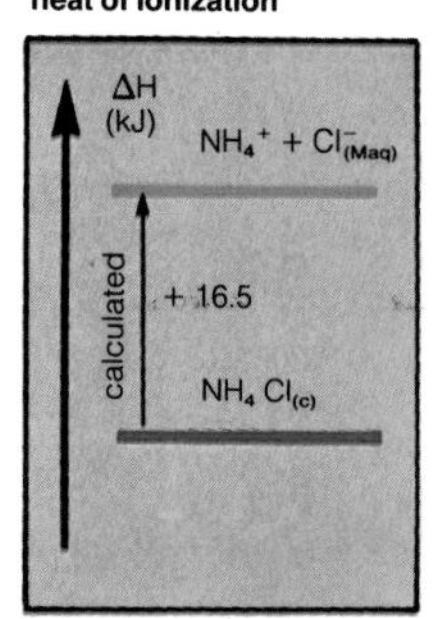

heat of ionization the heat change necessary to produce complete ionization (p.123) of a substance in an aqueous (p.88) solution.

thermochemical equation an equation showing the chemical reaction between whole numbers of moles of the reactants, giving information about the states of the reactants, the temperature of measurement, and the heat change, e.g.

$$CH_{4(g)} + 2O_{2(g)} = CO_{2(g)} + 2H_2O_{(l)}$$
$$\Delta H_{298} = -889\,kJ$$

The reactants are gaseous methane and oxygen; the products are gaseous carbon dioxide and liquid water. The reaction between 1 mole of methane and two moles of oxygen gives out 889 kJ of heat energy at 298 K (25°C). For crystalline solids, the symbol (c) (p.77) is written, e.g. $S_{(c)}$ for crystalline sulphur.

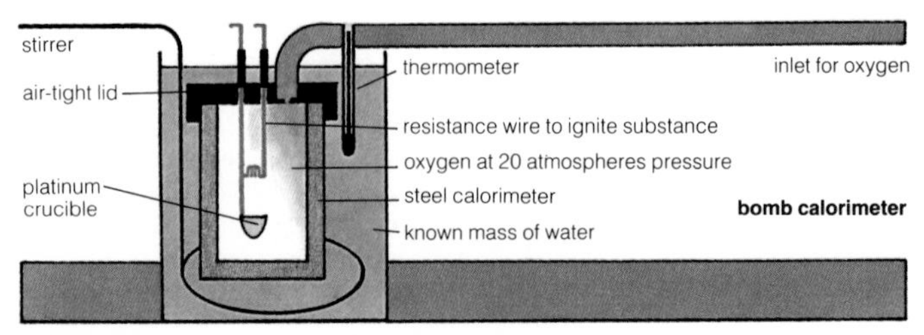

bomb calorimeter a device for measuring heats of combustion (p.146). It consists (p.55) of a thick-walled steel calorimeter, with an air-tight screw top. A known amount of substance is put in a platinum crucible and oxygen is pumped in to 20 atmospheres pressure. The substance is ignited (p.32) by a heated resistance wire. The heat evolved (p.40) heats the calorimeter, and thus the water. The temperature rise of the water is measured by a sensitive thermometer and the heat evolved is calculated, and hence the heat of combustion of the substance is found.

exothermic (*adj*) describes a reaction in which heat energy is given out.

endothermic (*adj*) describes a reaction in which heat energy is taken in.

Hess's law the heat energy given out, or taken in, in a chemical change is the same, no matter how the change takes place. For example, ammonium chloride solution can be made in two ways both starting with gaseous ammonia and gaseous hydrogen chloride:

(1) $NH_{3(g)} + aq \rightarrow NH_{3(aq)}$; $HCl_{(g)} + aq \rightarrow HCl_{(aq)}$
$NH_{3(aq)} + HCl_{(aq)} \rightarrow NH_4Cl_{(aq)}$
where (aq) is an aqueous solution.

(2) $HCl_{(g)} + NH_{3(g)} \rightarrow NH_4Cl_{(c)}$
$NH_4Cl_{(c)} + aq \rightarrow NH_4Cl_{(aq)}$

The changes of heat energy are shown in the diagram opposite. Both ways of preparation produce the same result.

thermochemistry (*n*) the study of the changes in heat energy that take place during a chemical reaction. Heat energy given out heats the products, while heat energy taken in cools the products.

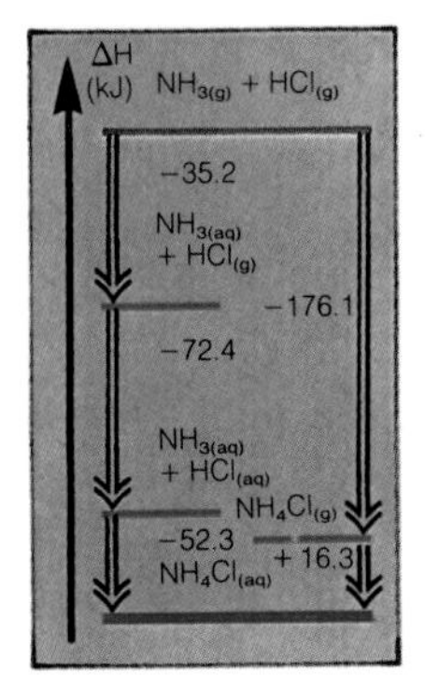

rate of reaction the rate of reaction is measured by the rate at which the reactants are used up or by the rate at which the products are formed.

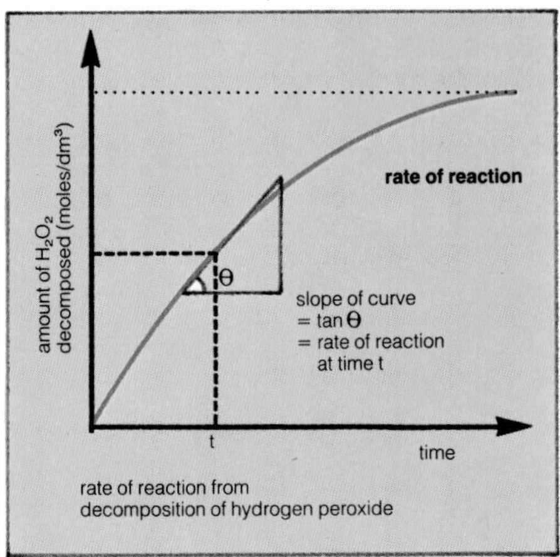

rate of reaction from decomposition of hydrogen peroxide

rate constant if an amount x of a substance is used up in a reaction and the initial (p.85) amount was a, then the rate of reaction ($\frac{dx}{dt}$) at a given time is: $\frac{dx}{dt} \propto (a - x)$
i.e. the rate is proportional to $(a - x)$.
This can be written as: $\frac{dx}{dt} = k(a - x)$
where k is the *rate constant*.
The rate constant is independent of pressure, but is changed by catalysis and temperature changes.

velocity constant another name for rate constant (↑).

law of mass action the rate of a reaction, at a constant temperature, is proportional to the product of the active masses of the reacting substances. In a reaction: $A + B \rightarrow$ products (A and B are two substances) the rate of reaction $\propto [A] \times [B]$, where [A] represents the active mass of A. For gases, the active mass is measured by the partial pressure (p.108) of the gas; for solutions, it is measured by the concentration in moles per dm^3. With solids, the active mass is difficult to measure.

equilibrium (*n*) in a reversible reaction (p.64), the products start to react the moment they are formed. The rate of reaction of the reactants (p.62) is high as their concentration is high, while the rate of reaction of the products (p.62) is low because their concentration is low; this is because the law of mass action (p.149) is obeyed (p.107). Eventually, the two rates of reaction will be the same, provided the products are not removed (p.215), and the reactants and products will both be present. This is a state of *equilibrium* and thereafter there is no change in concentration of reactants and products.

dynamic equilibrium a state of equilibrium (↑) in which the rate of change of two processes (p.157) is equal and opposite, e.g. (a) in a closed vessel (p.25) the rate of evaporation (p.11) of a liquid, and the rate of condensation (p.11) of its vapour (p.11), are equal at a constant temperature; the vapour exerts (p.106) its vapour pressure (p.103) for that temperature. (b) phosphorus pentachloride (PCl_5) decomposes (p.65) on heating to phosphorus trichloride (PCl_3) and chlorine (Cl_2). In a closed vessel, PCl_3 and Cl_2 combine to form PCl_5. At equilibrium (↑) all three substances are present, with the two opposite chemical processes being in dynamic equilibrium.

$$PCl_5 \rightleftharpoons PCl_3 + Cl_2$$

equilibrium mixture the relative (p.232) concentrations of reactants and products when equilibrium (↑) is reached in a reversible reaction (p.64).

equilibrium constant in a reversible reaction (p.64):

$$pA + qB \rightleftharpoons rC + sD$$

where A, B, C, D are substances and p, q, r, s are mole fractions, the equilibrium constant is given by:

$$K = \frac{[A]^p\,[B]^q}{[C]^r\,[D]^s}$$

[A], [B], [C], [D] represent the concentrations of the substances. The value of K changes with temperature.

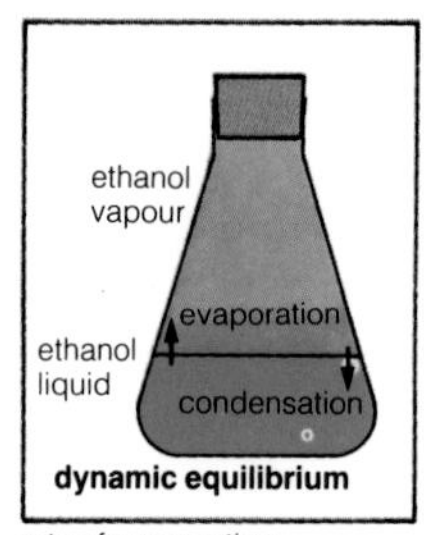

rate of evaporation
= rate of condensation

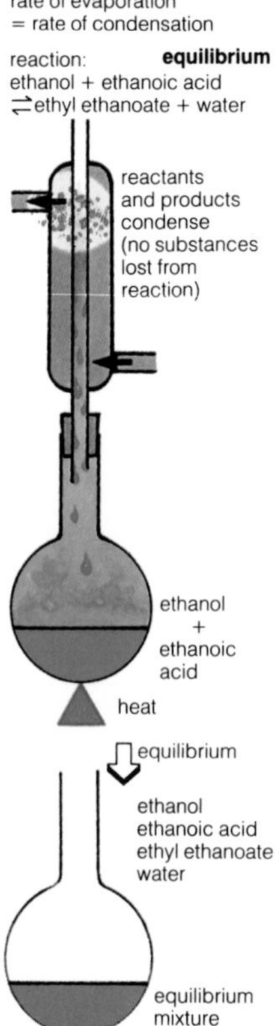

Le Chatelier's principle whenever changes take place in the external conditions of a chemical reaction *in equilibrium*, then changes occur (p.63), if possible, in the reaction which tend (p.216) to counteract (p.216) the effect of those external changes. The principle can be illustrated from the reversible reaction:

$$\underset{1\text{ vol.}}{N_{2(g)}} + \underset{3\text{ vols.}}{3H_{2(g)}} \rightleftharpoons \underset{2\text{ vols.}}{2NH_{3(g)}} \quad \Delta H = -50\text{ kJ}$$

The reaction between nitrogen and hydrogen is exothermic (p.148) with 50kJ of heat energy given out. The decomposition of ammonia is endothermic (p.148) with 50 kJ heat absorbed. One volume of nitrogen reacts with 3 volumes of hydrogen to produce 2 volumes of ammonia, hence there are 4 volumes of the reactants and 2 volumes of the products. If an equilibrium mixture has its temperature raised the mixture absorbs heat to counteract the rise in temperature; this favours (p.214) the endothermic reaction, i.e. the decomposition of ammonia. Raising of the temperature quickens both rates of reaction, so equilibrium is reached more quickly. Increasing the pressure favours a smaller volume of gases, so the equilibrium moves towards forming more ammonia. The effect of external conditions on equilibrium is summarized in the table below.

REACTION:	**A + B (larger volume)**	**C + D, (smaller volume)**	**+ heat given out**
EXTERNAL CHANGE	RATE OF REACTION	EQUILIBRIUM MIXTURE	EQUILIBRIUM CONSTANT
increase of temperature	increased	more A + B	changed
decrease of temperature	decreased	more C + D	changed
increase of pressure	increased for gaseous reactions	more C + D	unchanged
decrease of pressure	decrease for gaseous reactions	more A + B	unchanged
addition of catalyst	increased	unchanged	unchanged

energy level an electron possesses energy according to its distance from the nucleus (p.110) of an atom; the nearer it is to the nucleus, the lower is the energy it possesses. This energy is the energy level of an electron.

ground state the condition of an atom (p.110) when all its extranuclear (p.113) electrons are in their positions of lowest energy, i.e. at their lowest energy levels (↑).

excitation (*n*) the changing of an atom from its ground state (↑) by supplying energy in the form of heat, radiation, or bombardment (p.143) by sub-atomic (p.110) particles. This causes one or more electrons to move to orbitals further from the nucleus. Such electrons may then enter into covalent bonds (p.136), or with sufficient energy, be removed from the atom, forming an ion (p.123). **excitatory** (*adj*).

energy barrier a measure of the quantity of energy that must be supplied in order that a chemical reaction can take place. If a lesser quantity of energy is supplied, there is no reaction.

activation energy the additional energy that must be supplied to reactants before a reaction can take place. The atoms are in a state of excitation (↑). The activation energy is the energy barrier to a reaction; until the activation energy is supplied, no reaction can take place.

activated state the state of molecules, atoms, or ions in which they are able to react.

reaction profile a diagram which shows the relation between the ground states (↑) of reactions and products, and the energy barrier to a reaction.

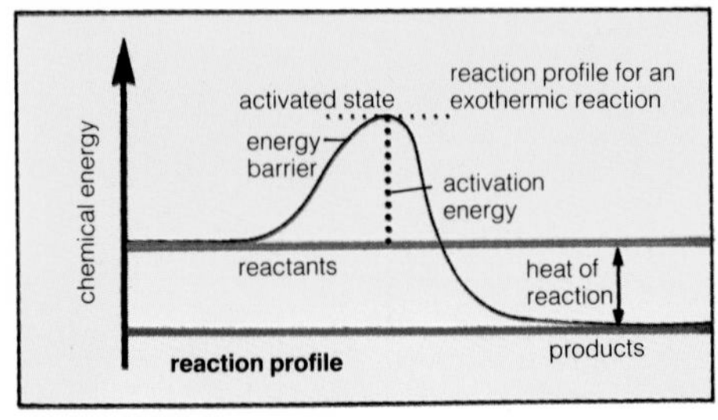

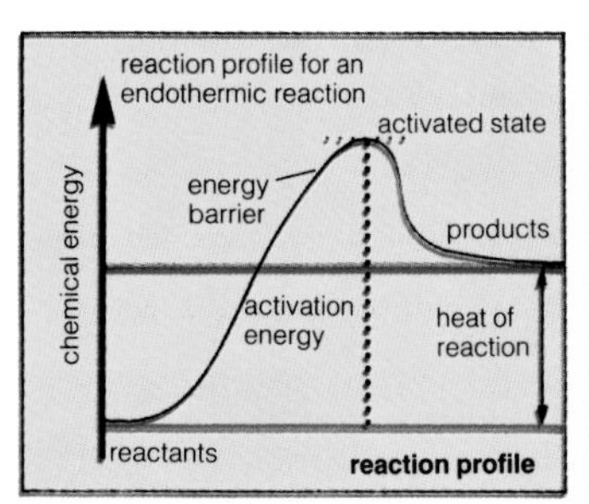

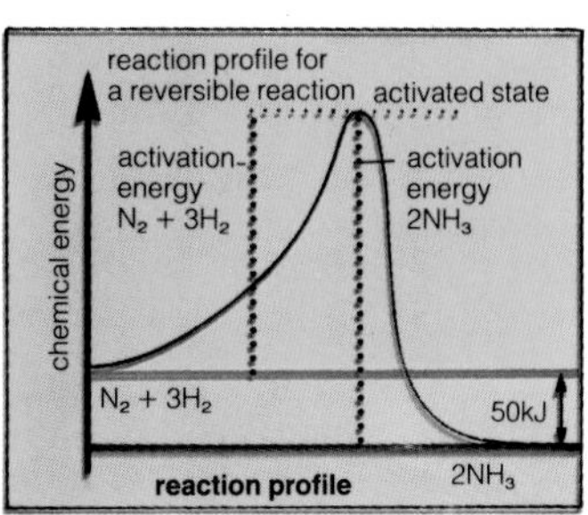

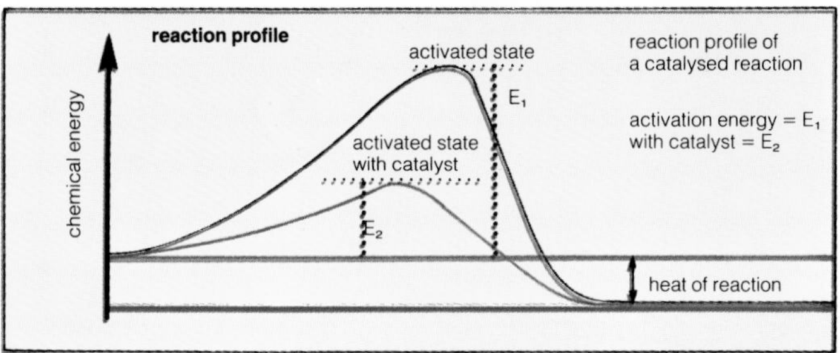

joule (*n*) the unit of energy and work. It is the work done when the point of application of a force of one newton is displaced through a distance of one metre in the direction of the force. It is also the work done when a current of 1 ampere flows between a voltage difference of 1 volt for 1 second. (A force of one newton gives a mass of 1 kg an acceleration of 1 ms^{-2}.) The symbol for joule is J.

calorie (*n*) a measurement of heat energy; it is that quantity of heat that will raise the temperature of 1 g of water through 1°C. The unit is no longer used. 1 calorie $\approx$ 4.18 joules.

electron-volt a unit of energy that is not included in S.I. units, but is useful in many chemical calculations on radiation (p.138) energy. It is the quantity of energy gained by an electron when it falls through a voltage difference of 1 volt.

1 electron-volt = 1 eV
$= 1.6 \times 10^{-19}$ coulomb $\times$ 1 volt
$= 1.6 \times 10^{-19}$ joule.

raw materials the materials which are used as the starting point of chemical processes (p.157). They include such materials as ores (↓), minerals (↓), limestone, common salt, coal, mineral oil.
mineral (*n*) any material which occurs (↓) naturally but does not come from animals or plants. Examples of minerals are: metal ores (↓), limestone (↓), rock salt (↓), coal (p.156), mineral oil (p.156). A mineral has a known chemical constitution (p.82) and definite physical and chemical properties.
ore (*n*) a mineral (↑) which is a compound (p.8) of a metal and from which the metal can be extracted (p.164).
deposit (*n*) a mass of a mineral found in the earth, in a large enough quantity to be worth mining (↓).
seam (*n*) a thick, fairly level, layer of a deposit (↑) of a mineral (↑), e.g. a seam of coal.
lode (*n*) a large, thin deposit (↑) of an ore (↑) which is between the walls of a deep crack in the Earth's structure.

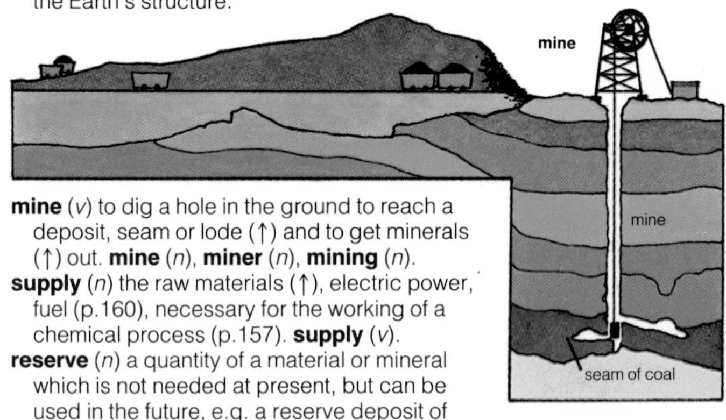

mine (*v*) to dig a hole in the ground to reach a deposit, seam or lode (↑) and to get minerals (↑) out. **mine** (*n*), **miner** (*n*), **mining** (*n*).
supply (*n*) the raw materials (↑), electric power, fuel (p.160), necessary for the working of a chemical process (p.157). **supply** (*v*).
reserve (*n*) a quantity of a material or mineral which is not needed at present, but can be used in the future, e.g. a reserve deposit of mineral oil which can be used when present supplies (↑) are finished.
occur[2] (*v*) to be in a particular place, e.g. petroleum (p.160) occurs in Saudi Arabia in great quantities. To contrast *occur* and *exist* (p.213): sulphur *occurs* in large deposits in North America; sulphur *exists* in two crystalline (p.15) allotropes (p.118).

assay (*v.t.*) to make chemical tests (p.42) to find the mineral content (p.85) of a material; in particular to determine (p.222) the relative amount of a metal in an ore (↑).

native (*adj*) describes an element found in the Earth and not combined in a mineral, e.g. gold, copper, and sulphur are found uncombined in deposits (↑); they occur (↑) native, and are native elements.

blende (*n*) a mineral which is a sulphide (p.51) of a metal, e.g. zinc blende which is zinc sulphide.

pyrites (*n*) a mineral which is a sulphide of a metal, e.g. iron pyrites is iron sulphide.

chalk (*n*) a soft white rock consisting (p.55) of calcium carbonate, formed from the shells of shellfish. The shells were deposited (↑) on the sea-bed over a long time and pressure formed the rock from the shells.

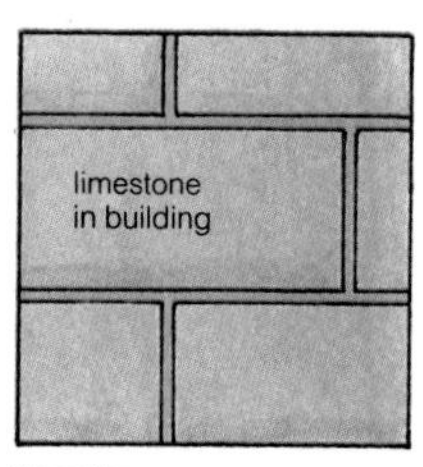

limestone

limestone (*n*) a rock consisting (p.55) mainly of calcium carbonate, but sometimes containing magnesium carbonate. It is harder than chalk and is used as a building material. On heating, it is converted (p.73) to lime (p.169).

marble (*n*) a hard, crystalline (p.15) rock, formed from limestone (↑) by heat and pressure within the Earth. It consists (p.55) of calcium carbonate. Marble is used as a building material.

dolomite (*n*) a white crystalline (p.15) mineral (↑) consisting of calcium and magnesium carbonates. The mineral is found in great abundance (p.231).

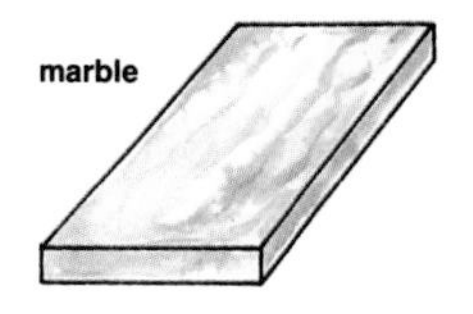

marble

common salt the substance sodium chloride; it occurs (↑) in sea water which has a salt content (p.85) of about 3%. Common salt is obtained by evaporating large quantities of sea water.

rock salt a mineral consisting of sodium chloride. Large deposits (↑) are found in many countries. It is the raw material (↑) for the manufacture (p.157) of chlorine and many sodium salts (p.46).

sulphur (*n*) a yellow solid element, occurring (↑) native (↑) in the earth, also found in many ores (↑) such as blendes (↑) and pyrites (↑). It is the raw material (↑) for the manufacture (p.157) of sulphuric acid, the most important industrial (p.157) chemical.

carboniferous (*adj*) describes any material having carbon in it, or being a form of coal (↓). Also describes any material producing carbon when heated, or any plant life which has formed coal (↓).

carbonaceous (*adj*) describes any mineral (p.154) containing carbon, but not carbonates, e.g. coal (↓) is a carbonaceous mineral.

coal (*n*) a black, fairly hard mineral which occurs in large deposits (p.154) and seams (p.154). Coal has been formed from plant life which existed (p.213) many millions of years ago. Chemical action, heat and pressure of the earth formed the mineral. It consists (p.55) mainly of carbon, but contains many other useful chemical compounds (p.8).

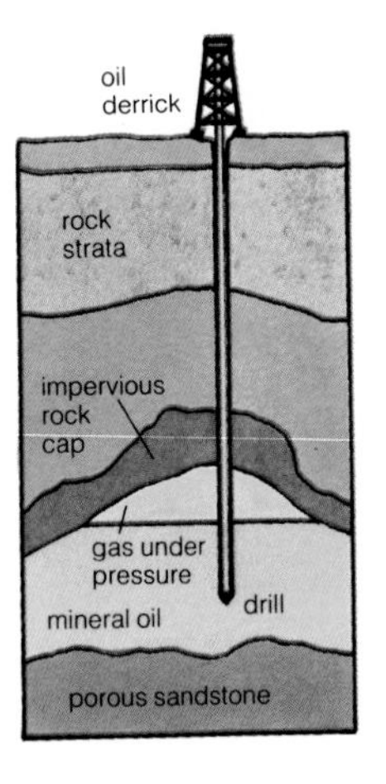

mineral oil a dark liquid mixture of various compounds (p.8) occurring (p.154) in deposits in many parts of the world; it varies widely in composition (p.82) depending on the place from where it comes. It is thought that mineral oil has been formed from decayed plants and animals. Deposits of mineral oil usually occur with natural gas.

bitumen (*n*) a naturally occuring black material which contains hydrocarbons (p.172). It is a solid or a very thick liquid. Bitumen is also obtained as a product of distillation from coal (↑).

charcoal (*n*) a form of carbon obtained by heating wood, or other plant materials or animal materials, when no air is present (p.217). Charcoal is black in colour and generally very porous (p.15). Charcoal absorbs (p.35) gases and decolorizes coloured liquids if the colour is from organic (p.55) materials.

coke (*n*) a solid material obtained by heating coal with no air present (p.217). It is mainly carbon.

liquid air air is cooled below its critical temperature (p.104) and then liquefied (p.11). Liquid air is the source (p.138) of nitrogen and oxygen.

silica (*n*) a very hard white substance, silicon dioxide. It occurs in many minerals, e.g. quartz, sand. It is the source of silicon, an element used in electronic circuits.

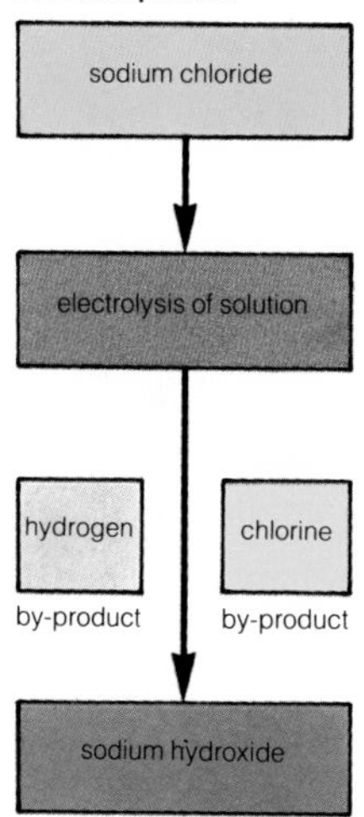

industrial process for the manufacture of sodium hydroxide

process (*n*) (1) in chemistry, a method of making a substance in large quantities; the method includes a description of the chemical reaction, the conditions (p.103) under which the reaction takes place and the plant (↓) needed for the process. (2) the different methods of preparing (p.43) and isolating (p.43) a substance, e.g. distillation, sublimation. (3) generally, a process is a number of events taking part in a connected and continuous change. **process** (*v.t.*).

plant (*n*) all the building, pipes, furnaces (p.164), machinery, special apparatus (p.23) and devices (p.23) used to make substances in large quantities for industry (↓).

industrial (*adj*) describes anything connected with the making of chemical substances, machinery, or materials in large quantities. Industrial chemicals are divided into heavy chemicals (used in other industrial processes), pharmaceutical chemicals (p.20) (used in curing diseases), fine chemicals (used in chemical analysis (p.82)), fertilizers and plastics. **industry** (*n*).

operate (*v.t.*) to make an industrial (↑) process work, e.g. to operate a process for the manufacture (↓) of ammonia is to provide the reactants and obtain the products using the necessary plant (↑). **operation** (*n*).

carry out to work on a process, an experiment or an investigation and to complete it.

manufacture (*v.t.*) to make chemical substances, materials, machinery by an industrial (↑) process (↑). **manufacture** (*n*).

main product the product for which a particular process (↑) is planned. In some processes it is the only product, e.g. sulphuric acid in the Contact process (p.166) is the only product.

by-product (*n*) a substance produced in the manufacture (↑) of a main product (↑); it has an industrial (↑) use, e.g. in the Castner-Kellner process (p.169) chlorine is a by-product.

waste product a substance that is produced in a chemical process and which has no industrial (↑) use, e.g. calcium chloride, produced in the Solvay process (p.169), has no industrial use.

end product a product formed after an original (p.220) substance has taken part in several reactions or processes, e.g. petrol is one end product from the refining (↓) of petroleum.
mill (*n*) (1) a machine for breaking large lumps (p.13) of solid materials and turning the material into a powder (p.13), e.g. a mill for making flour. (2) a work place where cloth is made, e.g. a cotton mill.
pulverize (*v.t.*) to turn pieces of solid into a powder (p.13) by hitting them repeatedly. A mill (↑) can pulverize a solid. **pulverization** (*n*).
spraying (*n*) the action of sending many small drops of water from small holes in a pipe so as to cover a large space. **spray** (*n*), **spray** (*v.t.*).

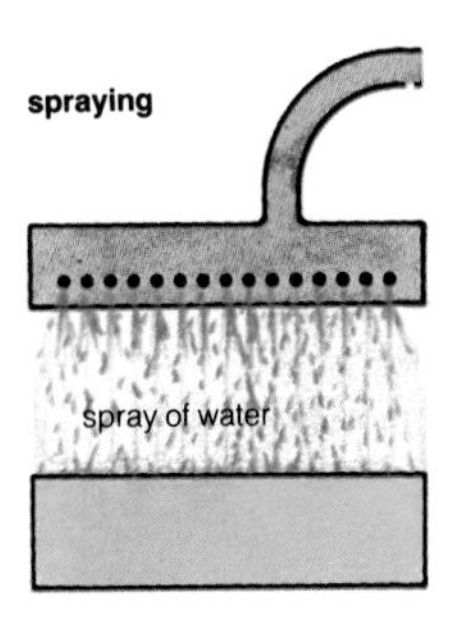

lixiviation (*n*) the process of removing one substance from a mixture of substances by using a suitable solvent (p.86), e.g. salts containing iodine are obtained from heated seaweed by lixiviation with water. **lixiviate** (*v*).
leaching (*n*) the removal (p.215) of soluble (p.17) substances by water from a mixture of solids when the water washes out the substance. To contrast *lixiviation* and *leaching*: if a mixture is treated (p.38) with water, the process is *lixiviation*; if water flows through the mixture to wash away a substance the process is *leaching*. Solvents other than water can be used in leaching. **leach** (*v*).
slaking (*n*) adding water to quicklime (calcium oxide) to form slaked lime (calcium hydroxide). **slake** (*v*).
scum (*n*) any solid substance, especially dirt or waste (p.170), which floats on the surface of a liquid, e.g. impurities floating on the surface of a molten (p.10) metal.
sludge (*n*) soft wet solid material, usually waste or unwanted material.
froth flotation a process for the separation (p.34) of ores (p.154) from earthy material. Oil and water form a froth (p.100) with particles of ore trapped in the liquid-air interfaces (p.18) of the bubbles (p.40). The foam is stabilized by frothing agents (p.63). The ore is removed with the froth.

scum/sludge

scum floating and sludge settling in a liquid

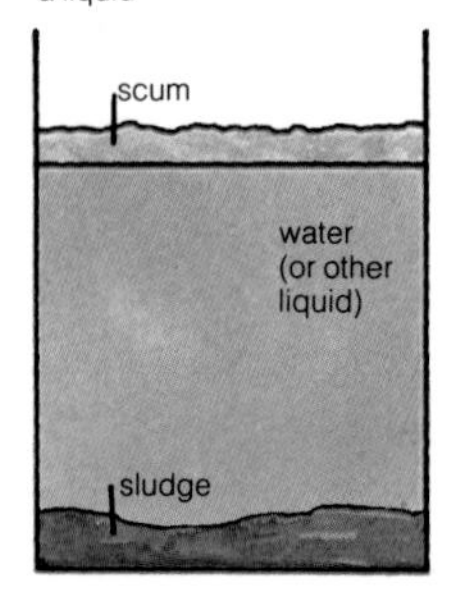

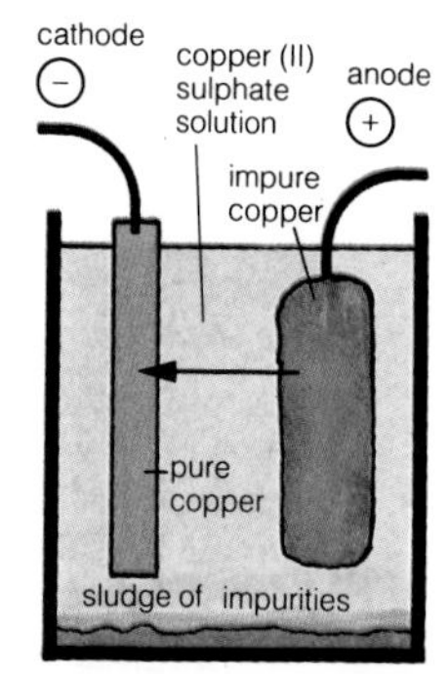

refining copper by electrolysis **refine**

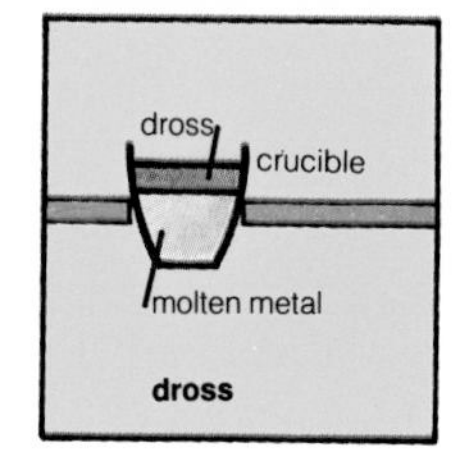

dross

stage (*n*) a length of space or time in a process, e.g. (a) the primary, intermediate and secondary stages of an education system (p.212); (b) the stages in the Bosch process (p.168) for manufacturing hydrogen are (i) mixing water gas and steam, (ii) passing the gases over a catalyst, (iii) removing carbon dioxide.

refine (*v*) (1) to remove impurities from a substance, e.g. to refine copper by electrolysis (p.122). (2) to separate the constituents (p.54) of a mixture to get pure specimens of some, or all, of the constituents, e.g. to refine petroleum (p.160) to get petrol, kerosene, etc. (3) to make a method, apparatus, or a technique more efficient or more suitable for its purpose, e.g. (a) a refined method of determining the equilibrium constant of a reaction; (b) a more refined apparatus for carrying out (p.157) chromatography. **refinement** (*n*).

liquation (*n*) a method of refining (↑) a metal in which a mixture of metals is heated until one metal melts (p.10) and flows away from the mixture.

dross (*n*) the impurities, or other waste material, that floats on top of molten (p.10) metal as a scum (↑). Dross is removed (p.215) in refining (↑) metals.

optimum (*adj*) describes a condition which is the most favourable (p.214) for a reaction, e.g. the optimum temperature for the Contact process (p.166) is 500°C. At this temperature the best yield (↓) is obtained.

yield (*n*) the quantity of a product obtained from an industrial (p.157) process or preparation (p.43). The actual yield is often compared, as a percentage, with the theoretical stoichiometrical (p.82) yield, e.g. the yield in the Haber process is only about 6%, i.e. only 6% of nitrogen and hydrogen combine to form ammonia. **yield** (*v*).

replenish (*v*) if a material is being used in a process, further quantities have to be added to replace (p.68) the losses. To replenish is to make these additions of the material. **replenishment** (*n*).

fuel (*n*) any material or substance that is burned to give heat, and through heat, to give power. Common fuels in industry are: coal, coke, kerosene, fuel oil, coal gas, natural gas.

petroleum (*n*) mineral oil (p.156) consisting (p.55) of hydrocarbons (p.172) and some compounds of sulphur and nitrogen. Petroleum is fractionally distilled (p.201) to obtain fractions (p.202), each of which has a particular use.

petrol (*n*) a volatile (p.18) liquid of low boiling point (p.12) obtained from petroleum (↑) by distillation (p.33). Petrol distils between 20°C and 150°C. It is a mixture of hydrocarbons (p.172) from hexane to decane. Petrol is highly inflammable (p.21); it is used in internal combustion engines.

gasoline (*n*) name for petrol in the U.S.A.

kerosene (*n*) a volatile liquid similar to petrol (↑) but with a higher boiling point. It is obtained from petroleum (↑) by fractional distillation; kerosene distils between 150°C and 250°C. It is used for heating, lighting and in jet engines.

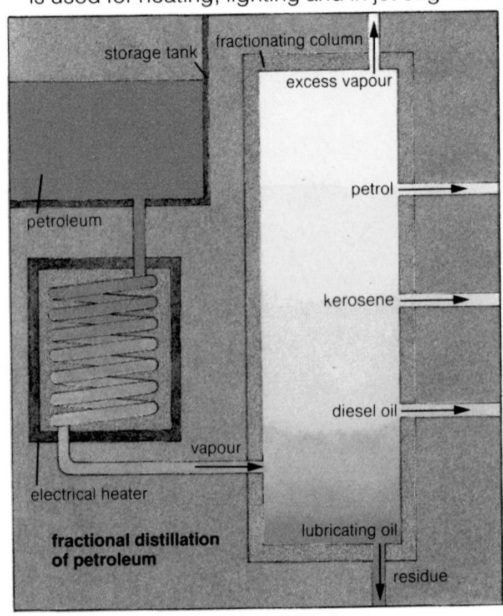

fractional distillation of petroleum

petrol	fuel in motor car engines
kerosene	for heating and lighting fuel in jet engines
gas oil	fuel in diesel engines
lubricants	lubricating moving parts of machinery
pitch	surfacing roads sealing roofs

asphalt (*n*) a black, sticky, solid material consisting (p.55) mainly of hydrocarbons (p.172). It occurs naturally in various places and is also found in some kinds of petroleum (↑). Asphalt is used for road surfaces and for making roofs waterproof.

naphtha (*n*) an organic (p.55) liquid obtained by dry distillation (p.203) of coal or wood. It can contain (p.55) a number of different compounds, but has no particular composition (p.82). Naphtha is an inflammable (p.21), volatile (p.18) liquid.

paraffin[1] (*n*) another name for kerosene (↑); paraffin oil is also used as a name.

gas oil a liquid obtained from petroleum (↑) by fractional distillation. It is less volatile than kerosene and distils between 250°C and 300°C. Gas oil is used in diesel engines.

lubricants (*n.pl.*) various kinds of lubricating oils. Lubricants are used to reduce friction. They are distilled from petroleum at between 300°C and 400°C. Lubricants with a high boiling point are soft solids such as petroleum jelly. Lubricants with lower boiling points are thick, sticky liquids.

doctor solution a solution which is used to remove bad-smelling compounds from petroleum (↑).

pitch (*n*) a black, sticky liquid left after fractional distillation of petroleum (↑). It is the same material as asphalt (↑).

tar (*n*) a thick, black, sticky liquid formed during the destructive distillation (p.203) of coal. It contains many different organic (p.55) substances from which many useful compounds are manufactured (p.157).

scrubber (*n*) a device for removing (p.215) ammonia and benzene from impure coal gas. The impure gas is passed up the scrubber, and a spray of water flows down the scrubber, washing out the ammonia and benzene. Ammoniacal liquor, containing these two substances, collects in the scrubber. **scrub** (*v*).

gasometer (*n*) a very large iron vessel (p.25) for the storing of coal gas.

vat (*n*) a large, open vessel (p.25) used for making wine or soap, dyeing cloth, etc.

dye (*n*) a substance used to give colour to cloth, plastics or paper. Some dyes are made from plants, but most are synthetic (p.200). **dye** (*v*).

pigment (*n*) a solid substance used to give colour to paint or varnish. A pigment is not soluble in water. Pigments can be either organic (p.55) or inorganic (p.55) substances.

mordant (*n*) a substance which is used with dyes (↑) that do not dye a material directly. The material is treated (p.38) first with the mordant; the treated material is then put in a vat (↑) with a dye. With acid dyes, the mordant is aluminium or tin hydroxide. With basic dyes, vinegar (ethanoic acid) or tannic acid is used as a mordant. The mordant causes the dye to dye the material.

soap (*n*) any salt of a metal and a fatty acid. The water-soluble soaps are the sodium and potassium salts; sodium forming hard soaps and potassium forming soft soaps. These soaps have a cleansing action, removing dirt from surfaces. Other soaps have different properties. The most common fatty acids in water-soluble soaps are stearic acid ($C_{17}H_{35}COOH$) and palmitic acid ($C_{15}H_{31}COOH$). **soapy** (*adj*).

salting out the addition of a concentrated solution of sodium chloride to a solution of an organic (p.55) compound in water or ethanol, in order to throw the organic compound out of solution.

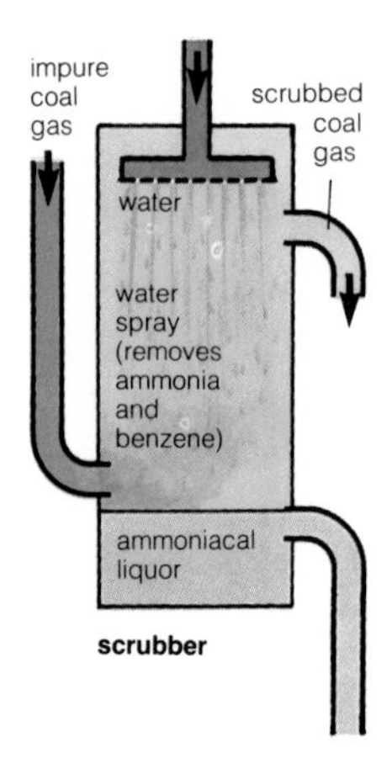

scrubber

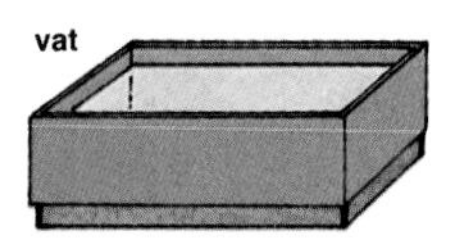
vat

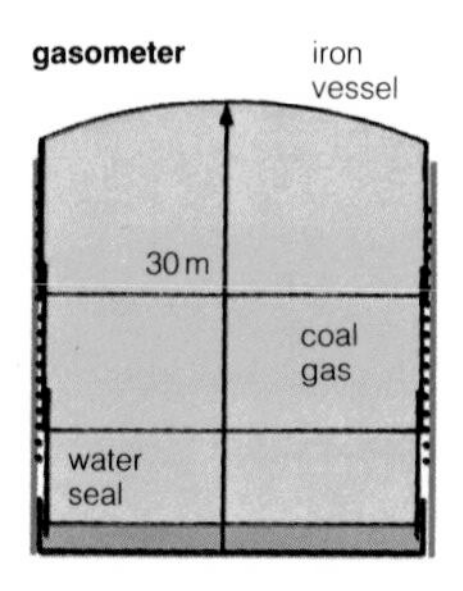

gasometer

blast furnace

blast furnace a vertical furnace (p.164) for obtaining iron from its ores. A strong current (a blast) of air is blown up the furnace. Coke (p.156) is mixed with limestone (p.155) and the iron ore. The furnace is kept at a temperature of about 1800°C in the hottest part. The coke is converted to carbon monoxide in two stages:

$$C + O_2 \rightarrow CO_2 \qquad CO_2 + C \rightarrow 2CO$$

The carbon monoxide reduces the iron ore to iron. Molten (p.10) iron runs out of the bottom of the furnace.

tuyere (*n*) a pipe, usually made of copper, through which air is blown into a blast furnace.

pig iron (*n*) iron, straight from a blast furnace (↑), which has cooled in moulds (p.210). It contains (p.55) 2–5% carbon and smaller quantities of other impurities making it hard and brittle (p.14).

cast iron pig iron which has been melted, mixed with steel scrap (p.171), and cooled in moulds (p.210) to give it a shape. Cast iron is impure iron and is hard and brittle.

wrought iron wrought iron is obtained from pig iron by heating it in a reverberatory furnace (p.164) with limestone and stirring the molten (p.10) mass with long iron rods. Wrought iron is almost pure, containing (p.55) less than 0.2% carbon. It is soft, malleable and easily welded.

puddling process cast iron is fused in a reverberatory furnace (p.164) with haematite (an ore of iron oxide) lining the hearth (p.164). The oxygen in haematite oxidises the carbon in cast iron and forms nearly pure wrought iron (↑).

slag (*n*) a waste product (p.157) from a blast furnace (↑). It consists (p.55) of calcium silicate formed by the reaction between limestone and the earthy parts of iron ore.

clinker (*n*) a hard mass of material which is not combustible (p.58), formed in furnaces and boiler fires. It consists of silicates formed by fusion (p.32) of earthy materials in fuels and ores.

roast (*v*) to heat metals or ores in air at a temperature too low for fusion. The roasting removes impurities (p.20) by oxidation (p.70) with atmospheric oxygen. Roasting removes (p.215) sulphur and sulphur dioxide from sulphide ores.

smelt (*v*) to separate a metal from its ore by heating the ore with a suitable reducing agent (p.71). The metal becomes molten (p.10) and impurities separate from the metal, e.g. iron ore is smelted in a blast furnace (p.163). **smelting** (*n*).

extract (*v*) to obtain an element from the Earth, either as a native element or by mining the ore and obtaining the element by chemical action, e.g. extracting aluminium by mining bauxite (aluminium oxide) and electrolyzing it to obtain aluminium. **extraction** (*n*).

furnace (*n*) a brick construction inside which great heat is produced by burning fuels (p.160) or by electricity; mainly used in the extraction of metals.

flue (*n*) a pipe leading hot air and smoke away from a fire.

reverberatory furnace a furnace (↑) with a low roof above the hearth (↓) so that the heat from the fire is directed onto the reactants on the hearth. Reverberatory furnaces are used for smelting ores.

hearth (*n*) the floor on which a fire burns or a reaction takes place in a furnace.

open-hearth furnace a shallow, rectangular furnace (↑) which is heated by burning a gaseous (p.11) fuel. The hot gases from the burning fuel pass over the hearth (↑) on which an ore is placed. The hearth is covered with limestone or dolomite (p.155).

ash (*n*) a residue (p.31) which is a powder, left after a material has been burned completely, e.g. the ash left after wood or plants have been burned. **ashen** (*adj*).

converter (*n*) a large iron vessel in which pig iron (p.163) is converted (p.73) to steel in the Bessemer process (↓).

metallurgy (*n*) the science of the extraction of metals from their ores (p.154), refining (p.159) the metals, and forming alloys (p.55). **metallurgist** (*n*).

weld (*v*) to join together two pieces of metal by (a) heating them so that they melt (p.10) and join or (b) by beating them with a hammer to make them soft enough to join together. **welding** (*n*).

reverbatory furnace

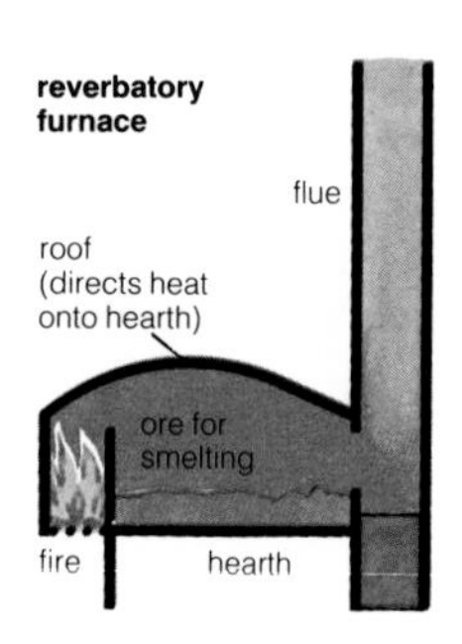

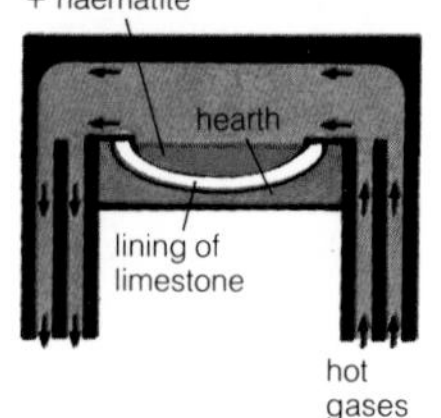

open-hearth furnace

steel (*n*) steel is obtained from pig iron by first oxidizing all the impurities away and then adding a known amount of carbon to the molten (p.10) iron. Other metals, such as manganese or chromium, can be added to make different kinds of steel. Steel contains between 0.15% to 1.5% carbon, depending on the kind of steel that is wanted. Steel is hard and elastic.

quench (*v*) to harden steel by heating it until red-hot, then quickly putting it into cold water or cold oil. **quenching** (*n*).

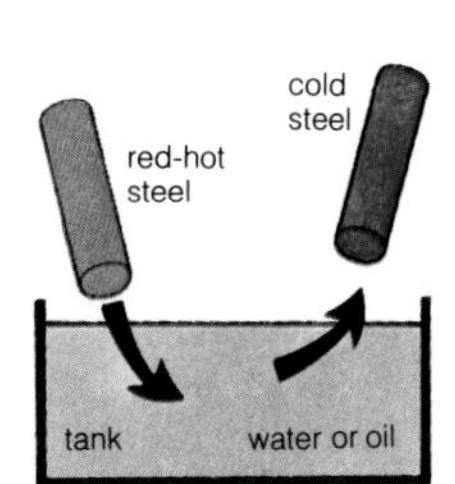

quenching steel

temper (*v*) quenched (↑) steel is heated to a temperature of 200°C-600°C, depending upon the kind of tempering needed. The steel is kept at that temperature for 30 minutes and then allowed to cool. Tempering makes a steel elastic as well as hard. **tempering** (*n*).

pickling (*n*) steel is put in vats with concentrated sulphuric acid. This removes rust (p.61) and any other surface impurities. The steel can then be galvanized (p.166), tinned (p.166), or painted. **pickle** (*v*).

Bessemer process a process to make steel (↑) from pig iron (p.163) or cast iron (p.163). A converter (↑) is lined with calcium and magnesium oxides; pig iron and scrap (p.171) steel are added as molten (p.10) metal. Air is blown through the metal oxidizing all the impurities. Carbon, as needed, is added to make the steel.

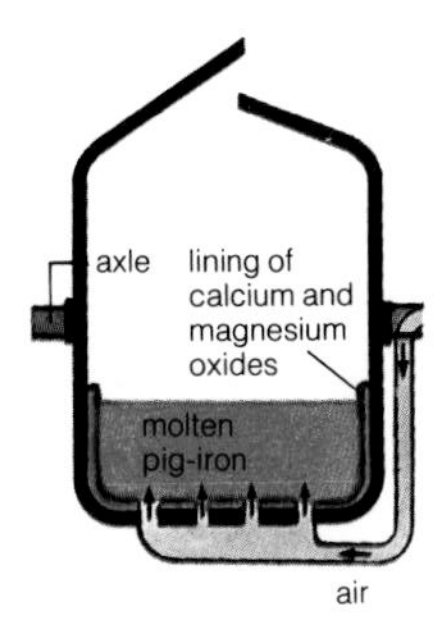

Bessemer converter

Siemens-Martin process pig iron (p.163), scrap (p.171) steel and haematite (iron (III) oxide) are heated in an open-hearth furnace (↑). The hearth (↑) is lined with limestone. The proportion of iron haematite and steel are calculated to give the correct proportion of carbon in the product.

open-hearth process another name for the Siemens-Martin process.

Linz-Donawitz process molten, impure pig iron (p.163) from the blast furnace is used to make steel. Oxygen is blown over the molten (p.10) iron causing the oxidation of carbon and other impurities. Lime powder is added to form a slag (p.163). This is a rapid process (it takes 10–20 minutes) and steel as pure as the steel from the open-hearth process is formed.

spelter (*n*) zinc that has not been refined (p.159); it contains other metals (e.g. lead) and some impurities.

galvanize (*v*) to cover an iron surface with a coat (p.127) of zinc, usually spelter (↑); a metal object is dipped into molten (p.10) zinc. **galvanized** (*adj*).

sherardize (*v*) to cover an iron surface with a coat (p.127) of zinc. A metal object is heated in a closed vessel containing zinc dust at a temperature just below the melting point of zinc.

tin (*v*) to coat (p.127) an iron article with a thin layer of tin. The iron is first pickled (p.165) and then dipped in molten (p.10) tin. **tinning** (*n*).

contact process a process for making concentrated sulphuric acid by converting (p.73) sulphur dioxide to sulphur trioxide and then converting the trioxide to sulphuric acid. The process is shown in the diagram below. Sulphur dioxide and oxygen (from air) are passed over a catalyst (p.72) of vanadium pentoxide at 500°C, forming sulphur trioxide. The sulphur trioxide is absorbed in concentrated sulphuric acid forming oleum (↓), which is diluted to form sulphuric acid of 98% concentration.

contact process

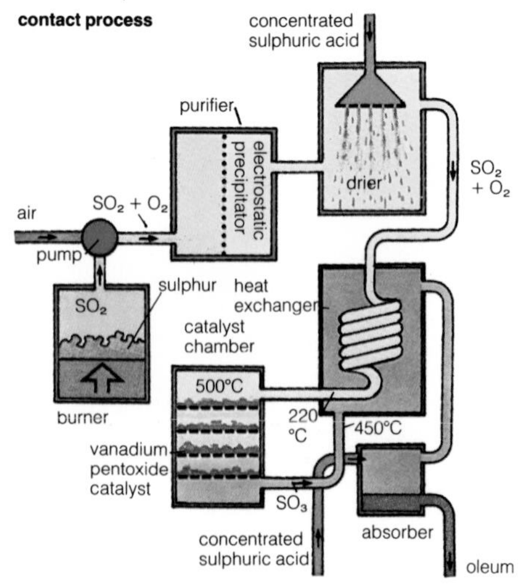

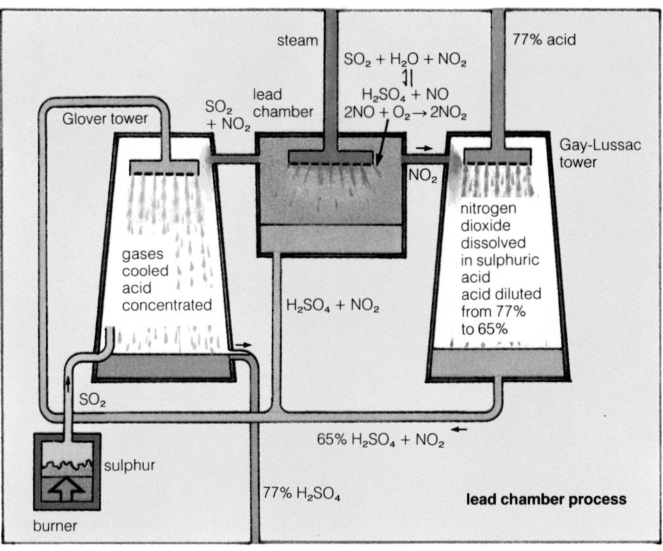

lead chamber process

lead-chamber process an older process for the manufacture of concentrated sulphuric acid. Sulphur dioxide is oxidized to sulphur trioxide by nitrogen dioxide: $SO_2 + NO_2 \rightarrow SO_3 + NO$ The nitrogen oxide (NO) is oxidized by oxygen to nitrogen dioxide; the nitrogen compounds act like a catalyst. The process is shown in the diagram. This process produces a 77% concentration of acid, not as concentrated and not as pure as the acid from the Contact process, but still useful for manufacturing processes.

oleum (*n*) concentrated sulphuric acid containing dissolved sulphur trioxide; also known as fuming sulphuric acid.

spent oxide iron (III) oxide is used to remove sulphur compounds from coal gas; the oxide eventually contains a high percentage of sulphur and is no longer of use for purifying (p.43) the coal gas; it then becomes spent oxide, i.e. it has been used up. Spent oxide is burned to form sulphur dioxide for use in the manufacture of sulphuric acid.

water gas a mixture of hydrogen and carbon monoxide formed by blowing steam over red-hot coke (p.156). The reaction cools the coke.

producer gas a mixture of nitrogen and carbon monoxide formed by blowing air through coke (p.156). The reaction heats the coke.

semi-water gas a mixture of hydrogen, carbon monoxide and nitrogen, formed by blowing steam and air alternately over coke (p.156). One reaction heats, and the other reaction cools, the coke, so its temperature is maintained.

Bosch process a process for making hydrogen from water gas (↑). Water gas and steam are passed over a catalyst of iron with traces (p.20) of chromium (III) oxide as a promoter (p.72). At 450°C, the reaction is:

$$H_2 + CO + H_2O \rightleftharpoons CO_2 + 2H_2$$

The carbon dioxide is removed by washing with hot potassium carbonate solution. Natural gas, containing methane, is now used in this process instead of water gas.

mercury cathode cell a method manufacturing sodium hydroxide and chlorine from sodium chloride. A solution of sodium chloride is electrolyzed (p.123) using carbon anodes (p.123) and a mercury cathode. Chlorine is taken away by pipes from the cell. The cathode forms a sodium amalgam (p.55), which is removed and taken to another tank where sodium hydroxide is formed and hydrogen set free.

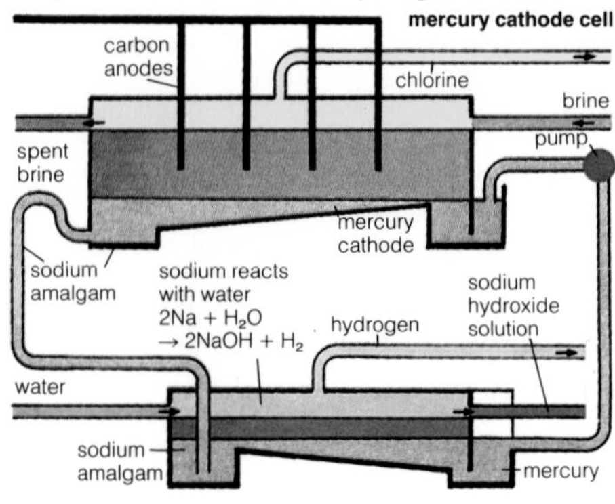

lime (*n*) lime is either **quicklime**, calcium oxide, or **slaked lime**, calcium hydroxide.

soda (*n*) soda is either **caustic soda**, sodium hydroxide, or **washing soda**, sodium carbonate.

brine (*n*) a solution of sodium chloride.

Castner-Kellner process another name for the mercury cathode cell (↑).

Kellner-Solvay process another name for the mercury cathode cell (↑).

Solvay process a process for the manufacture of sodium carbonate ($Na_2CO_3.10H_2O$) using sodium chloride and calcium carbonate as raw materials. Brine (↑) is saturated with ammonia and passed down a column, up which is passed carbon dioxide gas. The reaction which takes place is:
$NaCl + NH_3 + H_2O + CO_2 \rightarrow NaHCO_3 + NH_4Cl$
The sodium hydrogen carbonate ($NaHCO_3$) is heated to form sodium carbonate and carbon dioxide. The sodium carbonate is recrystallized to form washing soda ($Na_2CO_3.10H_2O$). The carbon dioxide and ammonia from the ammonium chloride are put back into the cycle (p.64) of operations (p.157) as shown in the diagram below. Calcium chloride is produced as a waste product.

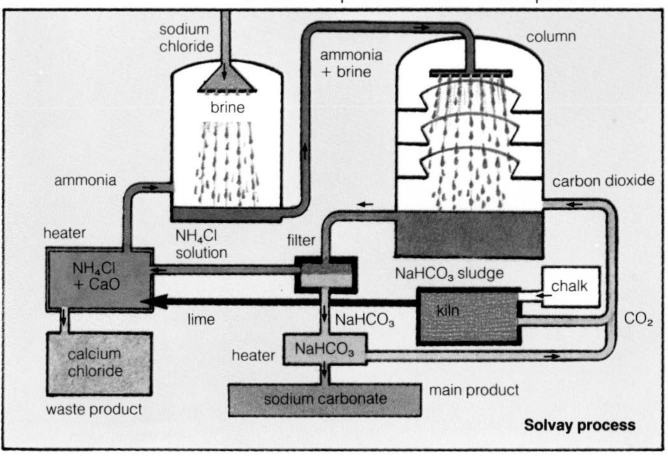

Solvay process

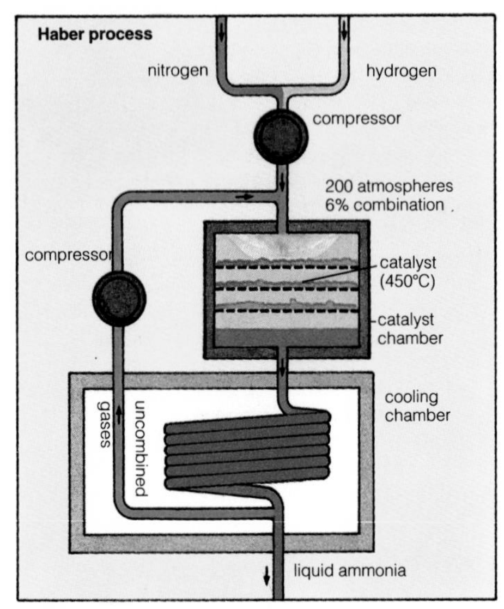

Haber process a process for the manufacture of ammonia by synthesis (p.200) from nitrogen and hydrogen. The two elements combine in the presence of a catalyst (iron with aluminium oxide as a promoter (p.72) under high pressure at 450°C – 500°C. The reaction is reversible (p.64) and only 6% of the elements combine. The ammonia is liquefied by cooling and then removed (p.215); the uncombined gases are passed back to the catalyst chamber.

kiln (*n*) a furnace (p.164) for making bricks, ceramics (↓) and heating chalk to form lime and carbon dioxide.

spent (*adj*) describes something which has been used and is finished; also something which has had an important constituent used, so it can no longer carry out its purpose, e.g. spent brine which is too dilute to be of use.

waste (*adj*) describes something that has been made but is not wanted, e.g. paper that is no longer wanted is waste paper, the paper is usable but not wanted for its original purpose.

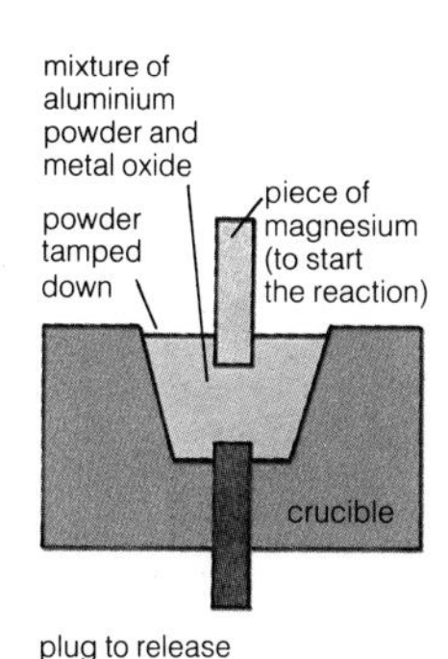

Thermit process

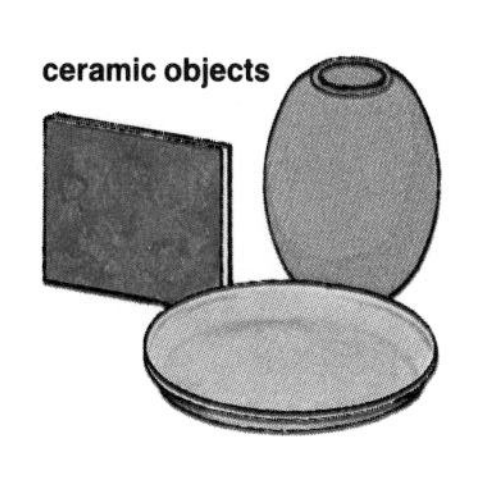
ceramic objects

obsolete (*adj*) describes something which is usable but has been replaced (p.68) by something which is better, and so is no longer used, e.g. retorts (p.28) are obsolete, as distillation flasks (p.28) have replaced them.

scrap (*n*) describes something which is no longer used, e.g. when a car is no longer used it becomes scrap. The steel in machines which are no longer used, is scrap steel. Metals are most commonly described as scrap.

Thermit process a process for the extraction (p.34) of chromium, manganese or tungsten from their oxides or for welding (p.164) iron objects. Aluminium powder is mixed with the metal oxide and a piece of magnesium is set alight to start the reaction. The aluminium reduces the oxide and, in the case of iron oxide, produces molten (p.10) iron to weld two iron objects together.

detergent (*n*) any substance used for removing (p.215) dirt. Soaps are detergents. Other substances used as detergents are made from sulphonic acids. Their molecules consist of hydrocarbon chains attached to acidic groups. *See sulphonate (p.193).*

vulcanization (*n*) the process for changing rubber from a weak material into a hard, strong material, e.g. the rubber used in car tyres has been hardened by vulcanization. Vulcanization is usually carried out by heating rubber with sulphur. **vulcanize** (*v*).

ceramics (*n*) the manufacture of earthenware and porcelain objects. **ceramic** (*adj*).

tamp (*v*) to push soft, powdery solids into a hole and to fill the hole completely, e.g. in the Thermit process, the aluminium powder and metal oxide is tamped into the crucible. **tamping** (*n*).

heavy chemical a chemical in great demand for industrial (p.157) processes and thus manufactured (p.157) in large quantities. Such chemicals are often not very pure (p.20). Examples of heavy chemicals include sulphuric acid, nitric acid, lime and sodium carbonate. *See fine chemicals (p.20).*

hydrocarbon (*n*) an organic (p.55) compound containing (p.55) only the elements carbon and hydrogen; all hydrocarbons are covalent (p.136) compounds. Hydrocarbons include alkanes (↓), alkenes (↓), alkynes (p.174) and benzene (p.179) compounds.

series [2] (*n*) a group of organic (p.55) compounds which all have similar chemical properties; physical properties showing a regular change with an increasing number of carbon atoms; can be prepared (p.43) by similar chemical methods; described by a general formula. The compounds are homologous (↓).

homologous (*adj*) describes structures which are alike without being exactly the same. In homologous series (↑), all the compounds have a general formula (p.181), the same functional group (p.185) and a gradation of properties (p.9), e.g. the alkanes (↓) form an homologous series; the carboxylic acids form an homologous series, each acid having the functional group – COOH. **homologue** (*n*).

alkane (*n*) a hydrocarbon (↑) with the general formula (p.181) $C_n H_{2n+2}$. The alkanes form an homologous (↑) series; the first four members are: CH_4 (n = 1); C_2H_6; C_3H_8; C_4H_{10} (n = 4). Alkanes are either straight chain (p.182) or branched chain compounds. The first four members of the alkanes are methane, ethane, propane and butane; thereafter they have a Greek number prefix for the number of carbon atoms, followed by *-ane*, e.g. hexane, C_6H_{14}. The alkanes have properties similar to methane (↓), becoming less reactive (p.62) with an increasing number of carbon atoms, and passing from gases through liquids to solids for members with a large number of carbon atoms. They are saturated (p.185) organic compounds.

methane (*n*) an alkane (↑) of formula CH_4. It is an odourless (p.15), colourless (p.15), inflammable (p.21) gas. The hydrogen atom can be replaced by halogens (p.117), otherwise methane is unreactive (p.62).

homologous series of alkanes

formula	melting point °C	boiling point °C
CH_4	−183	−162
C_2H_6	−184	−89
C_3H_8	−188	−42
C_4H_{10}	−138	−1
C_5H_{12}	−130	36
C_6H_{14}	−95	69
C_7H_{16}	−91	98
C_8H_{18}	−57	126

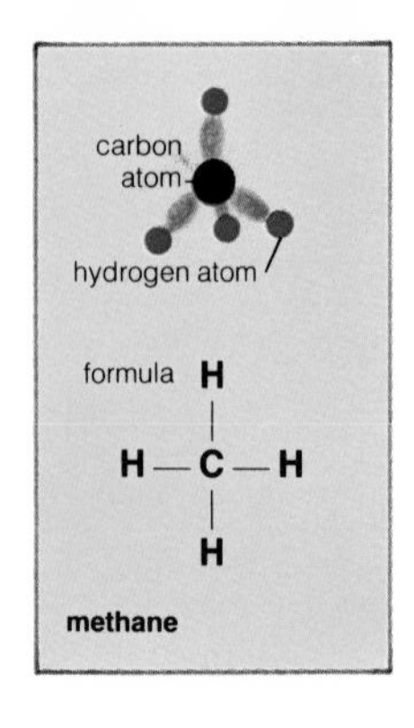

methane

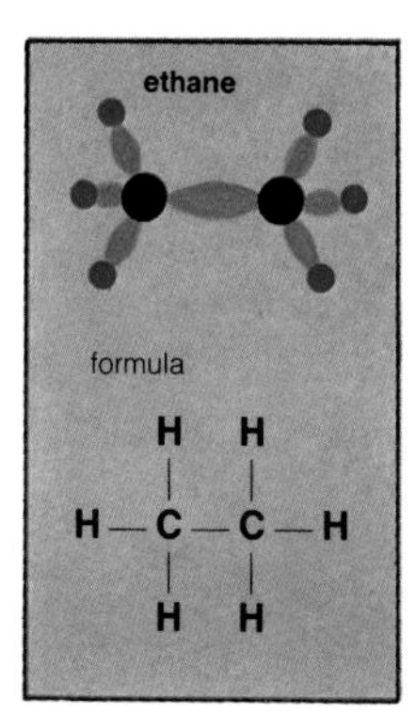

hydrogen atom
carbon atom

ethane (*n*) the second member of the alkanes (↓), formula C_2H_6. Its properties are similar to those of methane (↑), but it is less reactive, has a higher boiling point and a greater density.
paraffin[2] (*n*) traditional name (p.44) for an alkane (↑).

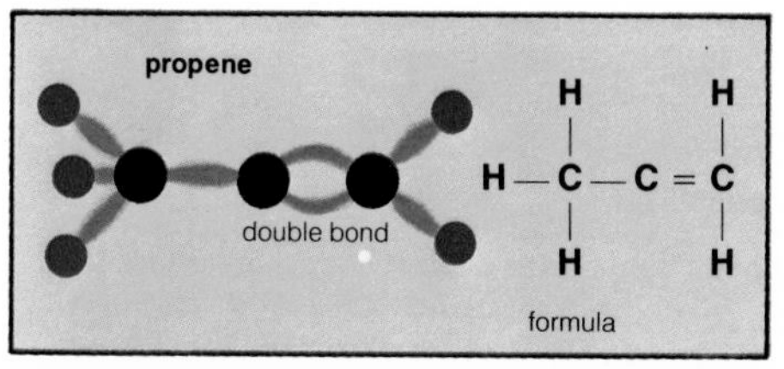

H H H H
H—C—C—C=C
H H H H
formula
$CH_3{\cdot}CH_2{\cdot}CH = CH_2$
a butene
but-1-ene

H H H H
H—C—C=C—C—H
H H
formula
but-2-ene
a butene
$CH_3{\cdot}CH = CH{\cdot}CH_3$

alkene (*n*) a hydrocarbon (↑) with the general formula (p.181) C_nH_{2n}. The alkenes form an homologous (↑) series; the first three members are: C_2H_4 (n = 2); C_3H_6; C_4H_8 (n = 4). Alkenes are either straight chain (p.182) or branched chain compounds. The first four members of the alkenes are: ethene, propene, butene and pentene; thereafter they have a Greek number prefix for the number of carbon atoms, followed by *-ene*, e.g. hexene, C_6H_{12}. The alkenes have properties similar to ethene (p.174), becoming less reactive with an increasing number of carbon atoms, and passing from gases through liquids to solids for members with a large number of carbon atoms. They are unsaturated (p.185) organic compounds.
olefine or **olefin** (*n*) the traditional name (p.44) for an alkene (↑).

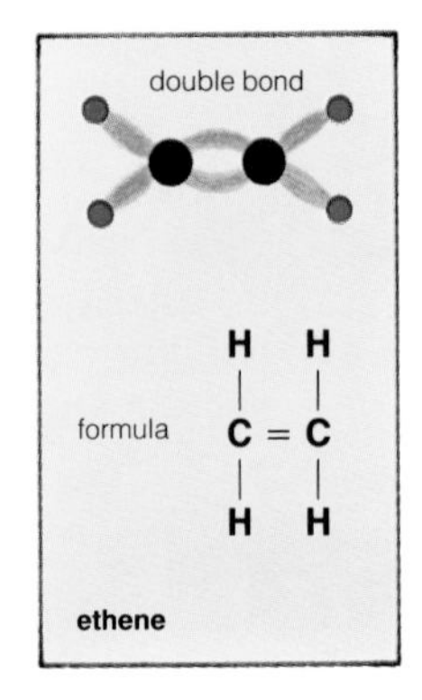

ethene (*n*) an alkene of formula C_2H_4. It is a colourless (p.15) gas with a sweet smell. The two carbon atoms are joined by a double bond (p.181) which makes ethene reactive (p.62). It undergoes (p.213) addition reactions (p.188), polymerization (p.207) and combustion (p.58).

ethylene (*n*) traditional name (p.44) for ethene (↑).

alkyne (*n*) a hydrocarbon (p.172) with the general formula (p.181) C_nH_{2n-2}. The alkynes form an homologous (p.172) series; the first three members are: C_2H_2 (n = 2); C_3H_4; C_4H_6 (n = 4). Alkynes are either straight chain (p.182) or branched chain compounds. The first three members of the alkynes are ethyne, propyne and butyne; thereafter they have a Greek number prefix for the number of carbon atoms, followed by *-yne*, e.g. hexyne, C_6H_{10}. The alkynes have properties similar to ethyne (↓), becoming less reactive with an increasing number of carbon atoms. They are unsaturated (p.185) organic compounds.

acetylenes (*n.pl.*) traditional name (p.44) for the alkynes (↑).

ethyne

triple bond

formula H—C≡C—H

ethyne (*n*) an alkyne (↑) of formula C_2H_2. It is a colourless (p.15) gas with a sweet smell. Ethyne burns with a bright, white flame (p.58) and is used for lighting. The two carbon atoms are joined by a triple bond (p.181) which makes ethyne very reactive (p.62). Ethyne undergoes (p.213) addition reactions (p.188) and polymerization (p.207). With alkali metals (p.117), acetylides (p.49) are formed.

acetylene (*n*) traditional name (p.44) for ethyne.

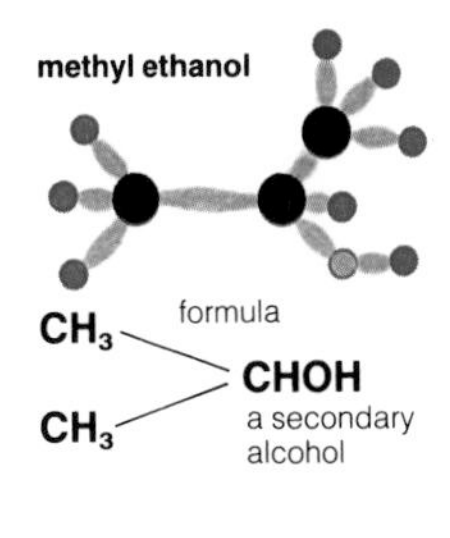

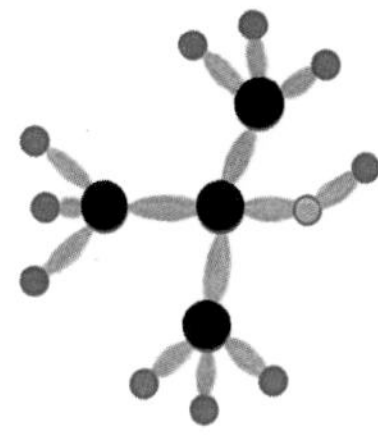

trimethyl methanol

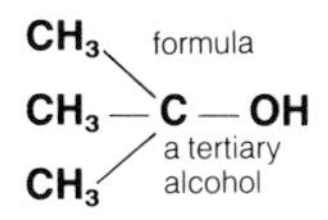

alcohol[1] (*n*) an organic (p.55) compound which contains one or more hydroxyl (p.185) groups. Alcohols are formed from alkanes (p.172) by substituting (p.188) a hydroxyl group for a hydrogen atom; they form homologous series (p.172) corresponding (p.233) to the alkanes. Alcohols are classified as primary, secondary or tertiary alcohols depending on the functional group (p.185). For primary alcohols, the functional group is $-CH_2OH$; for secondary alcohols it is $=CHOH$; for tertiary alcohols it is $\equiv C.OH$. Alcohols are also classified according to the number of hydroxyl groups in the molecule:

monohydric alcohol	CH_3CH_2OH	ethanol
dihydric alcohol	CH_2OH – CH_2OH	ethane-1,2-diol (glycol)
trihydric alcohol	CH_2OH – $CHOH$ – CH_2OH	propane-1,2,3-triol (glycerol)

Alcohols react with alkali metals (p.117) evolving (p.40) hydrogen and forming an alkoxide; they burn readily and are oxidized to aldehydes (↓), ketones (p.176) or carboxylic acids (p.176) depending on the alcohol and the strength of the oxidizing agent. With organic acids alcohols form esters (p.177). The hydroxyl group can be replaced by a halogen (p.117).

alcohol[2] (*n*) trivial name (p.44) for ethanol.

aldehyde (*n*) an organic (p.55) compound which contains the functional group $-CHO$. An aldehyde is formed as a first oxidation product (p.62) from the corresponding (p.233) alcohol (↑), e.g. ethanol is oxidized to ethanal. The name of an aldehyde is derived (p.106) from the corresponding alcohol by changing *-ol* to *-al*. Aldehydes are reduced to alcohols; they form addition (p.188) compounds with sodium hydrogen sulphite, hydrogen cyanide and other compounds.

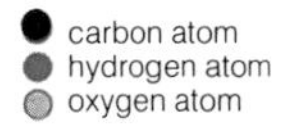

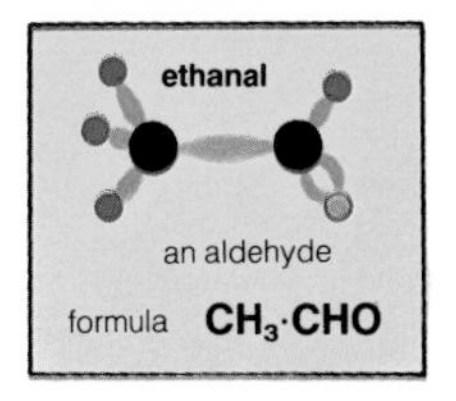

ketone (*n*) an organic (p.55) compound containing the functional group =CO. Ketones are prepared (p.43) from the corresponding secondary alcohol (p.175) by oxidation. They are good solvents for organic compounds and are less reactive than aldehydes (p.175), although their reactions are similar to those of aldehydes.

carboxylic acid an organic (p.55) compound containing the functional group –COOH. Carboxylic acids are prepared (p.43) from the corresponding alcohol (p.175) by complete oxidation; they are named from that alcohol, e.g. ethanol on oxidation forms ethanoic acid, the ending *-ol* is replaced by *-oic*. The

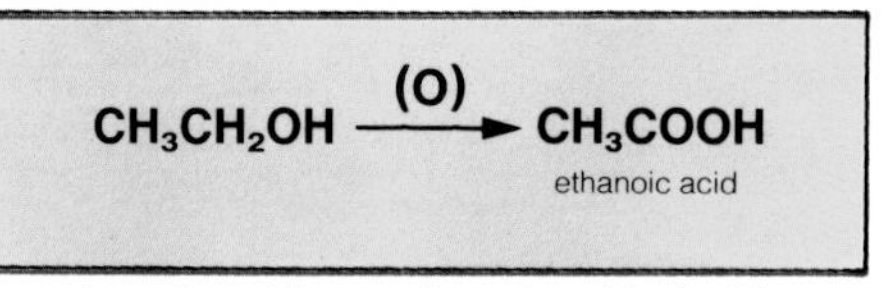

carboxylic acids are all weak acids (p.45). With alcohols, they form esters (↓); the hydroxyl group (p.185) can be replaced by a halogen (p.117); dehydration forms an acid anhydride (↓). The carboxylic acids form an homologous series (p.172).

dicarboxylic acid an organic (p.55) compound with two functional groups of –COOH. The simplest member is ethane di-oic acid (traditional name (p.44) oxalic acid). The dicarboxylic acids are stronger acids than the carboxylic acids; they are dibasic (p.46).

acid anhydride an organic (p.55) compound prepared from the corresponding (p.233) acid by dehydration or by the action of an acyl chloride (↓) on the sodium salt of the acid.

acyl chloride an organic (p.55) compound containing the functional group –COCl; it is the functional group –COOH with the hydroxyl group replaced (p.68) by chlorine. Acyl chlorides are prepared from the corresponding carboxylic acid (↑) by the action of phosphorus pentachloride. They are very reactive (p.62) compounds with reactions similar to those of the corresponding carboxylic acid.

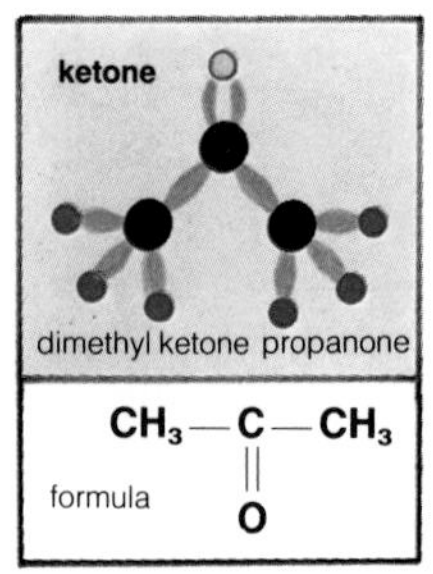

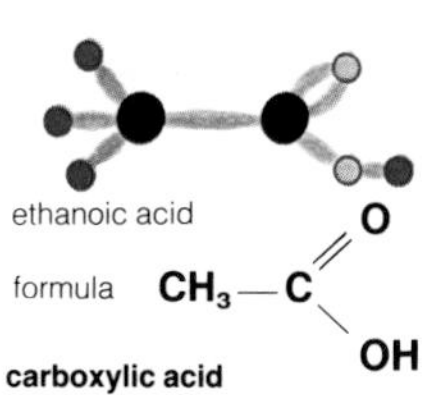

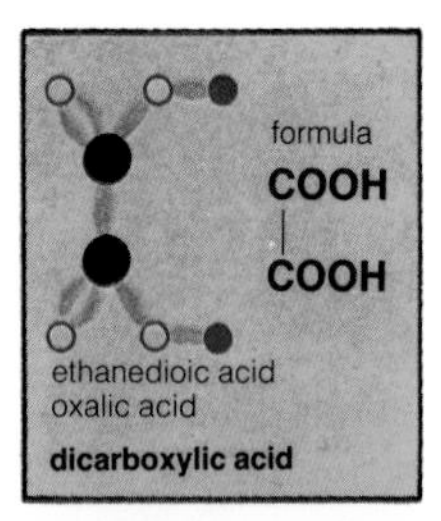

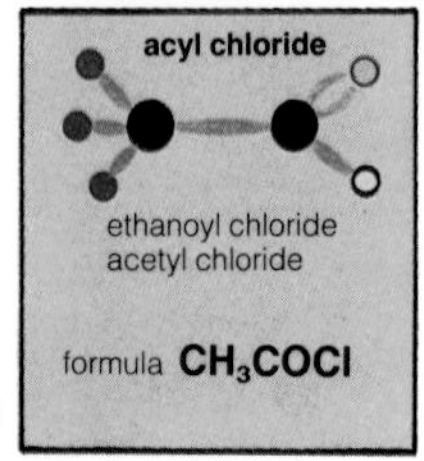

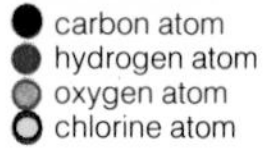

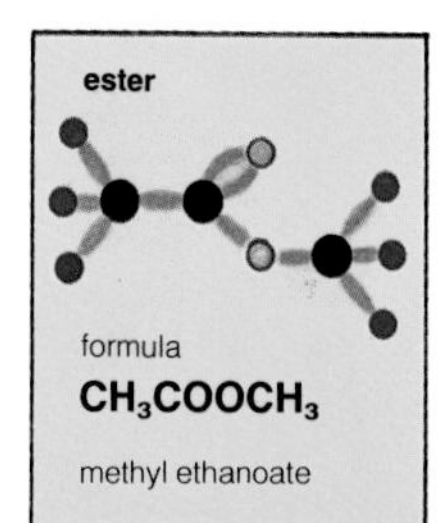

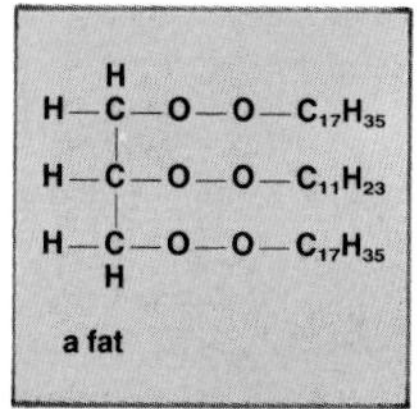

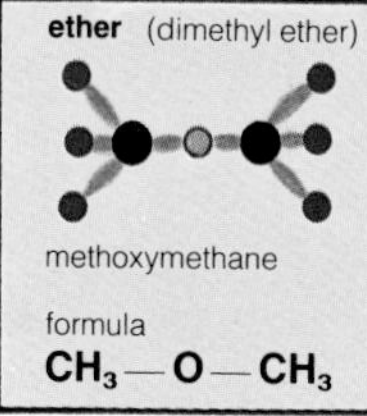

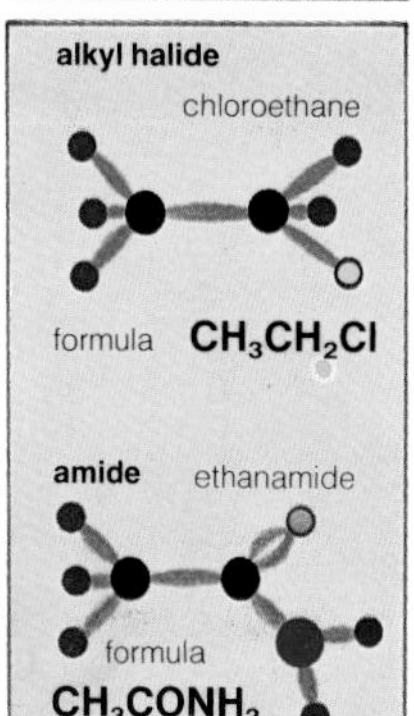

ester (*n*) an organic (p.55) compound formed when an alcohol (p.175) reacts with a carboxylic acid (↑). The hydrogen of the acid is replaced (p.68) by an alkyl (p.180) or any aryl (p.180) group. The reaction is reversible (p.64), e.g.

$CH_3CH_2OH + CH_3COOH \quad CH_3COOC_2H_5 + H_2O$

ethanol + ethanoic acid → ethyl ethanoate + water

An ester can also be formed with an inorganic (p.55) acid. Esters have fragrant (p.22) odours and are the odours of flowers and fruit; they are not very reactive, but undergo (p.213) hydrolysis (p.66).

fat (*n*) an ester (↑) of propane-1,2,3-triol (glycerol) and different carboxylic acids (↑). In fats, the acids are saturated (p.185), straight chain (p.182) compounds. In oils, the acids are unsaturated (p.185), straight chain compounds. The important acids in animal and plant fats are palmitic acid ($C_{15}H_{31}COOH$), stearic acid ($C_{17}H_{35}COOH$) and lauric acid ($C_{11}H_{23}COOH$).The structure of a fat is shown in the diagram opposite.

ether (*n*) an organic (p.55) compound with two hydrocarbon (p.172) groups, either alkyl (p.180) or aryl (p.180) joined to one oxygen atom. Ethers are very inflammable (p.21) but otherwise are unreactive (p.62).

alkyl halide an organic (p.55) compound formed when one hydrogen atom of an alkane (p.172) is replaced (p.68) by a halogen (p.117) atom. They are named from the halogen and the alkane, e.g. chloroethane, bromopropane, iodoethane. Alkyl chlorides are prepared from alcohols (p.175) by the action of sulphur dichloride oxide (thionyl chloride, $SOCl_2$). They are very reactive, and used in many organic preparations (p.43).

amide (*n*) an organic (p.55) compound containing the functional group $-CONH_2$, e.g. ethanamide CH_3CONH_2. Amides are named from the corresponding carboxylic acid (↑), e.g. propanoic acid forms propanamide. They are formed by dehydrating (p.66) the ammonium salt of the corresponding carboxylic acid.

amine (*n*) an organic (p.55) compound containing the functional group $-NH_2$. Amines are formed by the reaction between ammonia and an alkyl halide (p.177). The amines can be classified by the number of hydrogen atoms in ammonia that are replaced by alkyl (p.180) or aryl groups, e.g.

Primary amines	CH_3NH_2	methylamine	$R\text{-}NH_2$
Secondary amines	$(CH_3)_2NH$	dimethylamine	$R_2\text{-}NH$
Tertiary amines	$(CH_3)_3N$	trimethylamine	$R_3\text{-}N$

The amines are colourless (p.15) gases or liquids with a strong fishy odour; they are soluble in water forming weak bases, which form salts with inorganic (p.55) acids, e.g. methylamine reacts with hydrochloric acid to form methylammonium chloride $(CH_3NH_3)^+\ Cl^-$

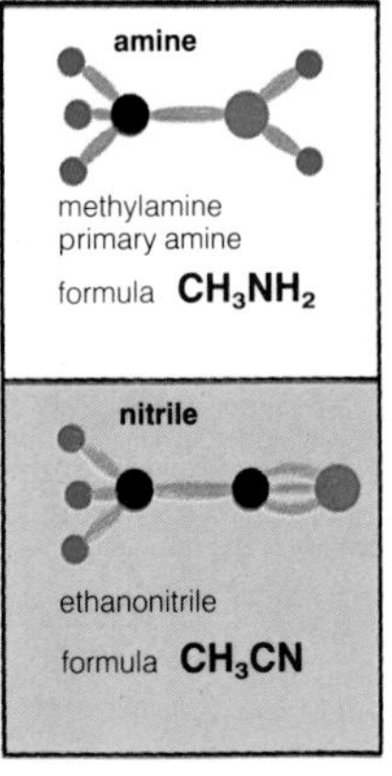

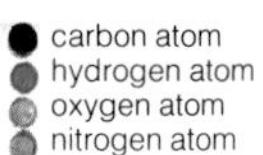

nitrile (*n*) an organic (p.55) compound containing the functional group $-CN$. The nitriles are named after the corresponding (p.233) alkane with the same number of carbon atoms, e.g. CH_3CH_2CN is propanonitrile. Nitriles are prepared by the action of sodium cyanide (NaCN) on an alkyl halide (p.177). They undergo (p.213) hydrolysis forming first an amide (p.177), then a carboxylic acid (p.176); they undergo reduction (p.193) to an amine (↑).

isocyanide (*n*) an organic (p.55) compound containing the group $-NC$; they are given names such as isocyanoethane (C_2H_5NC). Isocyanides are prepared by the action of silver cyanide (AgCN) on the corresponding (p.233) alkyl halide (p.177). They are colourless (p.15) liquids with a bad odour and are highly poisonous. Isocyanides are hydrolyzed (p.190) to primary amines and reduced (p.193) to secondary amines.

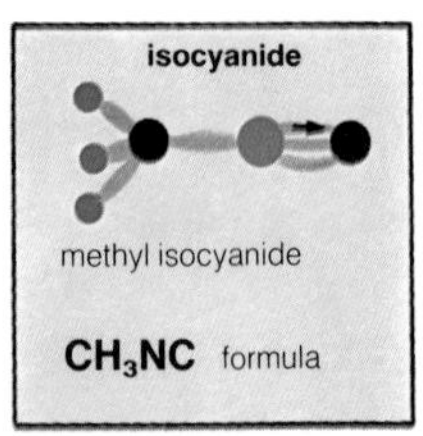

amino acid a carboxylic acid (p.176) with one hydrogen atom replaced by an amino group ($-NH_2$), e.g. $CH_3.NH_2COOH$, aminoethanoic acid, with a trivial name (p.44) of glycine. Amino acids in solution form both cations (p.125) and anions (p.125) and thus are amphoteric (p.46) in their reactions; they behave as carboxylic acids and as amines. Amino acids are the constituents (p.54) of proteins (p.209); they are capable of forming long chains by condensation (p.191).

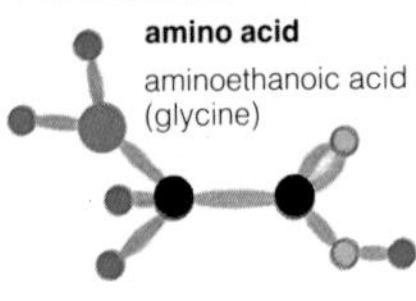

$CH_3 \cdot NH_2 \cdot COOH$ formula

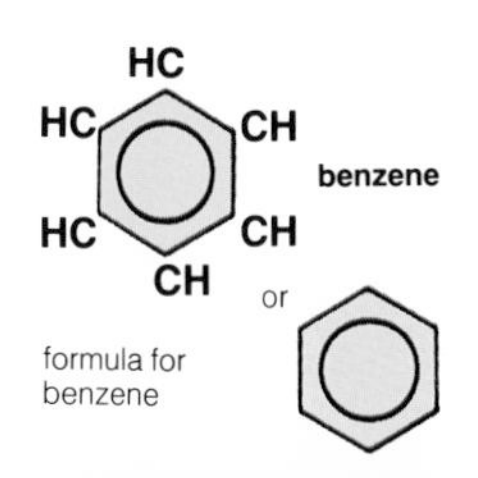

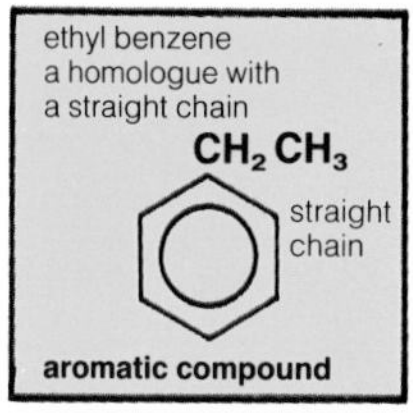

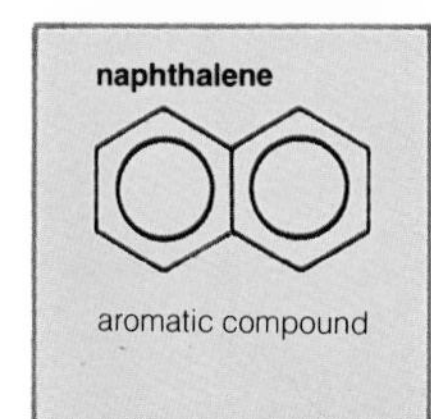

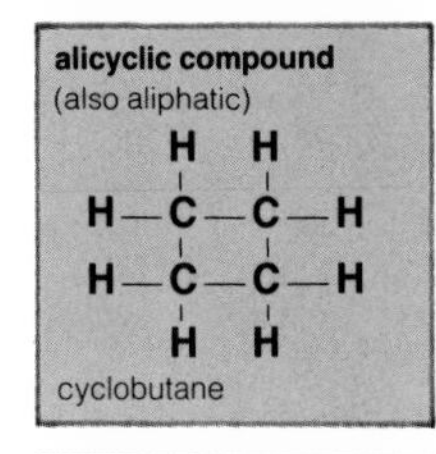

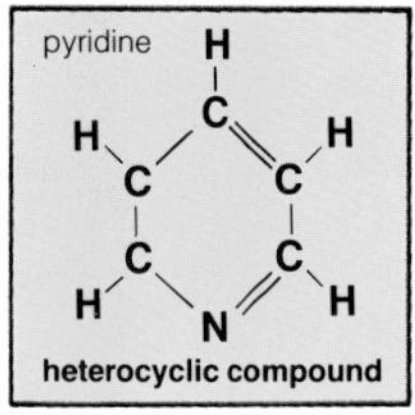

benzene (*n*) a hydrocarbon (p.172) of formula C_6H_6. The six carbon atoms are combined (p.64) in a ring structure, with all six bonds of equal length and equal activity. The bonds are neither single bonds (p.181) nor double bonds, but have a chemical character peculiar to themselves; this property is called **aromaticity**. Benzene is the first member of two homologous series (p.172); in one, chains (p.182) of carbon atoms are joined to the benzene ring; in the other, two or more benzene rings are combined. The formula of benzene is shown as a ring with a circle inside it; the circle represents 6 electrons which do not take part in any one bond, and are known as **delocalized electrons**. The delocalized electrons give benzene its aromaticity, or aromatic character. Benzene is a colourless (p.15) liquid with a pleasant (aromatic) smell, is inflammable (p.21), a good solvent for organic compounds and is chemically reactive (p.62).

naphthalene (*n*) a hydrocarbon (p.172) of formula $C_{10}H_8$. The ten carbon atoms are combined (p.64) in two benzene (↑) ring structures. Naphthalene is a white solid with a strong odour; it is a major (p.226) constituent (p.54) of coal tar and is chemically reactive (p.62). It is a homologue (p.172) of benzene.

aromatic (*adj*) describes a compound containing (p.55) a benzene (↑) ring. The compound has chemical properties similar to those of benzene.

aliphatic (*adj*) describes a compound consisting of straight chains (p.182) and branched chains, e.g. as in the alkanes, alkenes, and alkynes. Aliphatic compounds include alicyclic (↓) compounds.

alicyclic (*adj*) describes a compound consisting of single bonds, or double bonds, with the carbon atoms combined in a ring. Such compounds are aliphatic (↑) and not aromatic (↑), e.g. cyclobutane is an alicyclic compound.

heterocyclic (*adj*) describes a compound with an aliphatic (↑) ring structure, but including at least one atom which is not carbon, e.g. pyridine C_5H_5N.

alkyl (*adj*) describes an organic (p.55) group formed when one atom of hydrogen is taken from an alkane (p.172); examples of alkyl groups are: methyl (CH_3); ethyl (C_2H_5); pentyl (C_5H_{11}). The general formula of an alkyl group is C_nH_{2n+1}. **alkylation** (*n*).

acyl (*adj*) describes an organic (p.55) group formed when a hydroxyl group (−OH) is taken from a carboxylic acid (p.176). *See acyl chloride (p.176).*

aryl (*adj*) an aromatic (p.179) group formed when one atom of hydrogen is taken from an aromatic hydrocarbon (p.172). **arylation** (*n*).

phenol (*n*) an aromatic (p.179) compound of formula C_6H_5OH. It is a white crystalline (p.15) solid with a characteristic odour (p.15); it reacts (p.62) as a weak acid (p.45), forming **phenates**. Phenol is obtained from coal tar; it is used as a common antiseptic and is also important in the plastics (p.210) industry. Homologues (p.172) of phenol are called **phenols**.

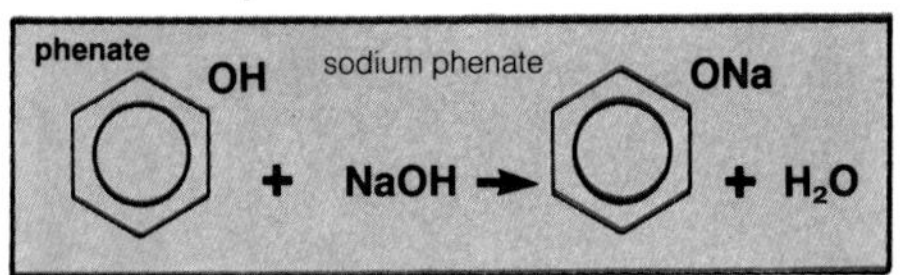

nitro compound a combination of an aryl (↑) group and a nitro group ($-NO_2$). Nitro compounds are important because (a) aromatic (p.179) hydrocarbons are easily nitrated by concentrated nitric acid, and (b) they are chemically reactive (p.62). Both nitro and dinitro compounds can be prepared. Alkyl nitro compounds also exist (p.213) but are less important.

diazonium salt a combination of an aryl (↑) group with the azo group $-N=N$ which forms a salt with an inorganic acid radical, e.g. $(C_6H_5-N=N)^+ Cl^-$ which is benzenediazonium chloride. Diazonium salts are prepared from primary aromatic (p.179) amines by the action of nitrous acid. The salts are unstable unless kept at a temperature below 0°C. They are important because many dyes (p.162) are manufactured from diazonium salts.

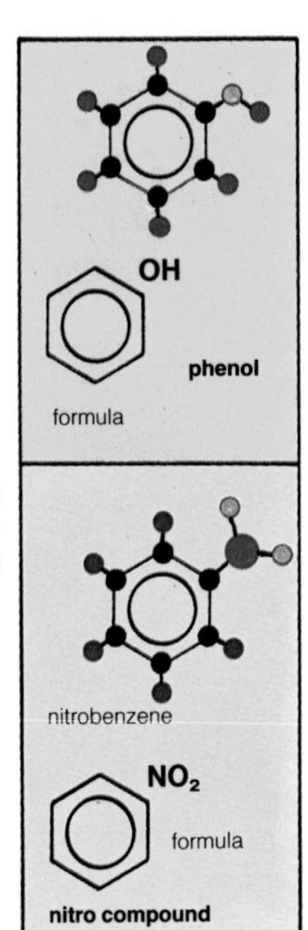

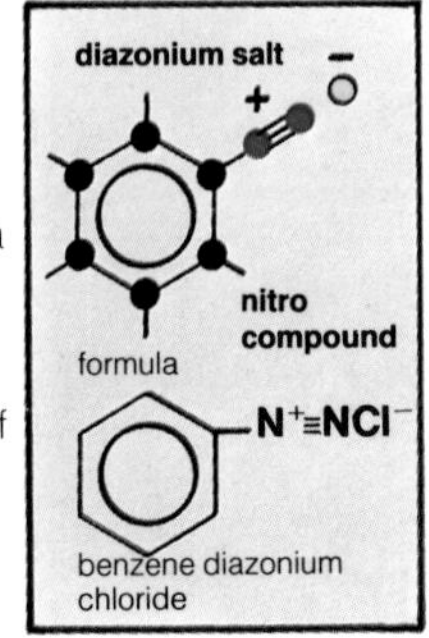

C:H:O = 1:2:1
by mass
empirical formula
CH_2O
for ethanoic acid
relative molecular mass for ethanoic acid is 60

molecular formula $C_2H_4O_2$

structural formula
CH_3COOH

carbon atom
hydrogen atom
oxygen atom

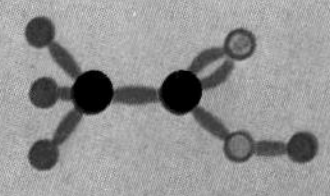

ethanoic acid

graphic formula

empirical formula a formula (p.78) which gives the simplest ratio of elements in a compound, e.g. the empirical formula of ethanoic acid is CH_2O.

molecular formula a formula (p.78) which shows the number of atoms of each element in a molecule of a compound. This formula is found from the relative molecular mass of a compound. For example, CH_2O is the empirical formula (↑) of ethanoic acid and of glucose (p.206). Their respective (p.233) relative molecular masses are 60 and 180. The relative molecular mass of CH_2O is 30, hence ethanoic acid is $C_2H_4O_2$ and glucose is $C_6H_{12}O_6$. These are molecular formulae.

structural formula a formula (p.78) which shows how the atoms in a molecule are grouped together, e.g. the structural formula of ethanoic acid is CH_3COOH. A structural formula of an organic (p.55) compound indicates (p.38) how the compound will react.

graphic formula a formula (p.78) which shows how the atoms in a molecule are oriented (p.93) in space, relative (p.232) to each other.

general formula a formula (p.78) which allows the formula for any member of an homologous series (p.172) to be written, e.g. C_nH_{2n} is the general formula for alkenes. Putting n = 2,3,4, etc. gives the formula for members of the series.

single bond a covalent bond (p.136), it is oriented (p.93) in space and allows atoms to turn relative (p.232) to each other.

double bond two covalent bonds (p.136) joining together two atoms; each covalent bond is formed by two atoms sharing two electrons. The double bond between two carbon atoms is important, it prevents the atoms turning about the bond and also can cause geometrical isomerism (p.184). A double bond makes an organic (p.55) compound reactive (p.62).

triple bond three covalent bonds (p.136) joining together two atoms; each covalent bond is formed by two atoms sharing two electrons. The triple bond between two carbon atoms is important. It has the same effect as a double bond (↑); but is much more reactive.

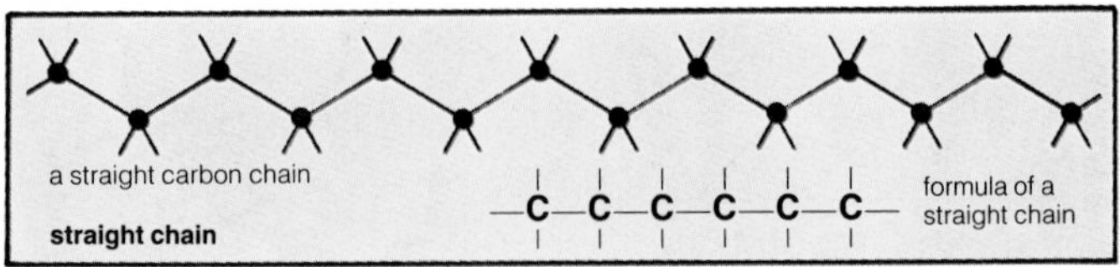

straight chain

chain (*n*) a structure in which each carbon atom is joined to the next carbon atom to form a chain, or line, of atoms. A chain may be straight (↓) or branched (↓).

straight chain a chain of carbon atoms in which any one carbon atom is not joined to more than two other carbon atoms, e.g. as in hexane.

branched chain a chain of carbon atoms in which one or more carbon atoms may be joined to 3 or 4 other carbon atoms. *See diagram.*

cyclic chain a chain of carbon atoms forming a circle in which the carbon atoms may be joined by either single or double bonds. The resulting (p.39) compound has an aliphatic (p.179) nature.

ring chain a chain of carbon atoms which can be cyclic (↑) or can be a benzene (p.179) ring with neither single nor double bonds.

isomerism (*n*) the property of having the same molecular formula (p.181), but different structural formulae, e.g. ethanol and methoxymethane (dimethyl ether) have the same molecular formula, but their structural formulae are $CH_3.CH_2OH$ and $CH_3.O.CH_3$ respectively (p.233). There are different kinds of isomerism. Isomerism is very common in organic (p.55) compounds. **isomer** *(n)*, **isomeric** *(adj)*.

isomer (*n*) one of two, or more, compounds which exhibit (p.221) isomerism (↑), e.g. ethanol and methoxymethane are isomers.

structural isomerism isomerism (↑) exhibited by two, or more, organic (p.55) compounds which have the same molecular formulae, but their different structures give them different physical or chemical properties, e.g. ethanol and methoxymethane are structural isomers; a branched chain and a straight chain alkane (p.172) with the same number of carbon atoms are isomers of each other. *See diagram.*

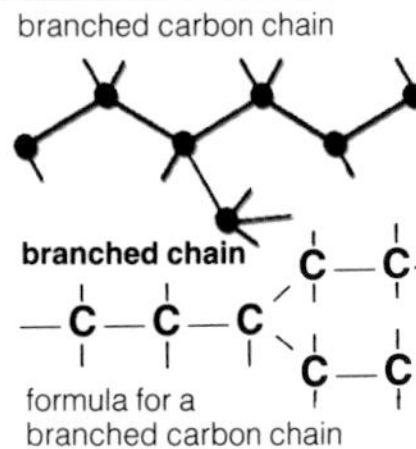

branched chain

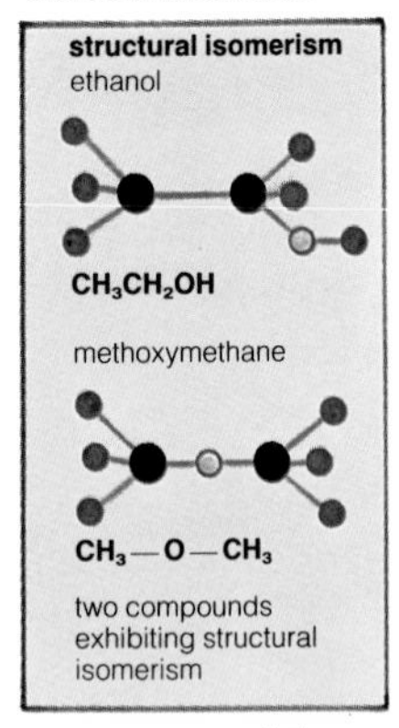

structural isomerism

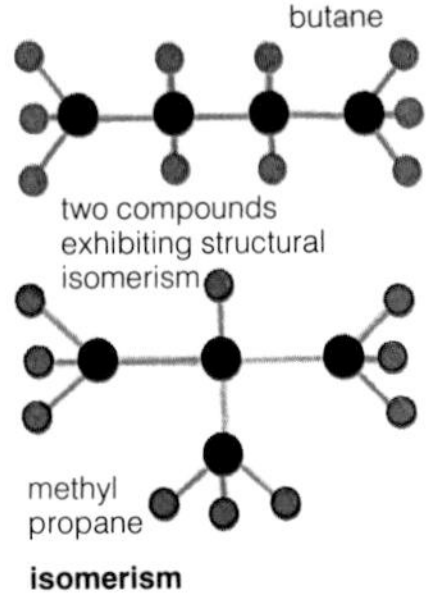

isomerism

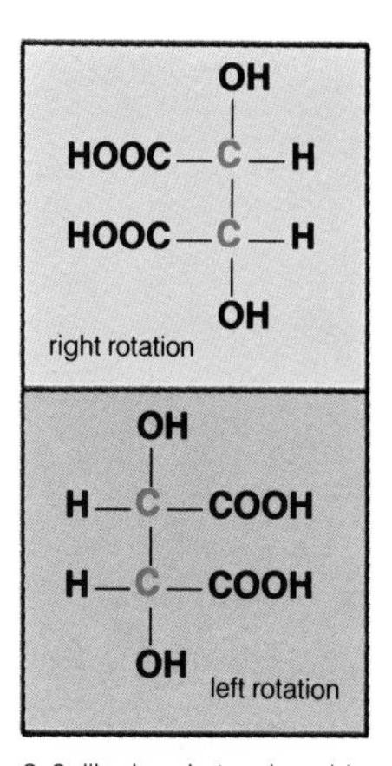

2, 3 dihydroxybutanoic acid
tartaric acid

(two asymmetric carbon atoms in the molecule) C

stereoisomerism (*n*) isomerism that results from a different arrangement in space of the atoms in a molecule. The structural formula does not show the different arrangements which are possible, but the graphic formula of such compounds can show the difference. Stereoisomerism results from the tetrahedral (p.83) directions of the four covalent bonds of carbon. There are two kinds of stereoisomerism: optical isomerism (↓) and geometrical isomerism (p.184).

optical isomerism optical isomerism exists when a molecule is not identical (p.233) with its image in a mirror. Two spatial (p.211) arrangements of the atoms are thus possible. The two structures are chemically identical, but one structure rotates (p.218) the plane of polarized light to the left and the other structure rotates it to the right; otherwise their physical properties are the same, e.g. lactic acid exhibits (p.221) optical isomerism.

enantiomorph (*n*) (1) a molecule exhibiting (p.221) optical isomerism (↑) which is one of two or more isomeric (↑) forms. (2) one of two crystalline forms when each crystal is the mirror image of the other. **enantiomorphic** (*adj*).

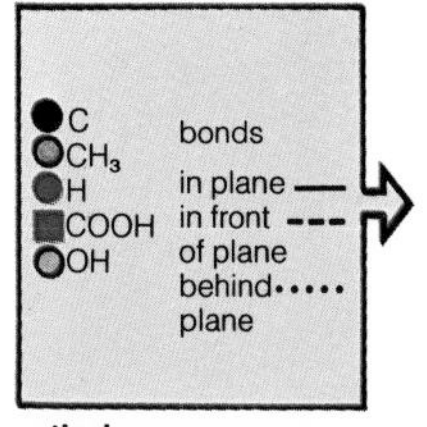

optical isomerism

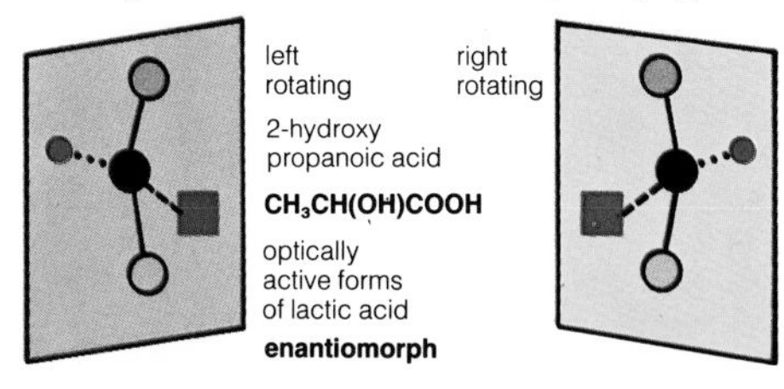

enantiomorph

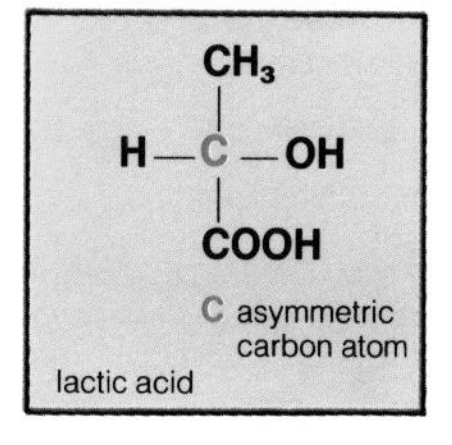

lactic acid

asymmetric carbon atom a carbon atom which has four different atoms or groups attached to it. It is the most common cause of optical isomerism (↑), e.g. lactic acid.

racemate (*n*) a physical mixture of two enantiomorphs (↑), one with rotation of the plane of polarized light to the right, and the other to the left. The result is optically inactive. When optically active compounds are synthesized, a racemate is produced. **racemic** (*adj*).

geometrical isomerism geometrical isomerism results from two conditions: (1) two carbon atoms joined by a double bond; (2) each carbon atom must have two different atoms or groups joined to it. A double bond prevents rotation (p.218) of the carbon atoms, so two spatial (p.211) arrangements of the atoms are possible. Geometrical isomers have very different physical properties, e.g. but-2-ene has one isomer with a melting point of −139°C and the other −106°C. Their chemical properties tend to be the same, but certain differences also exist, e.g. the difference between maleic and fumaric acids, which are isomers. **geometrical isomer** (*n*).

***cis*-configuration**

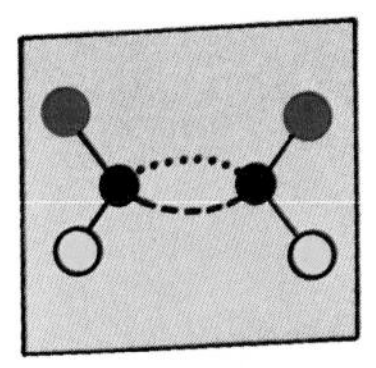

***trans*-configuration**

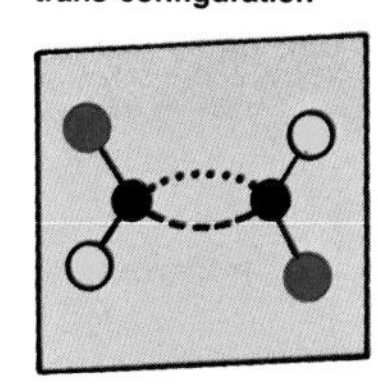

geometrical isomers

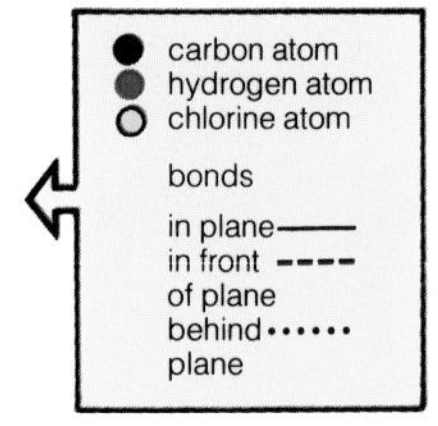

ClCH = CHCl

1, 2 dichloroethane

***cis*-configuration** (*n*) a geometrical isomer (↑) with two like groups on the same side of the double bond. *See diagram.*

***trans*-configuration** (*n*) a geometrical isomer (↑) with two like groups on opposite sides of the double bond. *See diagram.*

tautomerism (*n*) a state of dynamic equilibrium (p.150) between spontaneously convertible (p.73) isomers. Generally in the conversion of one isomer to another, a hydrogen atom changes its place in the molecular structure, e.g. as in ethyl 3-oxobutanoate (ethyl acetoacetate). The two isomeric forms in tautomerism are the 'keto' and 'enol' forms, shown in the diagram opposite.

tautomer (*n*) an individual isomer of tautomerism (↑); the 'keto' or 'enol' form of the tautomeric compound. **tautomeric** (*adj*).

methyl (*adj*) the alkyl (p.180) radical $-CH_3$.

ethyl (*adj*) the alkyl (p.180) radical $-C_2H_5$.

propyl (*adj*) the alkyl (p.180) radical $-C_3H_7$.

butyl (*adj*) the alkyl (p180) radical $-C_4H_9$.

pentyl (*adj*) the alkyl (p.180) radical $-C_5H_{11}$.

tautomerism

ethyl 3 oxobutanate (ethyl acetoacetate)

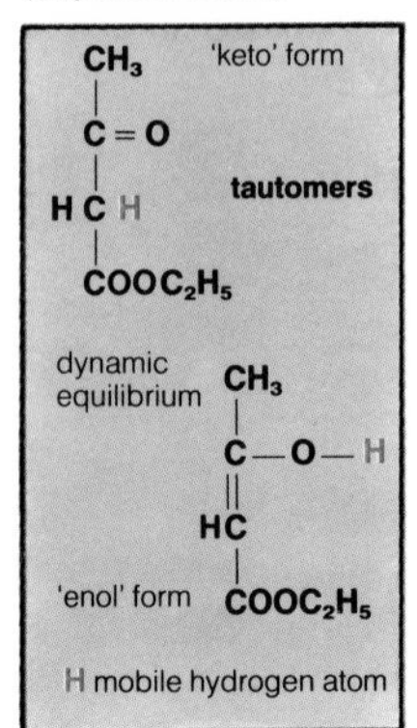

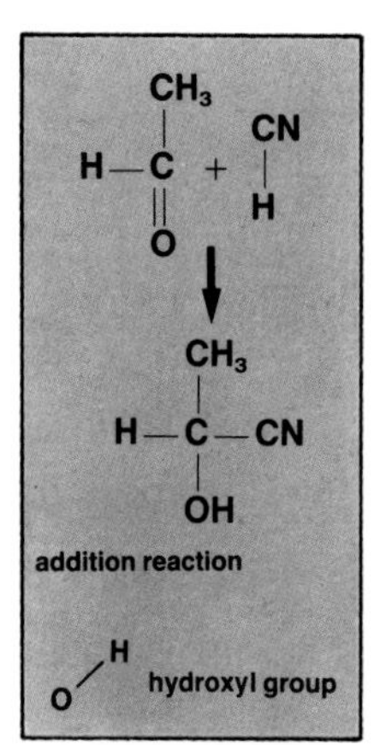

saturated[2] (*adj*) describes a carbon chain in which there are only single bonds connecting individual carbon atoms. A saturated chain is unreactive, as alkanes (p.172) are. **saturation** (*n*).

unsaturated[2] (*adj*) describes a carbon chain in which there is at least one double or triple bond. An unsaturated chain is reactive, particularly for addition reactions (p.188).

polyunsaturated (*adj*) describes a carboxylic acid containing more than two double bonds.

functional group an atom, or group of atoms, which gives an organic (p.55) compound its reactivity. A functional group is joined to an alkyl (p.180) or aryl group, which forms the homologous series (p.172) and the functional group gives the chemical characteristics. A long carbon chain with many carbon atoms reduces (p.219) the chemical activity of the functional group. Examples of functional groups are: hydroxyl, carboxyl, amino, chloro. In general formulae an alkyl group is represented by R and an aryl group by **Φ** or Ph, e.g. RCOOH is a general formula for a carboxylic acid; ΦNO_2 is an aromatic (p.179) nitro compound.

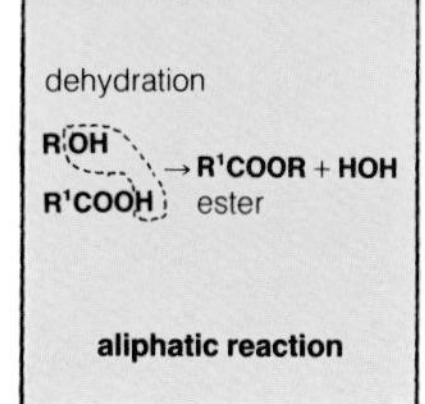

hydroxyl group the functional group (↑) –OH; it occurs (p.154) in alcohols (p.175) and phenols (p.180). Combined with alkyl groups, the hydroxyl group reacts with the hydrogen of acids to form water. Combined with aryl groups, the hydrogen of the hydroxyl group forms hydrogen ions, making the phenols weakly acidic.

monohydric (*adj*) describes an alcohol with one hydroxyl group. *See alcohol (p.175).*

dihydric (*adj*) describes an alcohol with two hydroxyl groups. *See alcohol (p.175).*

trihydric (*adj*) describes an alcohol with three hydroxyl groups. *See alcohol (p.175).*

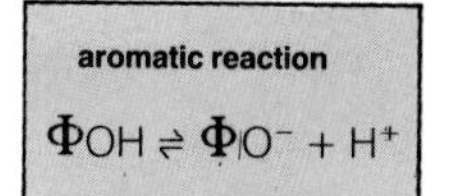

carbonyl group the functional group (↑) = CO; it occurs in aldehydes (p.175) and ketones (p.176). The carbonyl group has a double bond (p.181) between the carbon and oxygen. This double bond undergoes (p.213) addition reactions (p.188). A hydrogen atom joined to the carbon atom of the carbonyl group makes it more reactive, as in aldehydes.

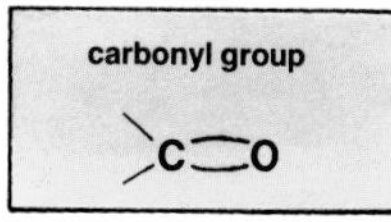

carboxyl group the functional group (p.185) $-COOH$. It forms weak acids and by neutralization it forms salts, e.g. sodium ethanoate. With alcohols (p.175) it forms esters (p.177).

sulphonate group the functional group (p.185) $-SO_2.OH$. It occurs almost entirely combined with aryl groups (p.180). Aromatic (p.179) sulphonic acids are prepared by the action of concentrated sulphuric acid on benzene (p.179) and its homologues; they form salts with alkalis, and the salts are used as detergents (p.171).

nitro group the functional group (p.185) $-NO_2$. Aliphatic (p.179) nitro compounds exhibit (p.221) tautomerism (p.184). They are of less importance than the aromatic (p.179) nitro compounds. Benzene and its homologues are converted to nitro compounds by nitration (p.193). Aromatic nitro compounds are reduced (p.70) to amines in several stages; some intermediate compounds are useful industrially. *See equations opposite*.

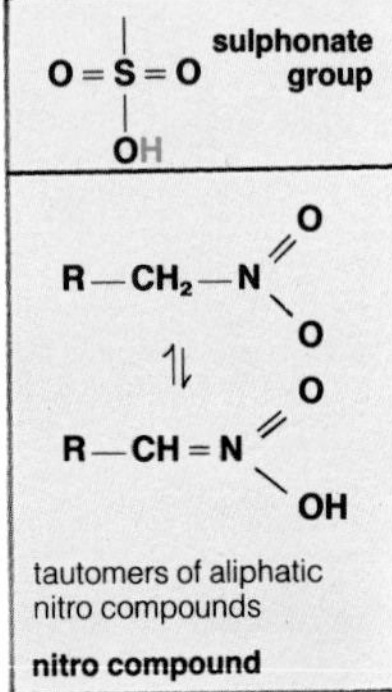

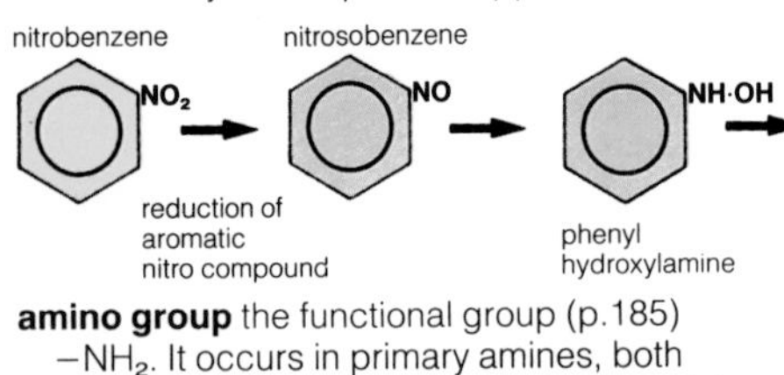

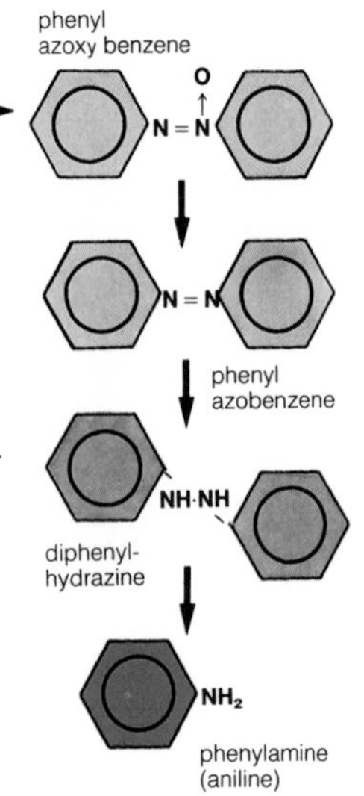

amino group the functional group (p.185) $-NH_2$. It occurs in primary amines, both aliphatic (p.179) and aromatic; in the amides as part of the amido (↓) group; and in the amino acids. The amino group causes amines and amino acids to have a basic (p.46) reaction.

amido group the functional group (p.185) $-CONH_2$. It occurs in amides and causes them to be amphoteric (p.46) in nature. The group is dehydrated to form a nitrile (p.178).

azo group the functional group (p.185) $-N{=}N-$. It occurs in diazo compounds (p.180) and is a highly reactive group.

cyano group the functional group (p.185) $-C{=}N$. It occurs in the nitriles (p.178). The cyano group is hydrolyzed (p.66) to the carboxyl group and reduced (p.70) to a primary amine.

chloro group the functional atom (p.185) –Cl. It occurs in the alkyl chlorides (p.177) and the aromatic chloro compounds. The alkyl chlorides are reactive, taking part in substitution (p.188) and elimination (p.189) reactions. The aromatic chloro compounds are relatively (p.232) inert (p.19). The presence of a nitro group (p.186) in the aromatic hydrocarbon increases the reactivity of the chloro group.

bromo group the functional atom (p.185) –Br. It occurs in the alkyl bromides, e.g. bromoethane CH_3CH_2Br, and in the aromatic (p.179) bromo compounds. The alkyl and aromatic bromo compounds are more reactive than the corresponding (p.233) chloro compounds, e.g. bromobenzene is more reactive than chlorobenzene.

bromo ethane
CH_3CH_2Br

iodo group the functional atom (p.185) –I. It occurs in the alkyl (p.180) iodides, e.g. iodoethane CH_3CH_2I, and in the aromatic (p.179) iodo compounds, e.g. iodobenzene C_6H_5I. The alkyl and aromatic iodo compounds are more reactive than the corresponding bromo compounds (↑).

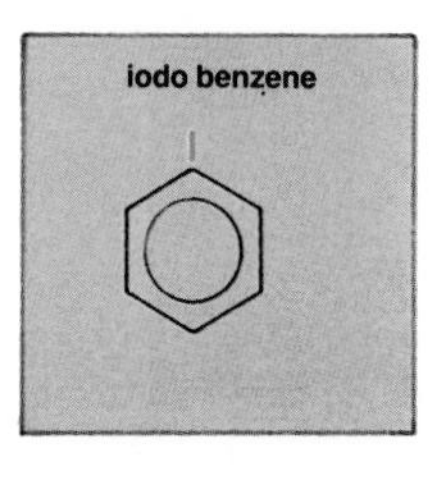

chromophore (*n*) any chemical group which causes a compound to have a distinctive (p.224) colour. In synthetic (p.200) organic dyes (p.162), such groups as the azo group cause the colour of a compound. **chromophoric** (*adj*).

auxochrome (*n*) a chemical group which deepens the colour of a compound, where red is taken as the deepest colour of the spectrum and violet as the least deep, e.g. the addition of an auxochrome can change a blue substance to a green substance. In addition, an auxochrome intensifies (p.230) the colour. The amino group (↑) is an example of an auxochrome. **auxochromic** (*adj*).

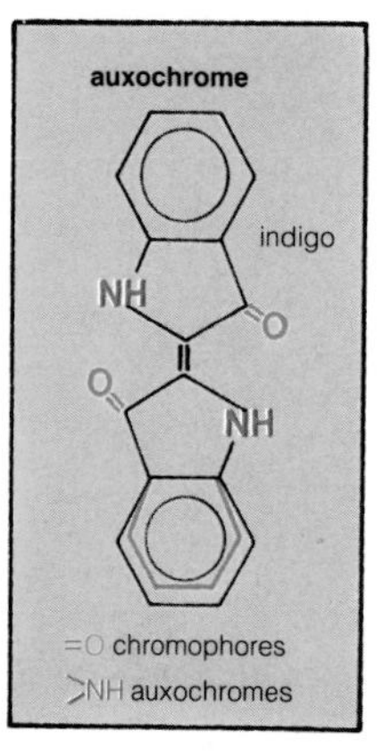

leuco base a compound formed from a dye (p.162) by reduction (p.70); it is colourless. An insoluble dye is converted to a leuco base which is soluble in alkalis; the alkaline solution is used for dyeing and the dye colour returned by oxidation, e.g. indigo is converted to a leuco base and the colour returned by subsequent oxidation.

leuco compound a name for leuco base (↑).

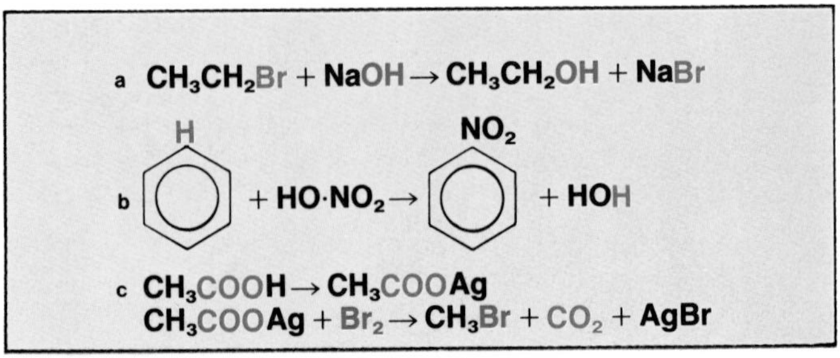

a $CH_3CH_2Br + NaOH \rightarrow CH_3CH_2OH + NaBr$

b $C_6H_5H + HO{\cdot}NO_2 \rightarrow C_6H_5NO_2 + HOH$

c $CH_3COOH \rightarrow CH_3COOAg$
$CH_3COOAg + Br_2 \rightarrow CH_3Br + CO_2 + AgBr$

substitution (*n*) a process in which an atom or a functional group in an organic (p.55) compound is replaced, directly or indirectly, by another atom or functional group. Substitution is one of the most important organic reactions. Examples of substitution are: (a) substituting a hydroxyl group for a halogen in an aliphatic (p.179) compound; (b) substituting a nitro group (p.186) for a hydrogen atom in benzene (p.179); (c) substituting a bromine atom for a carboxyl group in an aliphatic compound. **substitute** (*v*), **substitution** (*adj*).

addition (*n*) a process in which two substances react to form only one substance. In organic reactions, addition usually takes place across a double bond (p.181), e.g. hydrogen chloride adds across the double bond of ethene. **addition** (*adj*).

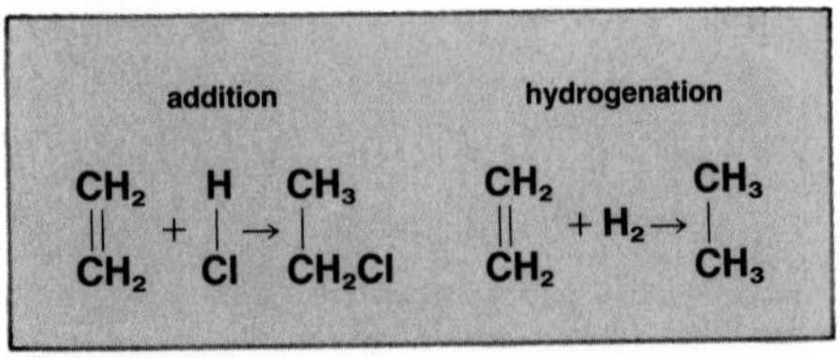

addition

$CH_2{=}CH_2 + HCl \rightarrow CH_3{-}CH_2Cl$

hydrogenation

$CH_2{=}CH_2 + H_2 \rightarrow CH_3{-}CH_3$

hydrogenation (*n*) an addition process (↑) in which hydrogen is added to a molecule in the presence of a suitable catalyst (often finely divided nickel) and at a suitable temperature and pressure. Hydrogenation converts unsaturated (p.185) compounds into saturated compounds. **hydrogenate** (*v*).

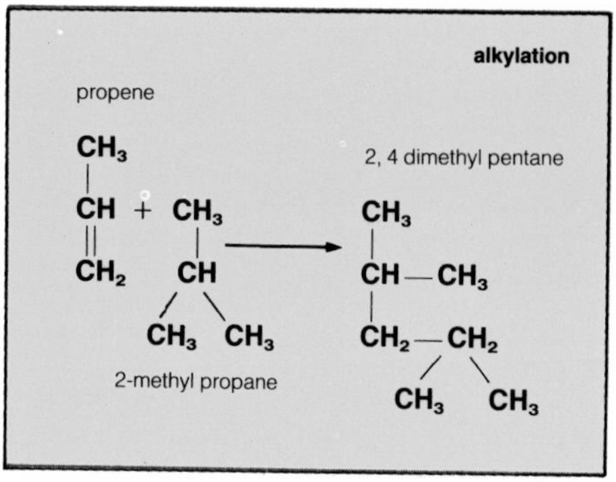

alkylation

propene

CH_3 | CH || CH_2 + CH_3 | CH / \ CH_3 CH_3 → CH_3 | $CH—CH_3$ | $CH_2—CH_2$ / \ CH_3 CH_3

2-methyl propane

2, 4 dimethyl pentane

alkylation (*n*) an addition process (↑) in which an alkane (p.172) undergoes (p.213) addition across the double bond of an alkene (p.173) to form a branched chain (p.182) alkane, e.g. propene and 2-methyl propane form 2,4,dimethylpentane. Alkylation is also the substitution (↑) of an alkyl (p.180) group in place of a hydrogen atom. *See Friedel-Crafts reaction (p.199).*

elimination (*n*) the removal (p.215) of atoms from a molecule; this usually results in the formation of a double bond, e.g. the removal of the elements of hydrogen and a halide (p.50) from an alkyl halide (p.177) to form an alkene. This is done by the action of a hot, concentrated solution of potassium hydroxide in ethanol (called alcoholic potash), e.g. bromoethane is converted to ethene. Other types of elimination include dehydrogenation (p.190) and dehydration (p.190).

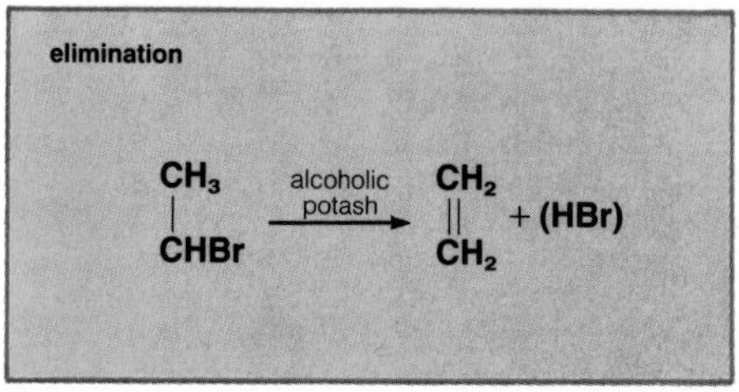

elimination

CH_3 | $CHBr$ —alcoholic potash→ CH_2 || CH_2 + (HBr)

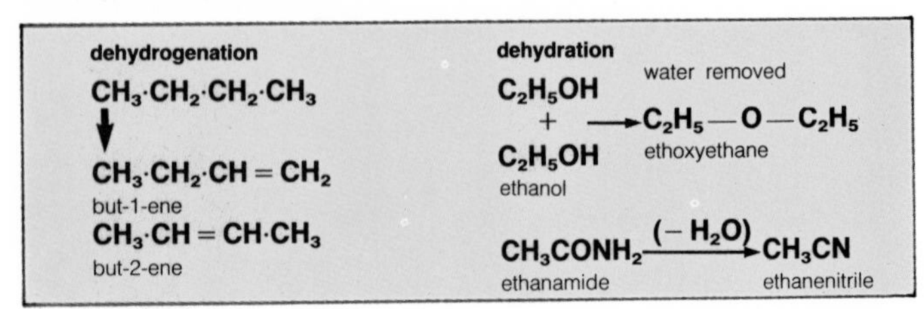

dehydrogenation (*n*) the removal (p.215) of combined hydrogen from an organic (p.55) compound; the process uses a suitable catalyst (e.g. aluminium and chromium oxides) at a high temperature (e.g. 600°C). For example, butane is dehydrogenated to butene. This is an elimination (p.189) process. **dehydrogenate** (*v*).

dehydration (*n*) the removal (p.215) of the elements of water from an organic (p.55) compound. For example, (a) the action of excess concentrated sulphuric acid on ethanol produces ethoxyethane (diethyl ether); (b) the action of phosphorus pentoxide on ethanamide produces ethanenitrile. This is an elimination (p.189) process. **dehydrate** (*v*).

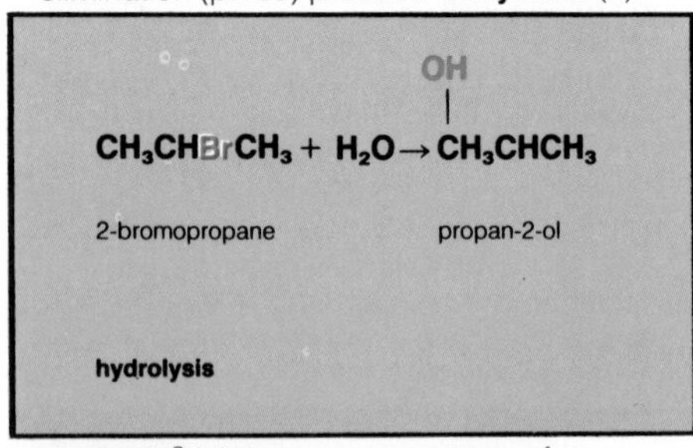

hydrolysis[2] (*n*) the reverse process of esterification (↓) and of condensation (↓), and also the substitution of a hydroxyl group into a compound. Hydrolysis of esters is catalyzed by hydrogen ions in acids or by hydroxyl ions in alkalis. Condensation products can be hydrolyzed by boiling with dilute acids. Water alone hydrolyzes alkyl halides (p.177). **hydrolyze** (*v*).

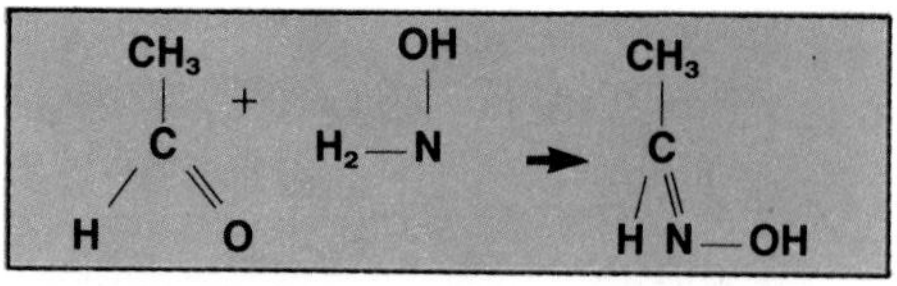

CH_3 + OH
C H_2—N → CH_3
H O C
H N—OH

condensation[3] (*n*) a reaction in which two organic (p.55) compounds combine to form one compound with the elimination of water, hydrogen chloride, ammonia, or other compounds with similar small molecules. The most common compound eliminated is water. Examples of condensation are (a) aldehydes and hydroxylamine condensing to aldoximes; (b) polymerization (p.207) with condensation, as in the urea-methanol condensation polymerization.

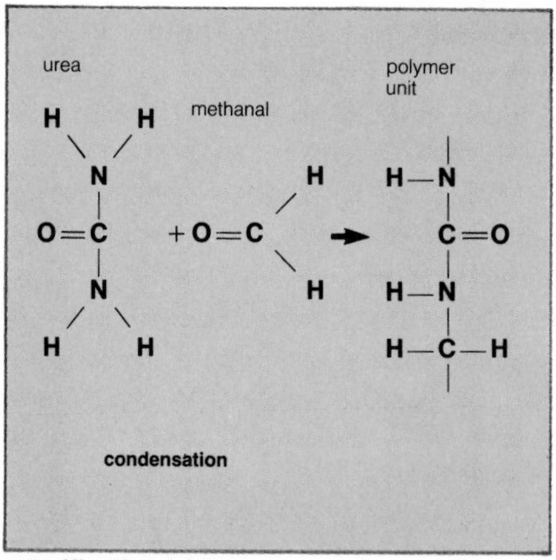

urea
methanal
polymer unit
H H
N H H—N
O=C +O=C → C=O
N H H—N
H H H—C—H
condensation

esterification (*n*) the process of converting an acid into an ester (p.177). The reaction is reversible, so the water formed in the reaction is removed by concentrated sulphuric acid to move the equilibrium mixture (p.150) in the direction of the ester. For example, the esterification of ethanoic acid with methanol:

$$CH_3COOH + H.CH_2OH \rightleftharpoons CH_3COOCH_3 + H_2O$$

This is a form of condensation (↑).

$$CH_3COOCH_3 + NaOH \rightarrow CH_3OH + CH_3COONa$$

methyl ethanoate → methanol + sodium ethanoate

saponification

saponification (*n*) hydrolysis (p.190) of an ester using boiling aqueous sodium or potassium hydroxide. The alkali combines with and removes (p.215) the acid from the products, and shifts the equilibrium mixture (p.150) towards the hydrolyzed product. **saponify** (*v*).

halogenation (*n*) the process of substituting (p.188) a halogen (p.117) atom for a hydrogen atom in a molecule of an organic (p.55) compound, e.g. the halogenation of phenol (p.180).

chlorination (*n*) halogenation (↑) when chlorine is the halogen. Chlorinating agents for aliphatic (p.179) compounds include phosphorus pentachloride, phosphorus trichloride and sulphur dichloride oxide (thionyl chloride, $SOCl_2$). For aromatic (p.179) compounds, a 'halogen carrier', such as aluminium chloride or powdered iron, is used and chlorine is passed through at room temperature, causing substitution in the benzene ring (p.179). **chlorinate** (*v*).

$$CH_3CH_2COOH + Cl_2 \rightarrow CH_3CHClCOOH + HCl$$

propanoic acid → 2-chloro propanoic acid

chlorination

bromination

$$CH_3{\cdot}CO{\cdot}CH_3 + Br_2 \rightarrow CH_3COCH_2Br + HBr$$

propanone → bromopropanone

bromination (*n*) halogenation (↑) when bromine is the halogen. For aliphatic (p.179) compounds, red phosphorus and bromine are used as brominating agents. For aromatic (p.179) compounds, a 'halogen carrier' is used as for chlorination (↑). **brominate** (*v*).

iodination (*n*) halogenation (↑) when iodine is the halogen. For aliphatic (p.179) compounds, red phosphorus and iodine are used as iodinating agents. For aromatic (p.179) compounds, iodine in the presence of mercury (II) oxide is used.

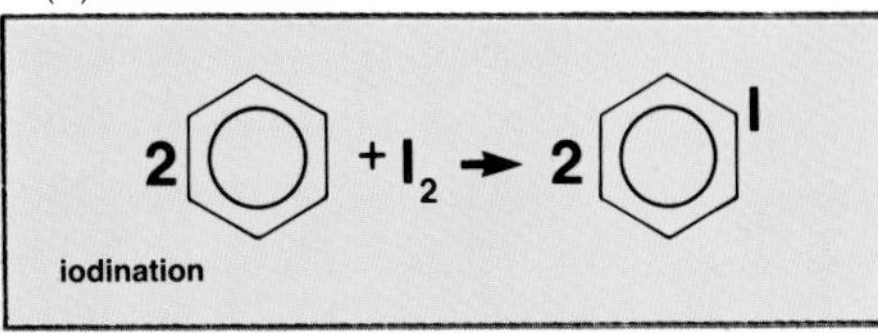

reduction² (*n*) the reduction of organic (p.55) compounds is either hydrogenation (p.188) or the removal of oxygen, e.g. the reduction of ethanoic acid to ethanol or to ethane.

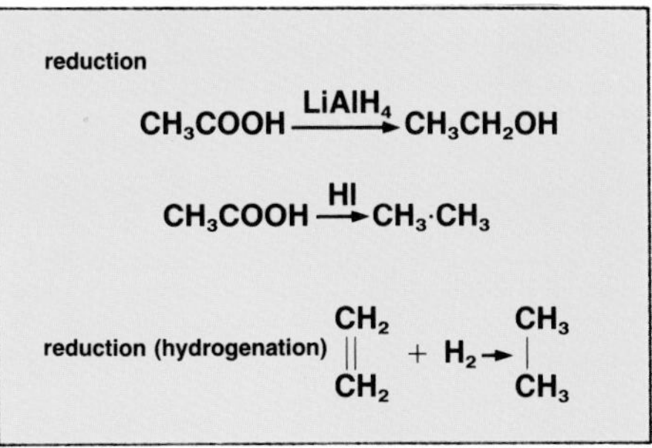

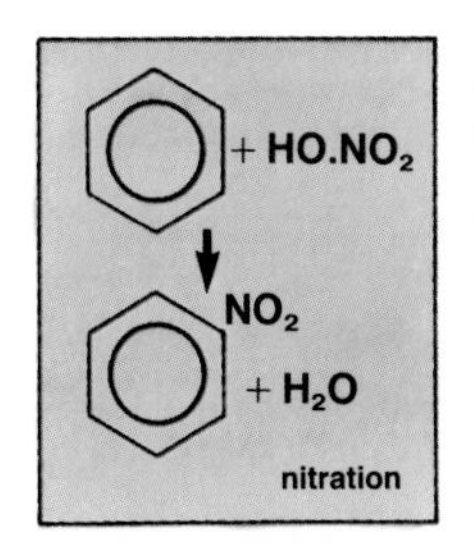

nitration (*n*) the substitution (p.188) of a nitro group (p.186) in an organic (p.55) compound, using concentrated nitric acid. With aliphatic (p.179) compounds, a high temperature is needed. With aromatic (p.179) compounds, the reaction takes place at much lower temperatures; concentrated sulphuric acid is used to quicken the reaction.

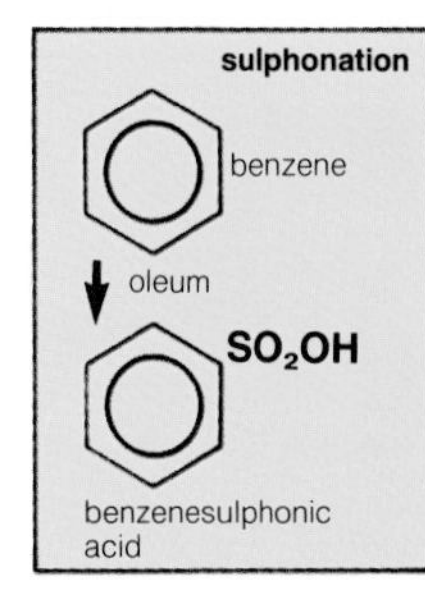

sulphonation (*n*) the substitution (p.188) of a sulphonate group (p.186) in place of a hydrogen atom in an organic (p.55) compound. Aromatic (p.179) compounds undergo (p.213) sulphonation readily, but aliphatic (p.179) hydrocarbons are extremely difficult to sulphonate. **sulphonate** (*v*).

$$C_2H_5COOH + CH_2N_2 \rightarrow C_2H_5COOCH_3 + N_2$$

diazobenzene **methylation**

methylation (*n*) a process in which a methyl group (p.184) is substituted for a hydrogen atom in an organic compound. Methylating agents include dimethyl sulphate, diazomethane and iodomethane. **methylate** (*v*).

cracking (*n*) an industrial (p.157) process in which hydrocarbons from petroleum fractions (p.160) are strongly heated under pressure; large molecules are broken up into smaller molecules, e.g. the higher fractions (p.202) are converted to petrol and kerosene. **crack** (*v*).

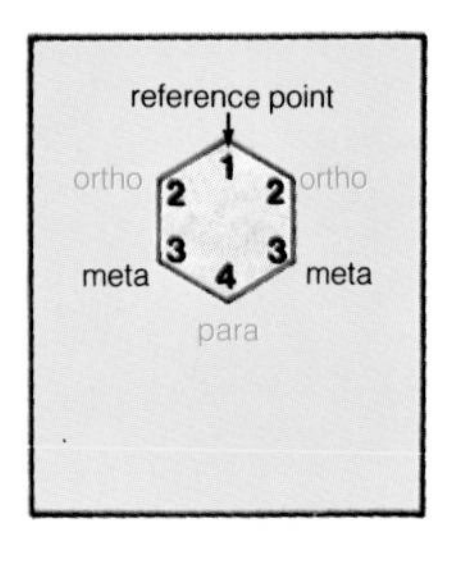

fermentation (*n*) the decomposition of carbohydrates (p.205) caused by enzymatic (p.72) action; the enzymes are produced by yeasts or by bacteria. **ferment** (*v*).

***ortho-para*-directing** (*adj*) a functional group (p.185) in a benzene ring (p.179) directs (p.233) the position of further substitution (p.188) in the ring. Some functional groups direct towards the *ortho* and *para* positions, and a second substitution will take place in these positions. An *ortho-para*-directing group activates (p.21) the benzene ring, so the second substitution is faster. The following groups are *ortho-para*-directing: alkyl, amino, hydroxyl, halogeno, methoxy.

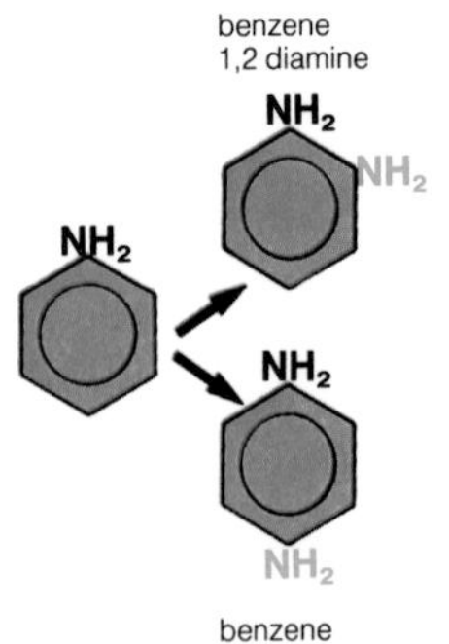

***ortho-para*-directing**

***meta*-directing** (*adj*) a functional group in a benzene ring directs (p.233) the position of further substitution in the ring. See *ortho-para* directing (↑). Some groups direct towards the *meta* position, these reduce (p.219) the activity of the ring so that the second substitution is slower and more difficult. The following groups are *meta*-directing: nitro, sulphonate, carboxyl, aldehyde, cyano.

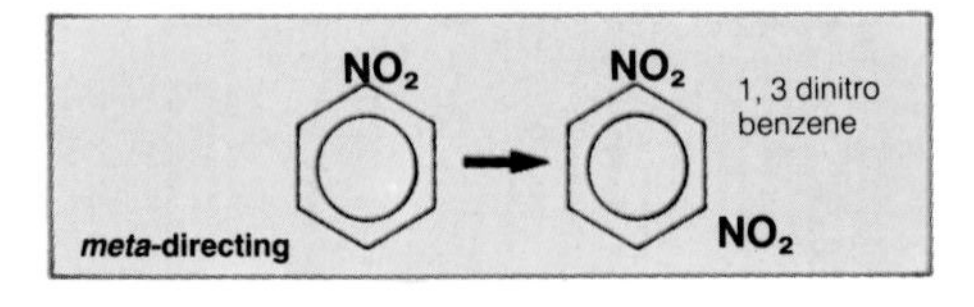

***meta*-directing**

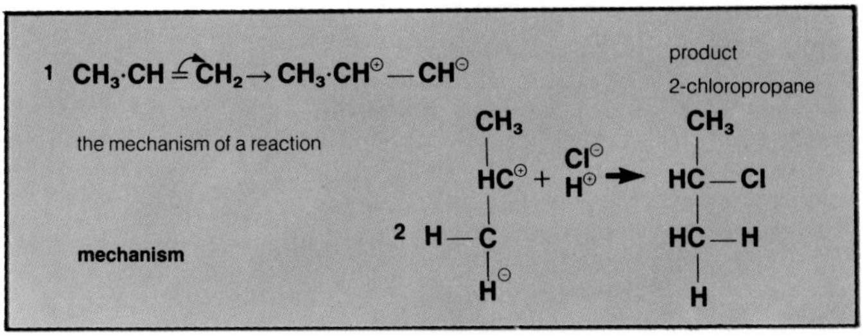

1 $CH_3{\cdot}CH = CH_2 \rightarrow CH_3{\cdot}CH^{\oplus} - CH^{\ominus}$

the mechanism of a reaction

product
2-chloropropane

$$\begin{array}{c} CH_3 \\ | \\ HC^{\oplus} \\ | \\ H-C \\ | \\ H^{\ominus} \end{array} + \begin{array}{c} Cl^{\ominus} \\ H^{\oplus} \end{array} \rightarrow \begin{array}{c} CH_3 \\ | \\ HC-Cl \\ | \\ HC-H \\ | \\ H \end{array}$$

2

mechanism

mechanism (*n*) the mechanism of a reaction gives an explanation of the way in which a reaction takes place. Reactivity occurs (p.63) because points of electron excess (p.230) or electron deficiency (p.232) appear in an organic (p.55) molecule and chemical attack takes place at these points. The reaction mechanism of the addition of hydrogen chloride to propene, shown above, explains how the reaction takes place and why the particular product is obtained.

ozonolysis (*n*) the process of passing ozonized oxygen (O_2 + O_3) through a solution of an alkene containing zinc dust and ethanoic acid. An ozonide, *see diagram*, is first formed by ozone adding across the double bond. This decomposes to form two compounds with a carbonyl group (p.185); these compounds may be aldehydes or ketones. Ozonolysis is used to determine (p.222) the position of a double bond in an unsaturated (p.185) molecule, by identifying (p.225) the products.

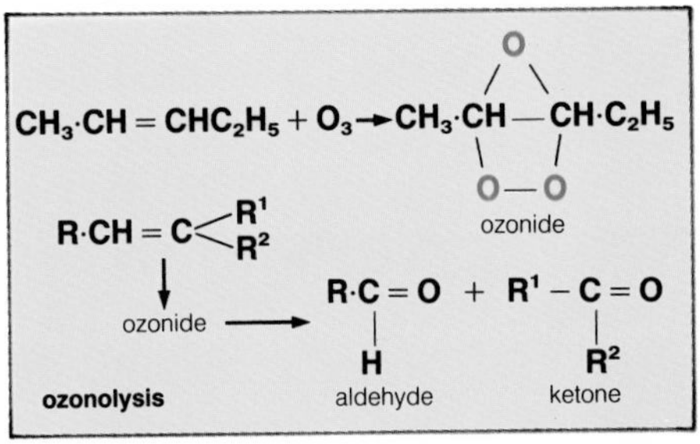

$$CH_3{\cdot}CH = CHC_2H_5 + O_3 \rightarrow CH_3{\cdot}CH \underset{O-O}{\overset{O}{\frown}} CH{\cdot}C_2H_5$$

ozonide

$$R{\cdot}CH = C \begin{array}{l} R^1 \\ R^2 \end{array}$$

↓

ozonide → $\begin{array}{c} R{\cdot}C = O \\ | \\ H \end{array} + \begin{array}{c} R^1 - C = O \\ | \\ R^2 \end{array}$

aldehyde ketone

ozonolysis

Schiff's reagent red fuchsine dye decolorized (p.73) by sulphur dioxide. When an aldehyde is added to Schiff's reagent, the red, or pink, colour returns. Ketones have no effect on the reagent, so the two kinds of compound can be distinguished (p.224).

Fehling's test a solution containing copper (II) sulphate, sodium potassium tartrate and sodium hydroxide is added to a solution to test for a sugar (p.205). If a red precipitate appears when the solution is boiled, a reducing sugar (p.206) is present (p.217). A red precipitate also indicates that an aldehyde (p.175) is present.

Benedict's test a solution containing copper (II) sulphate, sodium citrate and sodium carbonate is added to a solution to test for a sugar (p.205). If a red precipitate appears when the solution is boiled, a reducing sugar (p.206) is present (p.217). A red precipitate also indicates that an aldehyde (p.175) is present.

biuret test sodium hydroxide solution is added to a substance, followed by one or two drops of 1% copper (II) sulphate solution. A violet colour shows that protein (p.209) is present.

iodine test iodine solution is added to a substance. A deep blue colour indicates that starch (p.207) is present.

Lassaigne test a test to identify (p.225) the elements in an organic compound. The compound is fused with sodium metal in a test-tube. The tube is then broken and an aqueous solution of its contents formed. Carbon and nitrogen form sodium cyanide, tested for by iron (II) sulphate. Sulphur forms sodium sulphide, tested for as hydrogen sulphide. A halogen forms a sodium halide, tested for by silver nitrate. Thus C, N, S, Cl, Br, I are tested for.

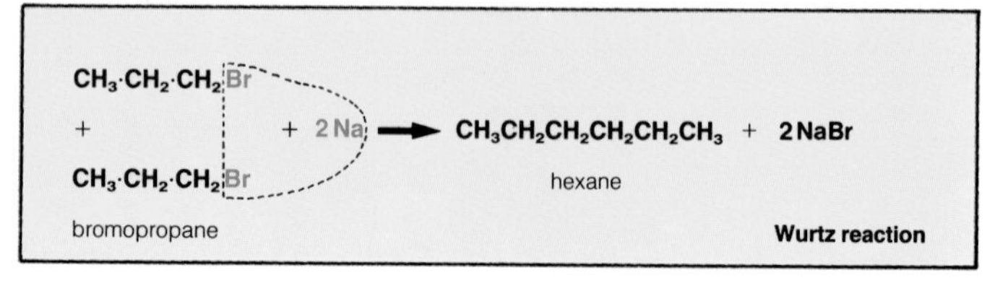

Wurtz reaction

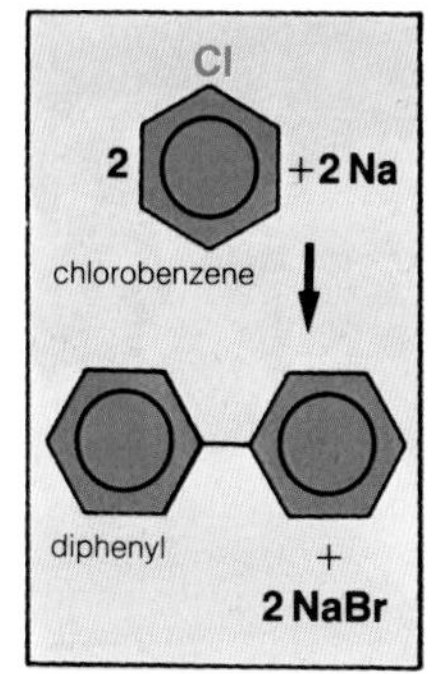

Fittig reaction

Wurtz reaction the production of an alkane (p.172) from an alkyl halide (p.177) by the action of metallic sodium, using an inert (p.19) solvent, e.g.

$$2C_2H_5Br + 2Na \rightarrow C_2H_5\text{–}C_2H_5 + 2NaBr$$

The general reaction is:

$$2RX + 2Na \rightarrow R\text{-}R + 2NaX.$$

Fittig reaction the production of a benzene homologue (p.172) from an aromatic (p.179) halide by the action of metallic sodium in ether.

Wurtz-Fittig reaction the production of an alkyl benzene from an aliphatic halide and an aromatic halide by the action of metallic sodium in ether. By-products are also formed.

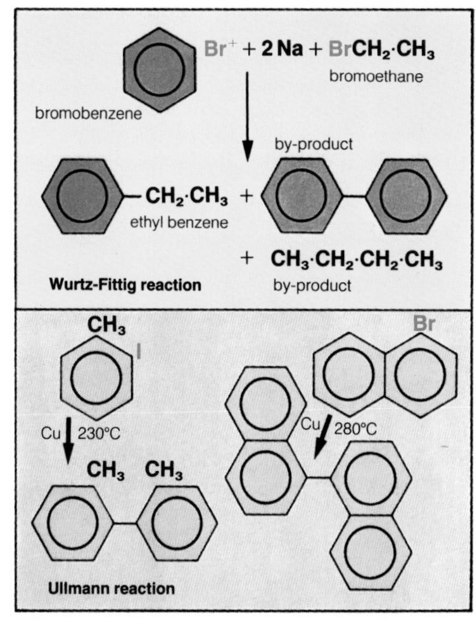

Wurtz-Fittig reaction

Ullmann reaction

Ullmann reaction the production of higher aromatic (p.179) homologues from a bromo or iodo compound by the action of copper powder. *See equation below*. The general reaction is $2\Phi X \rightarrow \Phi\text{–}\Phi$ where Φ is an aryl (p.180) group and X is iodine or bromine.

Williamson's synthesis a process for preparing simple or mixed ethers (p.177) from an alkyl halide (p.177) and the sodium derivative (p.200) of an alcohol, e.g.

$C_2H_5I + CH_3ONa \rightarrow C_2H_5-O-CH_3 + NaI$

Methoxyethane ($CH_3OC_2H_5$) is a mixed ether.

Grignard reagent a reagent which is prepared by the action of metallic magnesium on an alkyl halide in ether. An alkyl magnesium halide is formed, and this is the reagent, e.g.

$CH_3CH_2Br + Mg \rightarrow CH_3CH_2MgBr$

A Grignard reagent undergoes (p.213) many different reactions and the different reagents that can be prepared are used in the synthesis of many organic compounds. The symbol R is used in a Grignard reagent to represent any alkyl group (p.180) and the symbol X to represent a halogen (p.117). Grignard reagents can also be prepared using aryl (p.180) halides, e.g. $C_6H_5Br + Mg \rightarrow C_6H_5MgBr$ Only bromo and iodo compounds can be used for aryl halides.

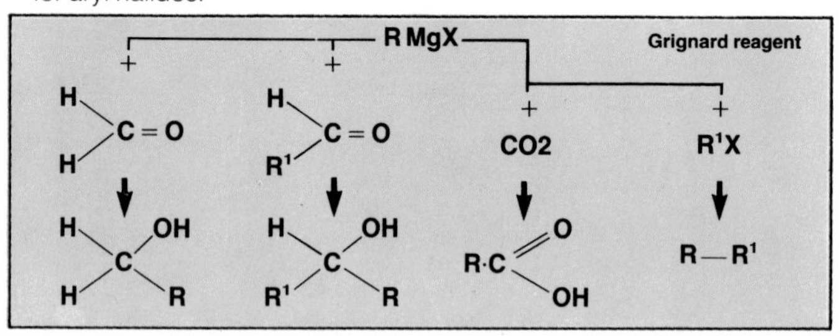

Sandmeyer reaction the use of copper (I) chloride in concentrated hydrochloric acid to convert an aqueous diazonium (p.180) chloride to a chloro-compound. A similar reaction takes place with copper (I) bromide and a diazonium bromide. A copper salt is unnecessary for the corresponding iodo compound.

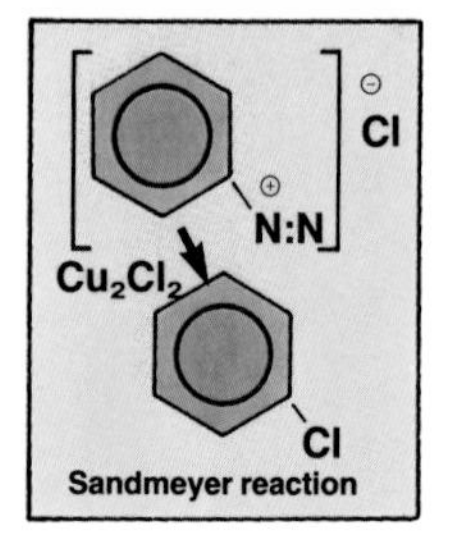

Sandmeyer reaction

Gattermann reaction[1] the use of copper powder to convert a diazonium salt to a halogeno compound, e.g. $C_6H_5-N=N-Cl \rightarrow C_6H_5Cl$ The yield (p.159) from this reaction is not as good as for the Sandmeyer reaction (↑).

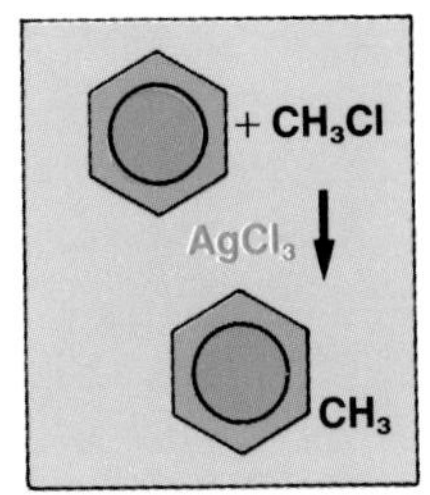

Friedel-Crafts reaction

Friedel-Crafts reaction a process for the alkylation (p.189) of benzene rings (p.179). Alkyl halides (p.177) react with aromatic hydrocarbons in the presence of anhydrous aluminium chloride.

Reimer-Tiemann reaction a reaction between phenols (p.180), aqueous sodium hydroxide and trichloromethane (chloroform), which puts an aldehyde group (p.175) on the benzene ring. The reaction does not produce a high yield (p.159) but is the easiest method of putting an aldehyde group on a benzene ring. The substitution generally takes place in the *ortho*-position.

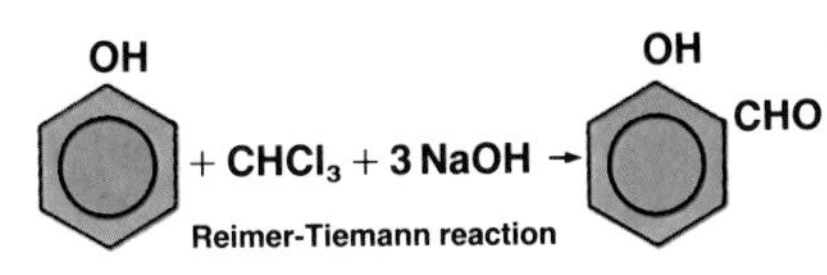

Reimer-Tiemann reaction

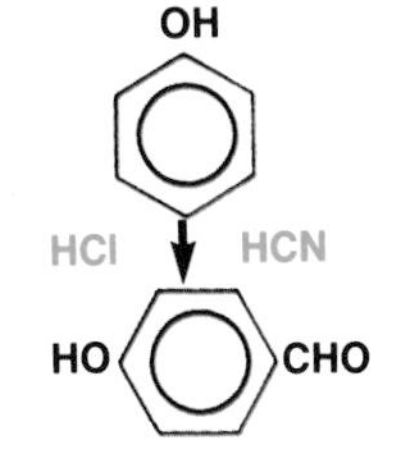

Gattermann reaction

Gattermann reaction[2] the combination of phenols with hydrogen chloride or hydrogen cyanide to give a product which is hydrolyzed by water to an aldehyde.

Cannizzaro reaction aldehydes (p.175) with no hydrogen atom on the second carbon atom in the aliphatic group, when treated (p.38) with cold concentrated aqueous sodium hydroxide, are both oxidized and reduced (p.70) at the same time; e.g. with methanol the reaction is:

$$2H.CHO + NaOH \rightarrow H.CH_2OH + H.COONa$$

The reaction also takes place with aromatic (p.179) aldehydes.

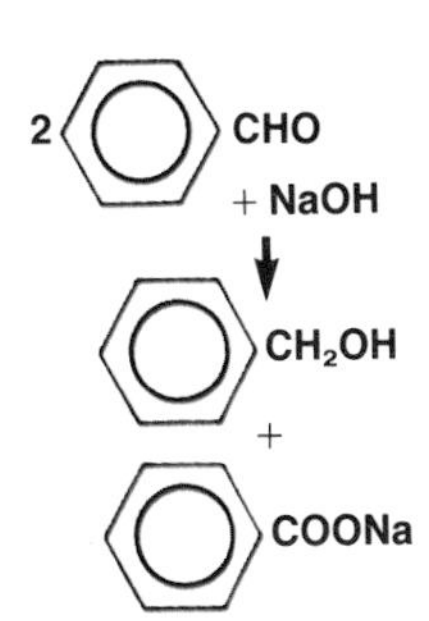

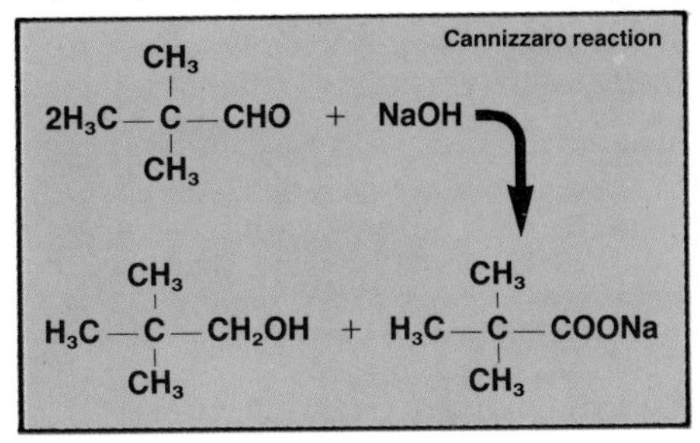

aldol addition a reaction in which aldehydes and ketones form addition compounds across the double bond (p.181) of the carbonyl group (p.185). The original reaction was:

$$CH_3CHO + CH_3CHO \rightleftharpoons$$
$$CH_3.CH(OH)CH_2CHO \text{ (aldol)}$$

The reacting molecules need not be identical. Equilibrium is largely to the left of the reaction, i.e. the addition product is formed only in small quantities. A trace (p.20) of potassium hydroxide catalyzes (p.72) the reaction.

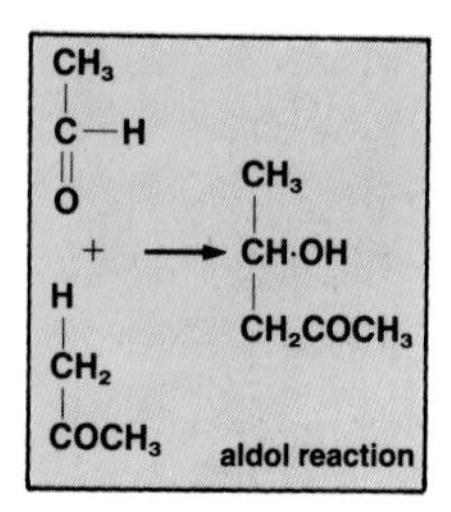

Kolbe electrolytic reaction a process for preparing alkanes (p.172) by the electrolysis (p.122) of the sodium salts of carboxylic acids. The carboxylic group must be at the end of a carbon chain. A cold concentrated aqueous solution of a salt is electrolyzed using platinum electrodes.

$$CH_3COONa \rightarrow CH_3\text{–}CH_3 + CO_2 + NaOH$$

Only alkanes with an even number of carbon atoms are prepared by this reaction.

derivative (*n*) a compound made by substitution of one or more atoms, or groups of atoms, in an original substance. Because the compound is a derivative of an original substance, it has the same kind of structure. For example, taking propane as an original substance, then its derivatives include propanol, propanal, propanoic acid and chloropropane. Derivatives are used to determine structure or composition. If the original structure is known, then the structure of the derivative is known. Ozonolysis (p.195) is frequently used to prepare derivatives. The derivatives are identified and from that the original structure of an alkene, including the position of the double bond (p.181), is found. **derive** (*v*).

derivatives

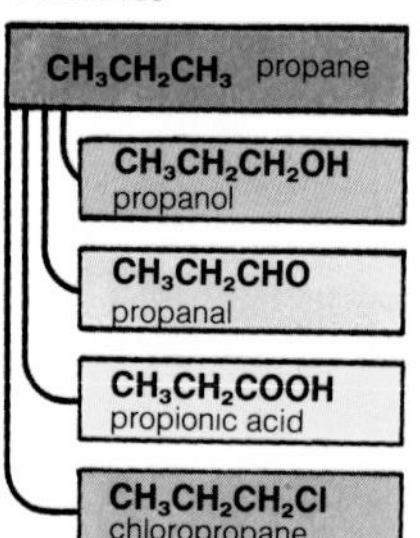

synthesis (*n*) (1) making a compound by chemical means from its elements. (2) making a compound by a series of chemical processes. e.g. the synthesis of ascorbic acid (vitamin C). A simple synthesis is shown in the equations opposite. **synthetic** (*adj*), **synthesize** (*v*).

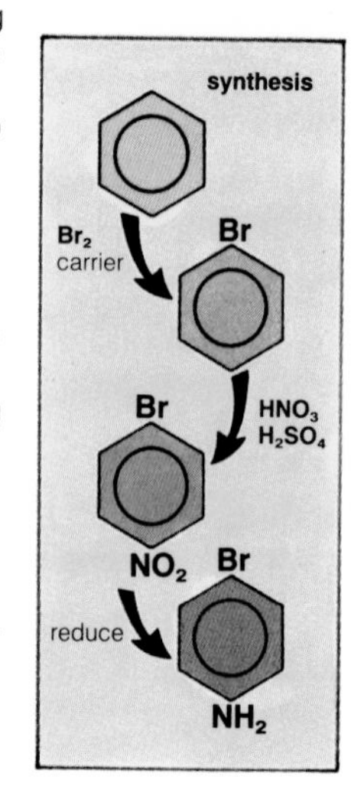

synthetic (*adj*) describes any compound made by synthesis (↑) from simpler compounds. A synthetic substance is one that generally replaces a naturally occurring (p.154) substance.

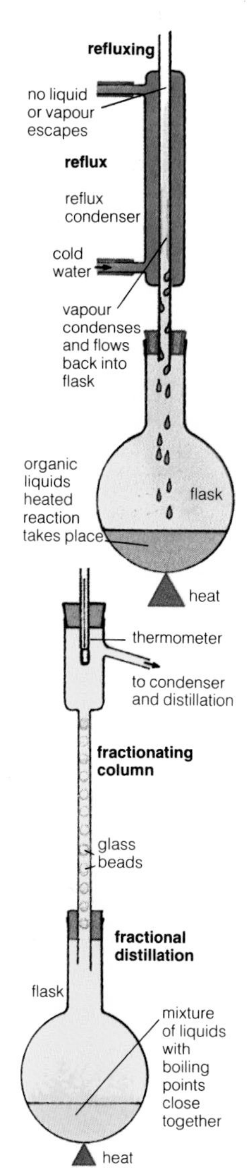

reforming (*n*) a process in which straight chain alkanes (p.172) are dehydrogenated (p.190) and form aromatic or branched chain hydrocarbons.

reflux (*n*) of a liquid or gas, the flowing back in an opposite direction from its original (p.220) direction. **reflux** (*adj*), **reflux** (*v*).

reflux condenser a condenser (p.28) fitted above a flask so that the vapour, formed by heating the flask, is condensed and flows back into the flask. This action prevents (p.216) the flask boiling dry and allows organic reactants sufficient (p.231) time to react while being heated, as the reactants do not escape from the flask.

still (*n*) an apparatus (p.23) used in industry for distillation (p.33). It is generally made of metal and produces (p.62) large quantities of distillate (↓).

distillate (*n*) the liquid obtained as the result of condensation from a process of distillation (p.33).

fractional distillation a process of distillation (p.33) in which the neck of a distillation flask (p.28) is extended (p.213) by a fractionating column (↓). Vapour from the flask condenses and falls back into the flask. Only vapour reaching the top of the column passes to a condenser (p.28) and is collected as a distillate (↑). This process is used to separate liquids which have boiling points close together. A column can have exits at different heights, allowing vapours which condense (p.11) at different temperatures to be separated and collected. This type of fractional distillation is used in industry (p.157), especially in the distillation of petroleum (p.160).

fractionating column an extension (p.213) of the neck of a distillation flask with an exit (p.215) at the top of the column, to pass the vapour to a condenser (p.28). An industrial fractionating column has several exits to take away vapour condensing to a liquid at a particular temperature. *See fractional distillation* (↑).

fraction (*n*) the distillate (p.201) collected at a particular temperature from a fractionating column (p.201), e.g. in the distillation of petroleum, four main fractions are obtained. These fractions can be fractionally distilled again to obtain a better separation (p.34) of the different liquids.

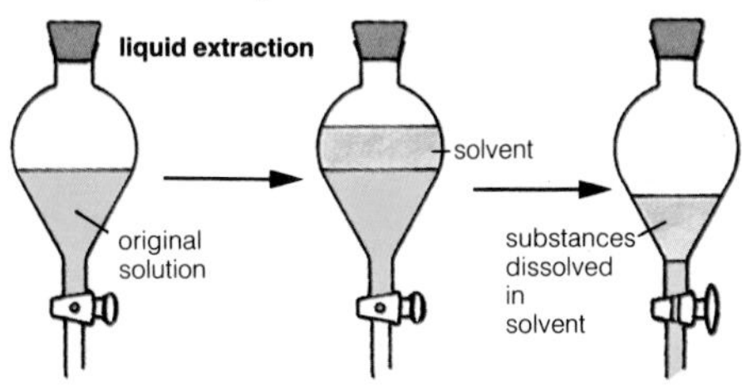

liquid extraction the removal (p.215) of a substance, dissolved in one solvent, by using a second solvent. The second solvent is added to the original (p.220) solution in a separating funnel (p.27). The two solvents must be immiscible (p.18). The solute (p.86) dissolves in both solvents and the two liquids are separated (p.34) by a separating funnel; e.g. ethanol is extracted from ethoxyethane (diethyl ether) by adding a concentrated solution of sodium chloride, and running off the aqueous layer in which the ethanol is dissolved.

steam distillation steam is blown through the heated mixture of the products of an organic (p.55) reaction. If a liquid product of the reaction is immiscible (p.18) with water, then the vapour of the product together with steam passes into a condenser and the distillate (p.201) contains water and the condensed liquid, e.g. phenylamine (aniline) can be steam distilled. The advantage of steam distillation is that the product is obtained at a temperature below its boiling point.

vacuum distillation distillation is carried out (p.157) at a low pressure. This method (p.221) of distillation is used with substances that decompose (p.65) at temperatures below their boiling points. At a lower pressure, a substance boils at a temperature below its boiling point.

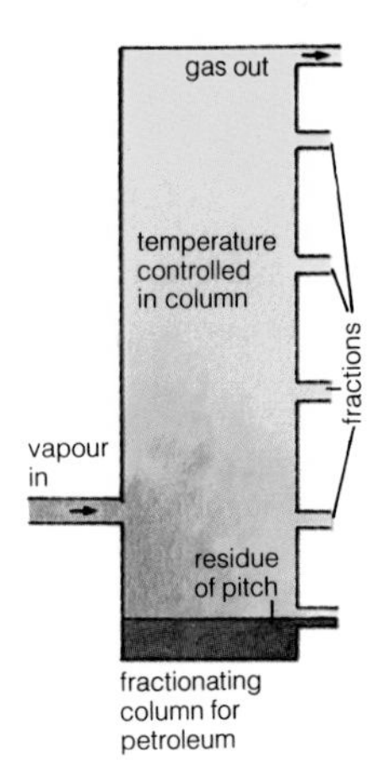

industrial fractional distillation

steam distillation

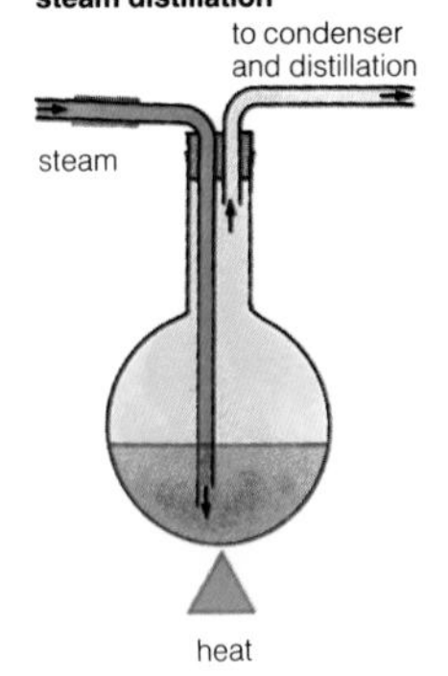

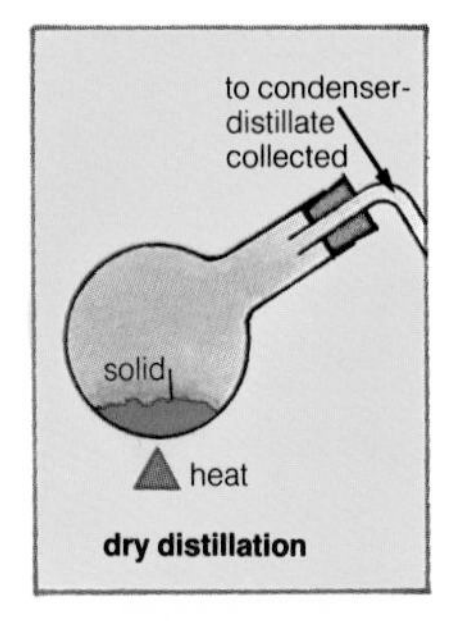

dry distillation

dry distillation the heating of a solid to make it give off (p.41) a vapour (p.11); the vapour is condensed to a liquid. For example, the dry distillation of solid calcium ethanoate (acetate) produces propanone (acetone), which is a volatile (p.18) liquid.

destructive distillation the heating of solid or liquid organic (p.55) materials to a temperature which is high enough to decompose (p.65) the material so that a residue and a distillate (p.201) are produced; a gas may be produced as well. Coal undergoes (p.213) destructive distillation. The residue is coke (p.156); the distillate consists of tar (p.162); and coal gas is also produced. Wood can also be destructively distilled.

destructive distillation of coal

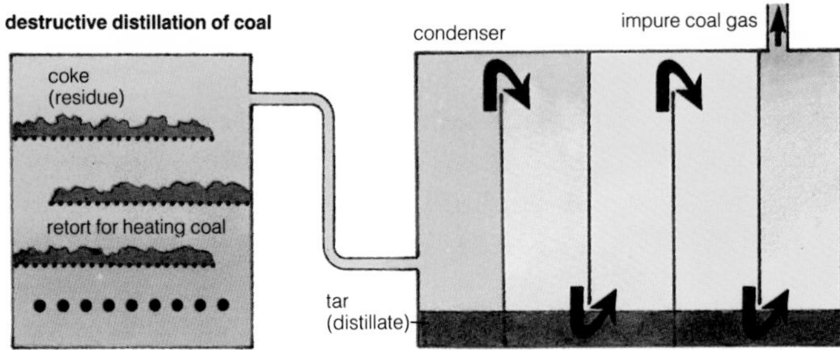

spirit (*n*) (1) a liquid consisting mainly of ethanol (ethyl alcohol) obtained by the distillation (p.33) of fermented (p.194) fruit, potatoes, or cereals. (2) any volatile (p.18) liquid obtained by distillation from natural products such as wood and petroleum; used for combustion (p.58), especially in internal combustion engines. (3) any solvent for fat, gums, paints.

methylated spirit a mixture of ethanol (ethyl alcohol) and methanol (methyl alcohol), usually with added colouring material to show the mixture is poisonous. Methylated spirit is used as a fuel (p.160) and as a solvent (p.86).

absolute alcohol ethanol (ethyl alcohol) from which all traces (p.20) of water have been removed. This is done by adding small quantities of calcium oxide (quicklime) which removes the water from 96% ethanol obtained by distillation.

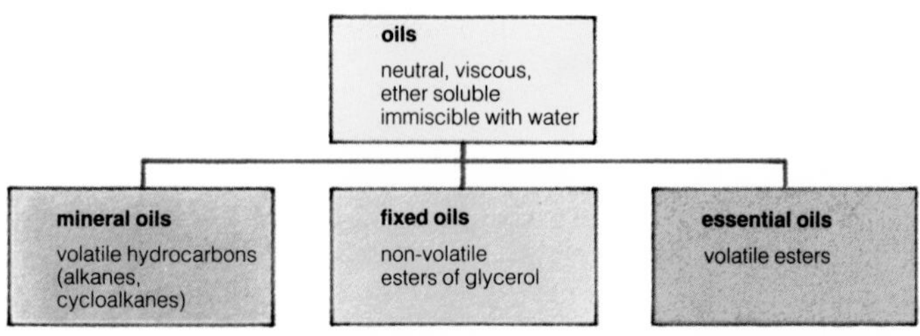

oil (*n*) (1) any substance, obtained from animals, plants or minerals, which is neutral (p.45), viscous (p.18), combustible (p.58) and soluble in ethanol or ether (ethoxyethane) but insoluble in water. The three main kinds of oil are fixed oils (↓), mineral oils and essential oils (↓). Mineral oils are petroleum (p.160). (2) a neutral fat (p.177) which is liquid below 20°C, is called an oil; such substances are fixed oils (↓). **oily** (*adj*).

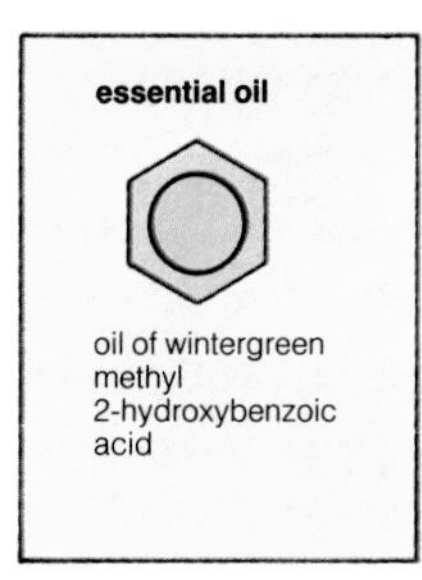

essential oil a volatile (p.18) oil produced from a plant and occurring (p.154) especially in the flowers of a plant. Essential oils give a plant its characteristic odour (p.15).

fixed oil a non-volatile oil occurring in plants. Fixed oils are generally edible and are used in cooking, e.g. coconut oil, peanut oil. The fixed oils are esters (p.177) of glycerol (a trihydric alcohol (p.175)) and unsaturated (p.185) or polyunsaturated (p.185) carboxylic acids, although they may contain a small proportion (p.76) of saturated (p.185) carboxylic acids.

hydrocarbon oil any oil obtained industrially (p.157) from petroleum (p.160).

flash point the lowest temperature at which a volatile (p.18) liquid, especially an oil (↑), gives off (p.41) enough vapour to produce a small flame, but not to catch fire, when touched by a small flame or hot object.

ignition point the lowest temperature at which a volatile (p.18) liquid, especially an oil, will catch fire, or burst into flames, when touched by a flame or hot object. **ignite** (*v*).

explosive (*adj*) describes any substance that ignites (p.32) so quickly that it causes an explosion (p.58).

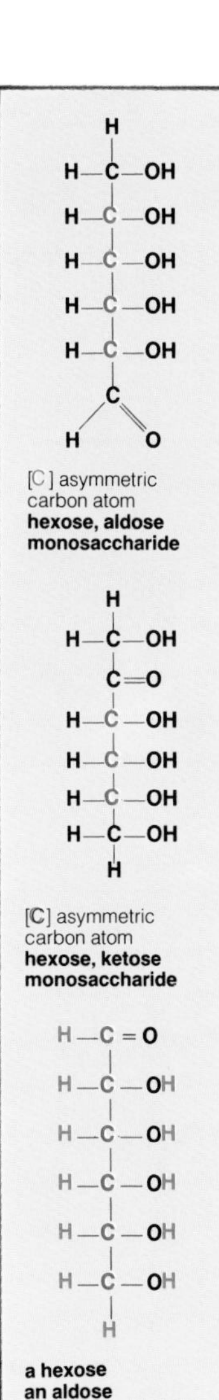

[C] asymmetric carbon atom
hexose, aldose monosaccharide

[C] asymmetric carbon atom
hexose, ketose monosaccharide

a hexose
an aldose
a monosaccharide

carbohydrate (*n*) an organic (p.55) substance with the general formula (p.181) $C_x(H_2O)_y$ and a complex molecular structure (p.83). The carbohydrates are divided into two groups, sugars (↓) and polysaccharides (p.207).

sugar (*n*) a colourless, crystalline, water-soluble solid with a sweet taste. Sugars are classified as monosaccharides (↓), disaccharides (p.206), trisaccharides, tetrasaccharides, etc.

monosaccharide (*n*) a sugar (↑) with the formula $C_n(H_2O)_n$, with the most common members of the group having n = 5 for a pentose (↓) or n = 6 for a hexose (↓). The two most common monosaccharides are glucose and fructose (p.206), they are both hexoses (↓). The monosaccharides exhibit stereoisomerism, the isomers are discussed under hexose (↓); monosaccharides are not hydrolyzed to simpler sugars.

pentose (*n*) a monosaccharide (↑) with a molecular formula (p.181) of $C_5H_{10}O_5$. The pentoses can be divided into aldoses (↓) and ketoses (p.206). The most common is ribose.

hexose (*n*) a monosaccharide (↑) with a molecular formula (p.181) of $C_6H_{12}O_6$. The hexoses can be divided into aldoses (↓) and ketoses (p.206). The carbon chains for the two kinds of molecule are shown in the diagram opposite in structural formulae. An aldose (↓) has an aldehyde (p.175) group and a ketose (p.206) has the carbonyl group (p.185) of a ketone (p.176). The hydrogen atoms and hydroxyl groups can have different spatial (p.211) directions, producing stereoisomers (p.183). Asymmetric carbon atoms (p.183) produce optical isomerism (p.183).

aldose (*n*) a monosaccharide (↑) with an aldehyde group (p.175); it is a reducing sugar (p.206). In a hexose, which is also an aldose, called an **aldohexose**, there are four asymmetric carbon atoms (p.183), none of which is combined with the same organic (p.55) group, hence each asymmetric atom will have two enantiomorphs (p.183). This produces 16 possible isomers; only three occur abundantly (p.231) in nature. Aldoses exhibit (p.221) the reactions of aldehydes and alcohols (p.175).

ketose (*n*) a monosaccharide (p.205) with a carbonyl group (p.185) giving the reactions of a ketone (p.176). A ketose is also a reducing sugar (↓). In a hexose which is also a ketose, called a **ketohexose**, there are three asymmetric carbon atoms (p.183); each asymmetric atom will have two enantiomorphs (p.183). This produces 8 possible isomers. Ketoses exhibit (p.221) the reactions of ketones and alcohols (p.175).

reducing sugar a sugar (p.205) which reduces Fehling's solution (p.196) and Benedict's solution (p.196). All monosaccharides are reducing sugars, but only some disaccharides.

glucose (*n*) an aldohexose. It is less sweet than sucrose (↓) and very soluble in water; the naturally occurring form is optically active (p.19), rotating the plane of polarized light to the right, and so is also known as 'dextrose'.

fructose (*n*) a ketohexose. It is the sweetest of all sugars and very soluble in water; the naturally occurring form is highly optically active (p.19), rotating the plane of polarized light to the left.

disaccharide (*n*) a sugar (p.205) which consists of two monosaccharides (p.205) chemically combined. On hydrolysis (p.66) one molecule of a disaccharide produces two molecules of monosaccharides; the monosaccharide molecules can be the same, or different. The most common disaccharides have a molecular formula (p.181) of $C_{12}H_{22}O_{11}$.

sucrose (*n*) a disaccharide (↑); on hydrolysis (p.66) it produces equal proportions of glucose (↑) and fructose (↑). A molecule of sucrose consists of a molecule of glucose and a molecule of fructose combined by a condensation (p.191) reaction. Sucrose is a non-reducing sugar (↓).

non-reducing sugar a sugar (p.205) which does not reduce Fehling's solution (p.196) or Benedict's solution (p.196).

maltose (*n*) a disaccharide (↑); on hydrolysis it produces glucose (↑). A molecule of maltose consists of two molecules of glucose combined by a condensation reaction (p.191). Maltose is a reducing sugar.

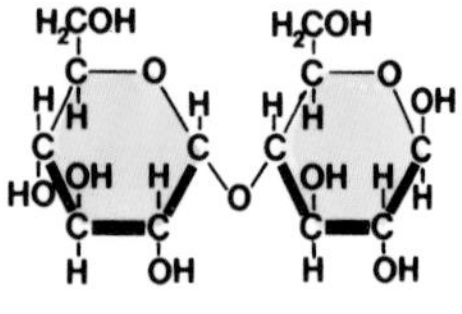

maltose disaccharide
(two hexoses combined)

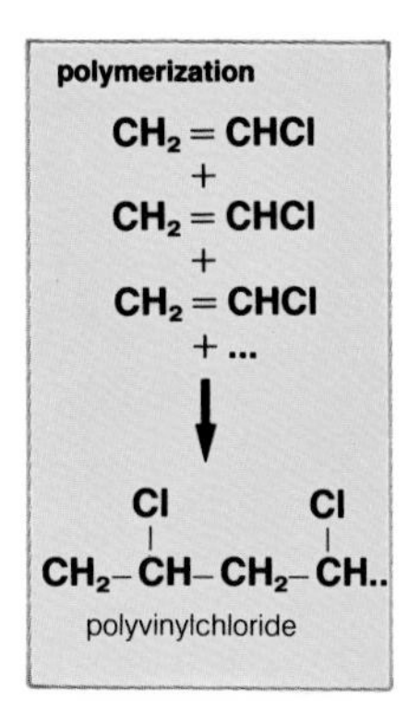

a polymer with 900 to 1300 molecules of the monomer combined **polymer**

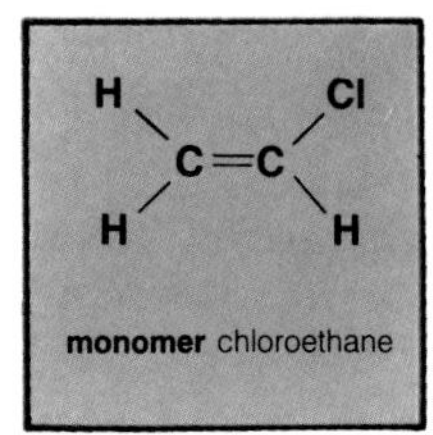

monomer chloroethane

polysaccharide (*n*) a carbohydrate (p.205) that is not sweet, is insoluble in water and is non-crystalline; most polysaccharides are colloidal (p.98) in nature. Polysaccharides formed by condensation reactions (p.191) from hexoses (p.205) have a general formula of $(C_6H_{10}O_5)_n$, where n is a very large number. Starch (↓) and cellulose (↓) are the most common polysaccharides.

starch (*n*) a polysaccharide (↑), which on hydrolysis produces glucose (↑). One starch molecule consists of 4000 to 30 000 glucose molecules chemically combined.

cellulose (*n*) a polysaccharide (↑), which on hydrolysis produces glucose (↑). One cellulose molecule consists of about 3000 glucose molecules.

polymerization (*n*) a chemical process in which molecules of the same compound combine together to form a molecule of high relative molecular mass (p.114). The formation of starch from glucose is a form of polymerization. There are two kinds of polymerization, addition polymerization (p.208) and condensation polymerization (p.208). **polymerize** (*v*), **polymer** (*n*), **polymeric** (*adj*).

monomer (*n*) a molecule or substance which can be polymerized; it has a molecule of low relative molecular mass (p.114), e.g. chloroethene (vinyl chloride) has a molecular formula (p.181) C_2H_3Cl; it is the monomer of polyvinyl chloride (P.V.C.) which contains between 900 to 1300 molecules of the monomer chemically combined.

dimer (*n*) a molecule or compound, formed by the chemical combination of two simpler molecules, e.g. nitrogen dioxide, NO_2, forms a dimer, dinitrogen tetroxide, N_2O_4. **dimerization** (*n*), **dimeric** (*adj*), **dimerize** (*v*).

polymer (*n*) a material with a molecule of high relative molecular mass (p.114) formed by polymerization (↑). For example, using the monomer (↑) chloroethane, a polymer is produced with a relative molecular mass of 50 000 – 80 000. The constitutions of the molecules of a polymer vary, it is described as being between a range of relative molecular masses.

$$CH_2 = CH_2 + CH_2 = CH_2 + CH_2 = CH_2 + \cdots$$

ethene (monomer)

$$\downarrow$$

$$-CH_2-CH_2-CH_2-CH_2-CH_2-CH_2-$$

addition polymerization polythene (polymer)

addition polymerization polymerization (p.207) in which monomers (p.207) combine together by an addition reaction, so that the polymer has the same empirical formula (p.181) as the monomer. For example, polyvinyl chloride is a polymer (p.207) with an empirical formula of $(C_2H_3Cl)_n$, where n is between 900 and 1300; the empirical formula of the monomer is C_2H_3Cl.Addition polymerization takes place with alkenes (p.173) and their derivatives (p.200). Chloroethene (C_2H_3Cl) is the monomer (p.207) of polyvinyl chloride.

condensation polymerization polymerization (p.207) in which monomers (p.207) combine together in a condensation reaction (p.191). For example, aminoethanoic acid (glycine) undergoes (p.213) condensation polymerization. Most condensation polymerizations are a kind of copolymerization (↓).

copolymerization (also a condensation polymerization)

methyl terephthalate ethane-1, 2-diol

$$CH_3OOC-C_6H_4-COOCH_3 + HO{\cdot}CH_2{\cdot}CH_2OH$$

$$\downarrow -CH_3OH$$

$$-OC-C_6H_4-COOCH_2{\cdot}CH_2OOC-C_6H_4-OC-$$

terylene

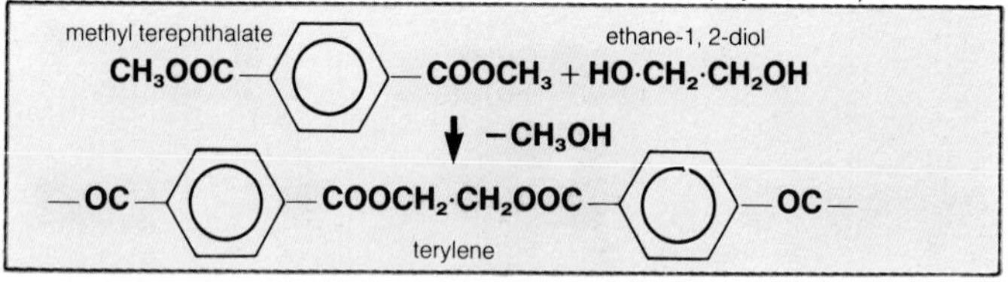

copolymerization (*n*) polymerization (p.207) using two, or more, monomers (p.207); e.g. the polymer terylene starts with two monomers: dimethyl benzene-1,4-dicarboxylate (methyl terephthalate) and ethane-1,2-diol. Condensation polymerization (↑) takes place with the elimination (p.189) of methanol.

polythene (*n*) a polymer (p.207) which has ethene (p.174) as a monomer (↑). Molecules of ethene in the presence of a catalyst undergo (p.213) addition polymerization (↑). Polythene is inert (p.19), a good electrical insulator, and can be moulded (p.210).

peptide (*n*) a polymer (p.207) formed by the condensation copolymerization (↑) of several amino acids (p.178) joined by a peptide bond (↓). On hydrolysis it produces amino acids.

peptide bond a condensation reaction (p.191) between an amino group (p.186) and a carboxyl group (p.186) with the elimination (p.189) of water. In hydrolysis the peptide bond is broken to form the original (p.220) amino acids.

protein (*n*) a polymer (p.207) formed from peptides (↑) by condensation polymerization (↑). A protein contains 50 or more amino acids (p.178) joined by peptide bonds (↑). On hydrolysis a protein is first decomposed to peptides and then to amino acids.

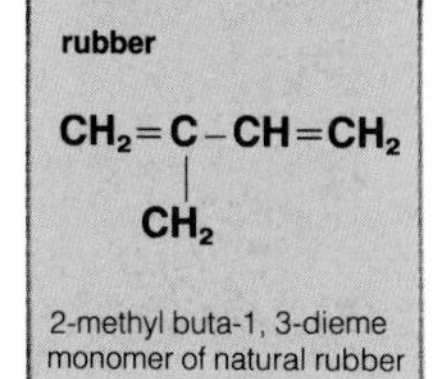

2-methyl buta-1, 3-dieme monomer of natural rubber

rubber (*n*) a naturally occurring (p.154) elastic material which is a polymerized (p.207) hydrocarbon (p.172). Destructive distillation (p.203) of natural rubber produces 2-methylbuta-1,3-diene. Synthetic rubber is generally manufactured (p.157) by the copolymerization of phenylethene (styrene) and buta-1,3-diene (butadiene). *See diagram.*

$CH_2=CH$ + $CH_2=CH-CH=CH_2$

phenyl ethene — buta-1, 3-dieme

↓

$-CH_2-CH-CH_2-CH=CH-CH_2-$

synthetic rubber

glasses (*n.pl.*) amorphous (p.15) solids composed of silicon dioxide (SiO_2) with the silicon and oxygen atoms forming a tetrahedral (p.83) structure. Cations of various metals form bonds with the atoms in the tetrahedral structure. Like polymers (p.207) glasses have molecules of very high relative molecular mass (p.114). Soda glass, the normal soft glass, is manufactured (p.157) by fusing together a mixture of sand, limestone and sodium carbonate. Glasses have no fixed melting point (p.12).

plastic[2] (*n*) a material manufactured (p.157) by the polymerization (p.207) of organic (p.55) substances. It is usually a hard material, or a threadlike material used for making cloth, and does not appear to be plastic (p.14) in the usual meaning of the word. A plastic material has its shape changed, either when hot or cold, and then retains (p.215) this shape. A manufactured plastic has the property of plasticity at some stage of its manufacture. There are two types of plastics, thermoplastics and thermosetting plastics.

thermoplastic (*n*) a kind of plastic which exhibits (p.221) plasticity when hot. The material is heated and pressed into shape in a mould (↓). On cooling it retains the shape of the mould. On heating again, its shape can be changed by pressure, e.g. polyvinyl chloride (P.V.C.) and polythene (p.208). These materials are soluble in organic solvents.

thermosetting plastic a kind of plastic which exhibits plasticity (p.14) when first heated; by heating, the structure of the polymer is changed and bonds are formed between molecules to produce a three-dimensional structure, making the plastic very strong when it has cooled. When the plastic is heated again, its shape cannot be altered, i.e. it has lost its plasticity, e.g. urea-formaldehyde resins and Bakelite.

mould (*n*) a vessel into which a hot liquid material is poured to become cold. When cold it solidifies (p.10) in the shape of the mould. Powders can be used instead of liquids and then pressure and heat are applied (p.232) to turn the powder into a solid of the same shape as the mould. This method is used with plastics (↑). **mould** (*v*), **moulding** (*n*).

plasticizer (*n*) a substance which is added to a polymer (p.207) to keep it plastic, e.g. a plasticizer is added to polyvinyl chloride to alter its properties. With a small quantity of plasticizer, a strong solid is produced; with a large quantity of plasticizer, an elastic solid is produced.

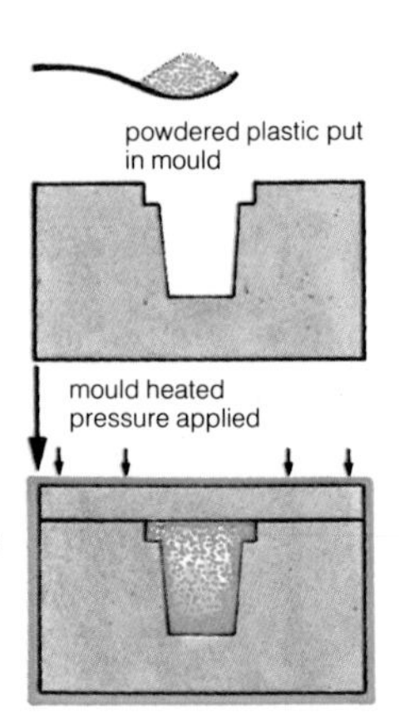

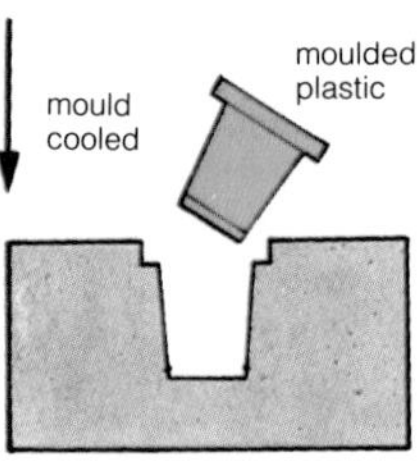

moulding plastic

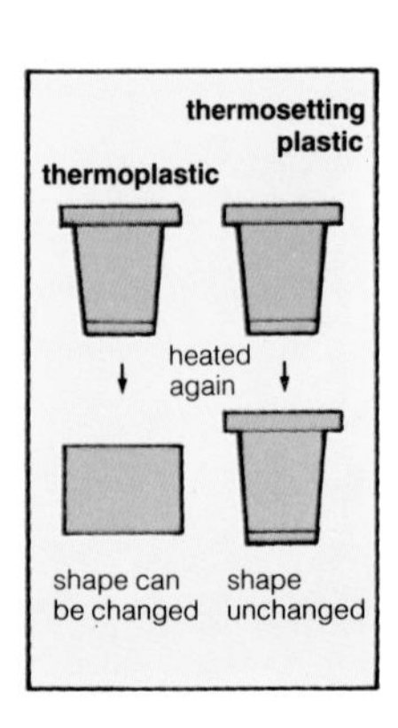

position of end

position

spatial

spatial arrangement of cross pieces

structure of a bridge

position (*n*) the position of an object is its place in space in relation (p.232) to other objects. For example, (a) the position of a table in a room is its place in relation to a door, a window, or chairs in the room; (b) the position of an element in the electrochemical series (p.130) shows the relation of that element to the other elements in the series, i.e. whether it has a greater or lesser electrode potential. **position** (*v*).

spatial (*adj*) describes an arrangement or a direction or an extent (p.213) in space, e.g. the spatial properties of covalent bonds (p.136) describe the direction and arrangement of the bonds in space. **space** (*n*), **space** (*v*).

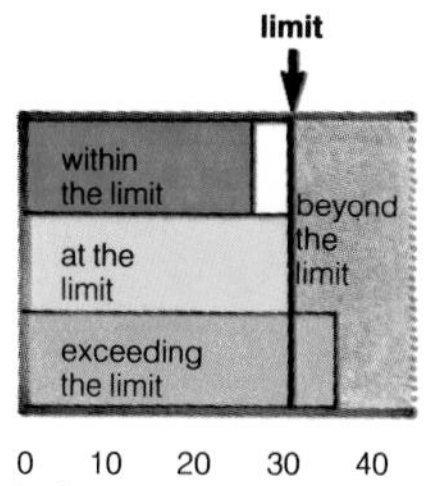

limit[1] (*n*) the value of a quantity (p.81) beyond which it is generally not possible to go. If a limit can be exceeded, *see diagram*, then different circumstances or different physical laws act on the quantity. A limit can be the largest or smallest possible value, e.g. the limit of solubility of a crystalline substance is the greatest amount that dissolves in boiling water.

structure[2] (*n*) the arrangement in space of the connected (p.24) parts of a whole object. For example, (a) the structure of a bridge is the arrangement in space of its different parts; (b) the structure of a molecule of ethanol (p.175) shows the arrangement in space of the atoms in the molecule. **structural** (*adj*).

structural (*adj*) describes anything to do with structure, e.g. structural isomerism (p.182) is a kind of isomerism which depends on molecules having different structures.

construct (*v*) to make a structure, e.g. to construct a model (p.223) is to make a model with a particular structure. **construction** (*n*).

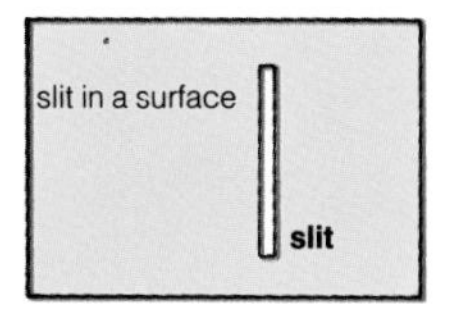

slit (*n*) a long, very narrow hole in a surface.

system (*n*) (1) a fixed way of carrying out (p.157) a process (p.157), e.g. the system of naming chemical substances. (2) a set of objects which obey physical laws, have an effect on each other and form a whole unit (↓), e.g. the substances in an equilibrium mixture (p.150) form a system. **systematic** (*adj*).

unit (*n*) (1) a value of a quantity which is accepted as a standard (p.229) for that quantity, e.g. the kilogram is accepted as the unit of mass. (2) a whole thing, made from different parts, which acts as a whole, e.g. a molecule (p.77) of a substance is made of different parts (the atoms) and it acts as a single object, it forms a unit. (3) each member of a series (p.172) is a unit.

circumstances (*n.pl.*) everything that may or may not have an effect on an object or a substance, together make the circumstances of that object or substance; e.g. the circumstances of a liquid are: the vessel containing it, the temperature, the atmospheric pressure, etc. Contrast *conditions*: *circumstances* that have an effect on a process involving the object or substance are its *conditions*.

local (*adj*) describes anything near an object or substance, e.g. the local conditions or the local circumstances. *Contrast* ambient (p.103) which describes anything surrounding an object.

general (*adj*) describes properties, qualities and nature possessed by all members of a set, e.g. the properties which all acids possess and their general properties. *Contrast* the additional properties of any one acid which are its **particular** properties. **generalize** (*v*), **generalization** (*n*).

special (*adj*) describes properties, qualities and nature which are very different for one member of a set and distinguish (p.224) it from all other members of the set.

universal (*adj*) (1) describes a statement which is always true and has no exceptions (p.230), e.g. a universal law, such as the attraction between unlike electric charges, has no exceptions. (2) describes a class of substances which have a characteristic action without exception, e.g. a universal solvent for organic substances.

use of general particular special

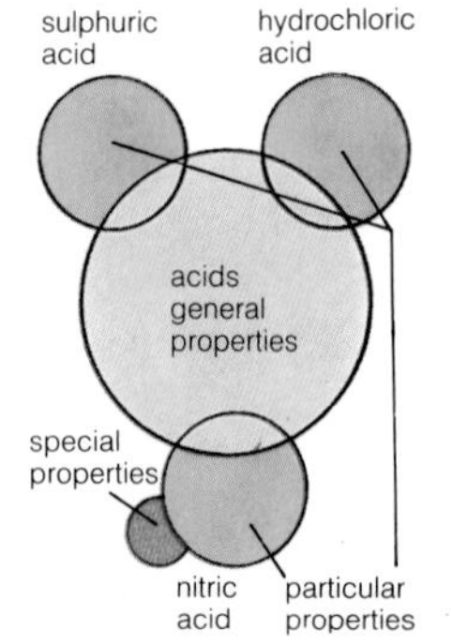

exist (*v*) to be, of objects, materials and substances, i.e. to be perceptible (p.42) to sight, hearing, feeling, tasting or smelling. For example, carbon exists in two crystalline forms. *See occur (p.154).* **existence** (*n*).

extend (*v*) (1) to take up space between two points or to cover an area, e.g. the coal seam (p.154) extends from a depth of 200 metres to a depth of 250 metres. (2) to take up time between two points in time, e.g. the whole process of separating the mixture by fractional distillation extended from 2.00 p.m. to 4.30 p.m. **extent** (*n*).

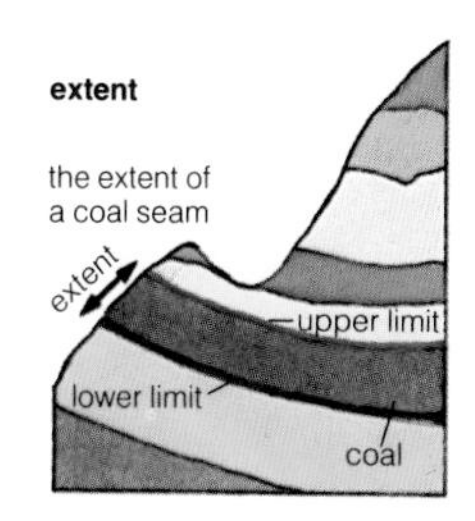

extent (*n*) (1) the space, time or activity between limits, e.g. the extent of the decomposition of phosphorus pentachloride into phosphorus trichloride and chlorine depends on the temperature and pressure (the temperature and pressure make the limits). (2) the space between limits over which something extends (↑). **extensive** (*adj*), **extensively** (*adv*).

accompany (*v*) to take place (p.63) or to exist (↑) at the same time, e.g. the reaction between metals and concentrated acids is usually accompanied by the evolution (p.40) of heat. **accompaniment** (*n*).

limit[2] (*v*) to make a limit (p.211) for time or activity, e.g. (a) refluxing (p.201) the mixture was limited to half an hour; (b) pressure limits the formation of ammonia from nitrogen and hydrogen.

effort (*n*) the use of force or energy, e.g. an effort is needed to keep the reactants in a chemical process under pressure.

undergo (*v*) to take part in an action or a chemical reaction because of an outside agent (p.63), e.g. iron undergoes rusting in damp air (the damp air is an outside agent causing the rusting).

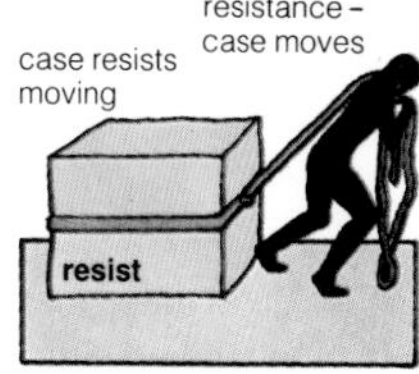

overcome (*v*) to make an action take place when there is resistance (↓) to the action, e.g. sufficient energy has to be supplied (p.154) to reactants to overcome the energy barrier (p.152).

resist (*v*) to try to prevent an action taking place, e.g. a chemical reaction does not take place until the energy barrier (p.152) is overcome (↑); the energy barrier resists the reaction taking place. **resistance** (*n*).

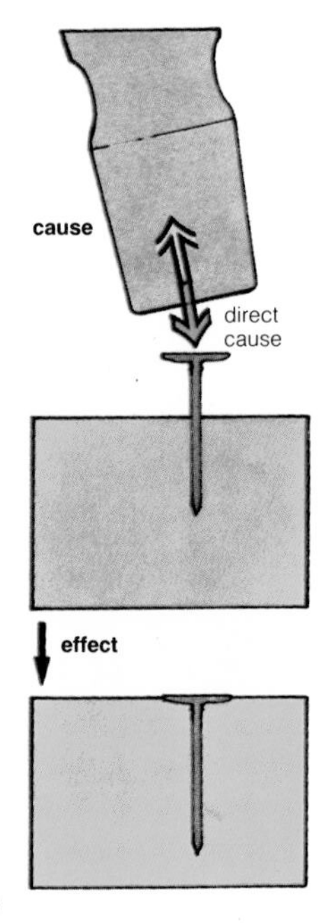

cause (*n*) a substance, a form of energy, or an event which makes a change, a process, or another event take place. For example, (a) oxygen in the air is the cause of iron rusting; (b) a flame is the cause of many explosions of methane in coal mines; (c) the escape of coal gas was the cause of the explosion. **cause** (*v*).

effect (*n*) the change, process or event produced by a cause (↑), e.g. heat causes water to boil, the boiling of the water is the effect of the heat.

effective (*adj*) describes anything that produces an effect (↑). For example, (a) the hydroxyl group in an alcohol (p.175) is the effective part of the molecule; (b) propanone (acetone) is a very effective solvent for organic substances (it always produces the effect of dissolving the solute). **effectiveness** (*n*).

efficient (*adj*) describes a device, a piece of apparatus, or a process which produces a result with little or no waste of energy or material. For example, (a) a fractionating column is an efficient piece of apparatus for separating a liquid mixture; (b) the Solvay process (p.169) is the most efficient way of manufacturing sodium carbonate. **efficiency** (*n*).

facilitate (*v*) to make an action or a process take place more easily, e.g. a catalyst facilitates the chemical combination of nitrogen and hydrogen in the Haber process (p.170). **facility** (*n*).

favour (*v*) to provide the conditions (p.103) so that a chemical reaction takes place more easily or more quickly, e.g. high pressure favours the combination of nitrogen and hydrogen to form ammonia; high pressure is the condition, it does not cause the reaction to take place more easily, but the pressure offers less resistance (p.213) to the change. **favourable** (*adj*).

terminate (*v*) to stop a process, or a chemical reaction, before it has had time to finish, e.g. the distillation was terminated when the flask cracked. **termination** (*n*).

duplicate (*v*) to carry out the same action twice, e.g. the experiment was duplicated so that the two sets of results could be compared (p.224). **duplicate** (*n*), **duplicate** (*adj*).

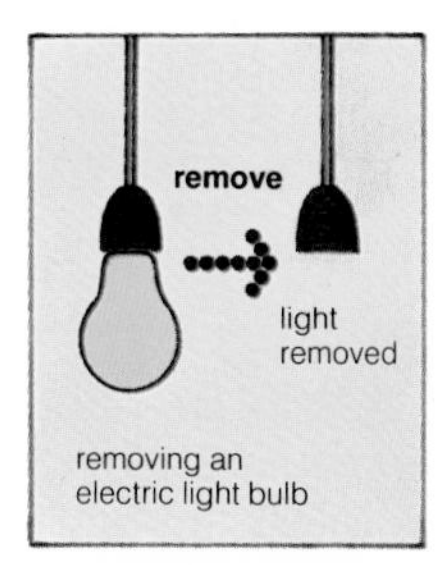

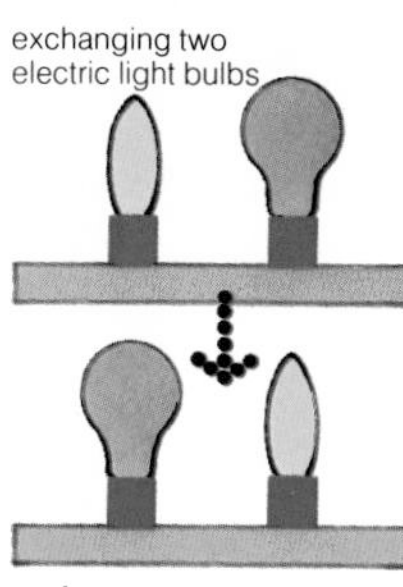

exchange

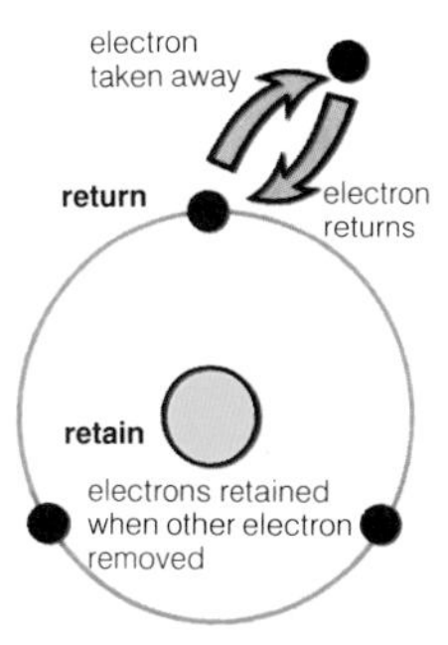

remove (*v*) to take an object, a material, or a substance from a place, e.g. (a) to remove a residue (p.31) from a filter paper; (b) to remove impurities from a metal. **removal** (*n*), **removable** (*adj*).

substitute (*v*) to put one object in the place of another object, when the two objects are not similar and do not have the same properties, e.g. to substitute a chlorine atom for a hydrogen atom in an organic compound (the two atoms are not similar and do not have the same properties). **substitution** (*n*).

exchange (*v*) to put one object in the place of another object when the two objects are not similar but do have the same use or property, e.g. when a solution of ammonium sulphate passes through soil, calcium sulphate solution passes out of the soil; the ammonium ions have been exchanged for calcium ions (the two ions are not similar, but they have the same properties of being cations (p.125). **exchange** (*n*).

interchange (*v*) to put one object in the place of another object when the two objects are identical, e.g. when water is in equilibrium (p.150) with its vapour in a cloud vessel, the molecules in the water interchange with the molecules in the vapour. **interchange** (*n*).

retain (*v*) to continue to possess an object, energy, substance or property when conditions try to remove (↑) the object, energy, substance or property. For example, (a) after distillation, the flask always retains some liquid; (b) after purification of pig iron to wrought iron (p.163) the iron still retains some impurities. **retention** (*n*).

revert (*v*) to go back to an original (p.220) state, e.g. on heating rhombic sulphur slowly it changes to monoclinic sulphur at 96.5°C. On cooling, monoclinic sulphur slowly reverts to rhombic sulphur.

return (*v*) to go back to an original (p.220) place, e.g. if an electron is removed from an atom, a positive ion is formed; if an electron returns to the ion, the atom is formed.

exit (*n*) a place through which something goes out, e.g. in a fractionating column there are several exits, one for each fraction. **exit** (*v*).

prevent (*n*) to cause an action or a process not to take place or not to happen. An agent is needed as the cause. The presence (↓) of the agent, or the action of the agent, is the cause. The agent can be an object, substance, condition or a person. **prevention** (*n*).

interfere (*v*) to cause a process to become slower, or to stop, or make it difficult to observe (p.42). The process is usually one that is wanted, e.g. (a) the presence of propane-1,2,3-triol (glycerol) interferes with the decomposition of hydrogen peroxide, the process is slowed down; (b) the presence of a sodium salt interferes with the flame test for other metals (makes the test difficult to observe). **interference** (*n*).

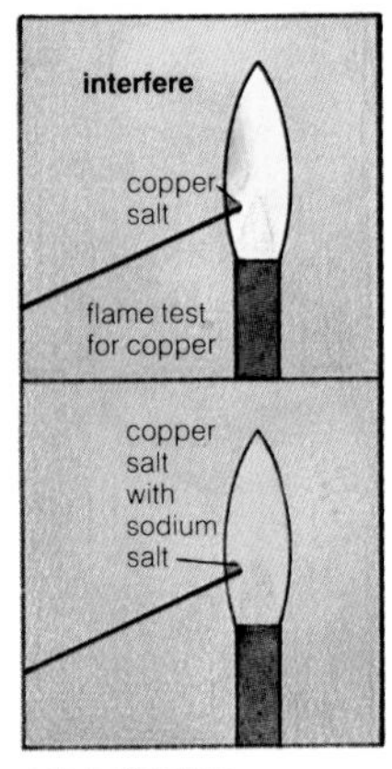

colour of sodium flame interferes with test for copper

counteract (*v*) to act against a process so that the process is slowed down or stopped, or made to go in the reverse (↓) direction. The process is usually, but not necessarily, one that is not wanted, e.g. substances are added to rubber to prevent atmospheric oxygen destroying its properties, these substances counteract the effect of atmospheric oxygen. **counteraction** (*n*).

reverse (*v*) to make a process go in the opposite direction, e.g. heating a liquid causes it to vaporize (p.11), cooling the vapour reverses the process and the liquid is condensed. **reverse** (*n*), **reverse** (*adj*).

tend (*v*) to have a possible action or behaviour, with the action taking place slowly, or not taking place if conditions (p.103) are not suitable; e.g. a solution of sodium hydroxide tends to absorb carbon dioxide from the air, this can be counteracted (↑) by using a tight-fitting rubber bung (p.24). **tendency** (*n*).

trend (*n*) the general direction of change in a set of related facts, e.g. (a) in homologous (p.172) series the trend is for a decrease in activity with an increase in the number of carbon atoms (the general direction of related change); (b) in the set of alkali metals (p.117) the trend is for increasing ease of ionization with increasing atomic number.

present[1] (*v*) to produce a thought in an observer, e.g. (a) making volatile (p.18) compounds react when heated presents a difficulty to an observer which he overcomes (p.213) by refluxing (p.201) the compounds; (b) finding a use for waste products presents a problem to a manufacturer.

present[2] (*adj*) describes an object or substance which is in a particular, named place. For example, (a) the iodine test shows whether starch is present in food; (b) the presence of chloride ions (p.123) is tested by silver nitrate solution. **presence** (*n*).

recur (*v*) to occur (p.63) time after time, usually at a definite interval (p.220) of time, e.g. conditions for the pollution of the atmosphere recur every year when the temperature is suitable for the formation of oxides of nitrogen.

flow (*v*) to move, of a liquid or gas, along a pipe or over a surface. Electric current and heat flow along a conductor. **flow** (*n*).

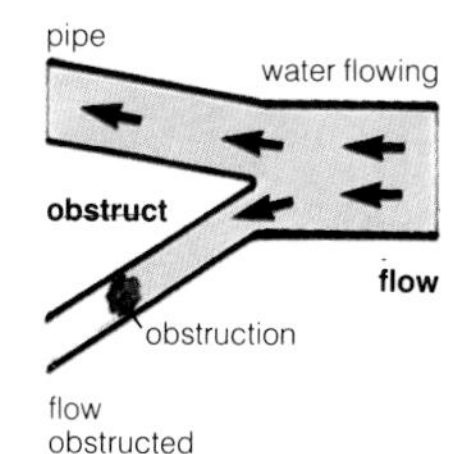

obstruct (*v*) to prevent (↑) the flow of a liquid or gas, or the flow of a stream of particles (p.110), e.g. a deposit of calcium carbonate in a water-pipe obstructs the flow of water. **obstruction** (*n*).

stoppage (*n*) the state of a flow being stopped, e.g. a stoppage in the delivery tube connecting a distillation flask to a condenser means that the flow of vapour has stopped and an explosion is likely.

capture (*v*) to attract and to hold an object by the use of force. To catch and to keep an escaping object. For example, (a) an atomic nucleus (p.110) captures a neutron (p.110) during bombardment (p.143) by neutrons (the neutron is held in the nucleus by nuclear forces); (b) a steam trap (p.29) captures any water escaping with the steam. **capture** (*n*).

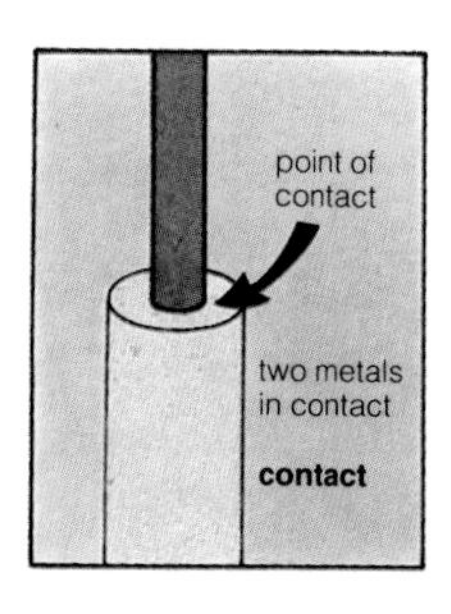

contact (*n*) the state of two, or more, objects being in touch with each other. For example, (a) an electrical contact is formed when two conductors (p.122) touch each other; (b) when sulphur dioxide and oxygen are both in contact with a platinum catalyst, the two gases combine (the two gases touch each other and the catalyst). **contact** (*n*).

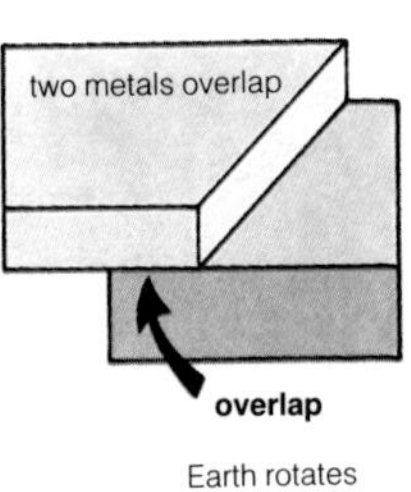

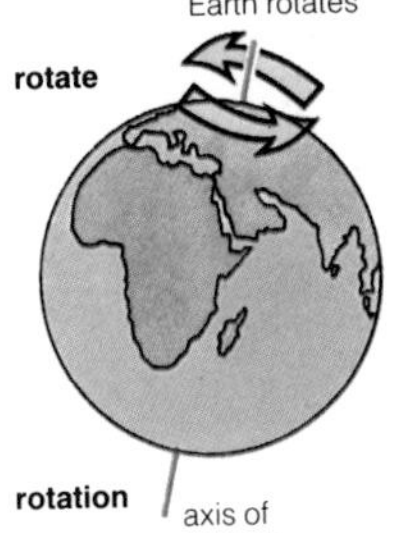

overlap (*v*) to cover part of a flat object by putting another flat object above it, e.g. the pieces of wood on the side of a boat overlap each other. **overlap** (*n*).

rotate (*v*) to turn, of an object, if the object turns about an axis going through the body, e.g. the Earth rotates about its axis. **rotation** (*n*).

vary (*v*) to change in detail (p.226) only, of a quantity (p.81), a quality (p.15), or a shape, e.g. (a) atmospheric pressure varies from day to day; (b) the electric current passing through an electrolyte (p.122) varies with the voltage (p.126) applied to the electrodes (the process of electric current being conducted by the electrolyte remains the same, only the detail of the value of the current varies). **variation** (*n*), **variable** (*adj*), **variable** (*n*), **varying** (*adj*), **varied** (*adj*).

variation (*n*) (1) the amount of change which takes place if a quantity or quality varies (↑); e.g. the variation in atmospheric temperature from the highest to the lowest temperature of the day, (2) the action of varying (↑).

variable[1] (*n*) a quantity, such as temperature, pressure, humidity, concentration, etc. which can change in value or can be changed in value, e.g. the vapour pressure (p.103) of a liquid is a variable because it varies (↑) with temperature.

variable[2] (*adj*) describes a quantity, quality, or shape which varies (↑) or can be varied, e.g. atmospheric pressure is variable as it changes from day to day.

maximum (*n*) (maxima *n.pl.*) the greatest possible, or the greatest recorded (p.39), value of a variable (↑), e.g. (a) 41°C was the maximum recorded for atmospheric temperature in the country; (b) the maximum pressure the chamber will withstand is 200 atmospheres.

minimum (*n*) (minima *n.pl.*) the least possible, or the lowest recorded, value of a variable (↑), e.g. 630 mm of mercury was the minimum recorded for atmospheric pressure in the country; (b) the minimum pressure recorded in the vacuum distillation was 50 mm of mercury.

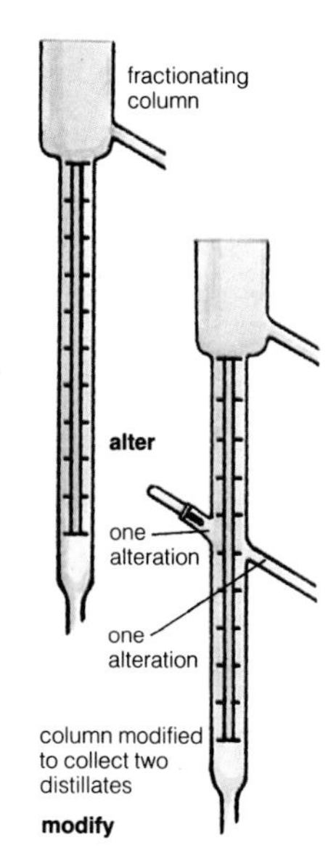

alter (*v*) to make one variation (↑) in a quantity, shape, or condition, e.g. (a) to alter the pressure of a gas from 1 atmosphere to 2 atmospheres; (b) to alter the condition in a reversible reaction by the addition of a catalyst (change of conditions). **alteration** (*n*).

modify (*v*) to alter (↑) a process, or an object, to make it more suitable for a particular purpose, e.g. a fractionating column (p.201) can be modified to produce two fractions (p.202) instead of one (both a process and an object are altered). If a process or an object is modified, its purpose is not changed. **modification** (*n*).

decrease (*v*) to become, or to cause to become, less, of a quantity (p.81) or number. For example, (a) the rate of reaction (p.149) between ethanol and ethanoic acid is decreased by lowering the temperature of the reactants (p.62); (b) the volume of a gas decreases as the pressure on the gas increases. **decrease** (*n*).

reduce (*v*) to cause a quantity (p.81), or number, to become less, e.g. the concentration of the solution was reduced from 2.0 M to 1.0 M by adding a suitable volume of solvent. To contrast *decrease* (↑) and *reduce*: lowering the temperature by 80°C *decreases* the volume of the gas by 160 cm^3. Lowering the temperature of the gas by 80°C *reduces* the volume to 5.30 dm^3. *Decreasing* describes a continuous change; *reducing* describes a change from one state, or degree (p.227), to another. **reduction** (*n*).

accumulate (*v*) to increase a quantity (p.81) or an amount by addition over a period of time, e.g. (a) during electrolysis (p.122) the deposit on an electrode accumulates as the process goes on; (b) as a suspension settles, the sediment accumulates at the bottom of the vessel. **accumulation** (*n*).

accelerate (*v*) to increase speed with time, e.g. some catalysts accelerate the rate of reaction more than others do, i.e. they increase the rate of reaction.

event (*n*) a change that takes place (p.63) at a particular time and usually at a particular place, e.g. (a) the appearance of a precipitate is an event; (b) an explosion is an event. **eventual** (*adj*), **eventually** (*adv*).

interval (*n*) the distance between two points or the length of time between two events.

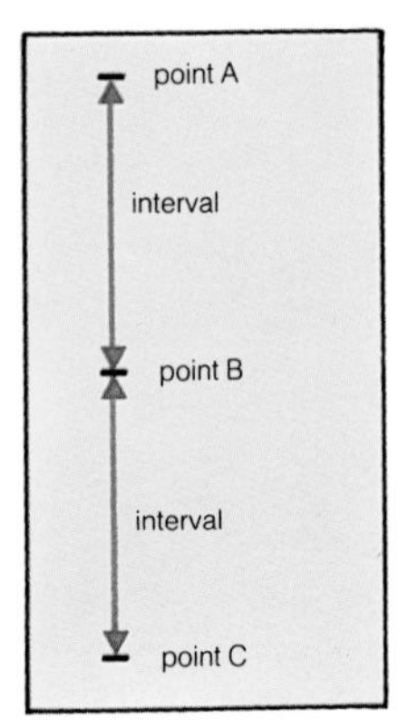

interval

duration (*n*) the length of time in which a process takes place, e.g. (a) the duration of a process of distillation; (b) the duration of a chemical reaction. To contrast *duration* and *extent* (p.213): The *duration* of the distillation is one hour, the *extent* of the distillation is from the time of starting to heat the flask until the last drop of distillate is collected, i.e. *duration* gives the time for the process, while *extent* gives the limits of the process. **durable** (*adj*), **endure** (*v*).

precede (*v*) to come immediately before an event in time, or before a unit (p.212) in a series (p.172), e.g. number 7 precedes number 8 in the series 5, 6, 7, 8, 9, 10. Numbers 5, 6, 7, are the preceding terms for number 8, while 7 is the preceding term. **preceding** (*adj*).

previous (*adj*) describes events which come before a named event, or a named time, or the present time, e.g. in a previous experiment, which took place last year, the results were different. To contrast *previous* with *preceding*: the *preceding* (↑) experiment is one immediately before the present experiment; a *previous* experiment is one which took place some time ago. **previously** (*adj*).

preliminary (*adj*) describes the first stage of a process, e.g. preliminary tests in analysis indicate (p.38) the subsequent (↓) stages that have to be carried out.

original (*adj*) describes the very first event, action or process, e.g. Becquerel carried out the original experiments on radioactivity.

subsequent (*adj*) describes events which come after a named event, or a named time, or the present time, e.g. the results of the present experiment are not sufficient for a conclusion, subsequent experiments may give a better result. **subsequently** (*adv*).

event

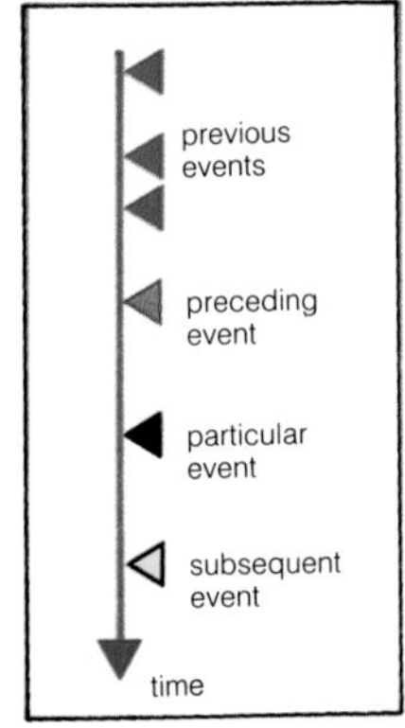

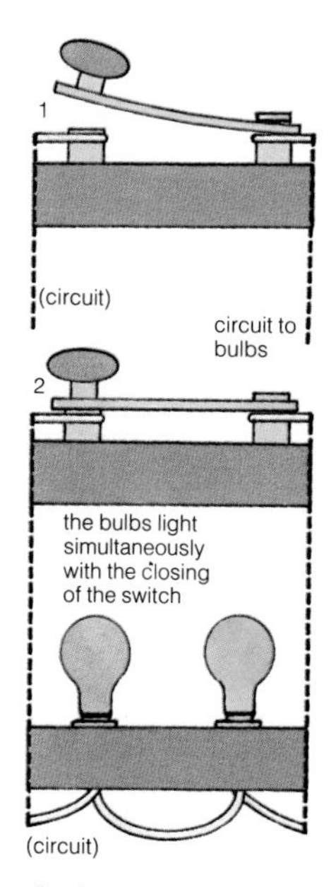

simultaneous

simultaneous (*adj*) describes two or more events which take place at the same time, e.g. the closing of a switch in an electrical circuit and the flow of electric current in the circuit are simultaneous. **simultaneously** (*adv*).

immediate (*adj*) describes one event which is next in time to another event without anything happening between the two events, e.g. precipitation is immediate when silver nitrate solution is added to the solution of a chloride. **immediately** (*adv*).

order (*n*) the arrangement of parts of a whole, or of members of a series (p.172), so that a pattern (p.93) is formed; or the arrangement follows particular rules; or the arrangement occurs (p.63) in time. For example, (a) the molecules in a giant structure are arranged in a particular pattern which has a definite order; (b) the stages of distillation follow each other in order; (c) words can be arranged in an alphabetical order. **ordered** (*adj*).

method (*n*) a particular way of carrying out a process, e.g. (a) the method of separating a liquid mixture by fractional distillation; (b) copper (II) sulphate can be prepared by digesting copper (II) oxide in sulphuric acid or by digesting copper (II) carbonate in sulphuric acid; these are two different methods.

control (*v*) to start or stop a process, to increase or decrease the rate of a reaction; to vary a quantity for a purpose; e.g. to control the manufacture of ammonia by using suitable conditions of temperature and pressure and adding suitable amounts of reactants (p.62). **control** (*n*).

progress (*v*) of a process, to go successfully from one stage (p.159) to the next stage in the correct order (↑), e.g. as the distillation progressed, the distillate (p.201) accumulated (p.219) in the receiver. **progress** (*n*).

exhibit (*v*) to allow a property (p.9) to be observed when it is not always open to observation (p.42). For example sodium carbonate crystals exhibit efflorescence (p.67), i.e. they can be observed to effloresce only under particular conditions.

statement (*n*) a description of related facts in language or in symbols, e.g. a statement of Boyle's law in language or in symbols. The related facts are the mass of the gas, the pressure, its volume and its temperature. **state** (*v*).

information (*n*) the knowledge given by a statement (↑), e.g. information on the change in volume of a gas under particular conditions can be obtained from a statement of the gas laws (p.109). **inform** (*v*).

determine (*v*) (1) to find a value of a quantity if more than one measurement has to be made and the final value obtained by calculation, or to find a very accurate value of a quantity by careful measurement, e.g. to determine the boiling point of a substance using a very accurate thermometer and special apparatus. **determination** (*n*). (2) if one variable (p.218) depends on another variable, then the second variable determines the value of the first variable, e.g. the volume of a gas depends on its pressure, so its pressure determines its volume.

deduce (*v*) to come to a conclusion (p.43) using known facts, e.g. fact 1: substance X neutralizes an alkali; fact 2: acids neutralize alkalis; deduce that substance X is an acid. **deduction** (*n*).

verify (*v*) to carry out an experiment and to record results using materials and substances, when similar experiments have been carried out previously (p.220), to show that the previous results, and the deductions (↑) from the results, were true. For example, to carry out an experiment to verify the statement (↑) of Boyle's law. **verification** (*n*).

assume (*v*) to take a statement (↑) or a fact, without verification (↑) to be true, e.g. in a calculation on relative vapour density (p.12) we assume the gas laws. **assumption** (*n*).

clarify (*v*) to make a statement (↑) clear enough to understand it, e.g. a statement about the direction of the covalent bonds of a carbon atom is clarified by making a drawing of the atom and bonds. **clarification** (*n*).

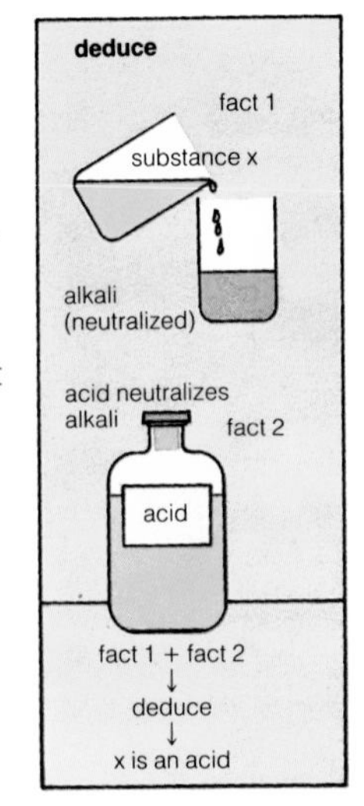

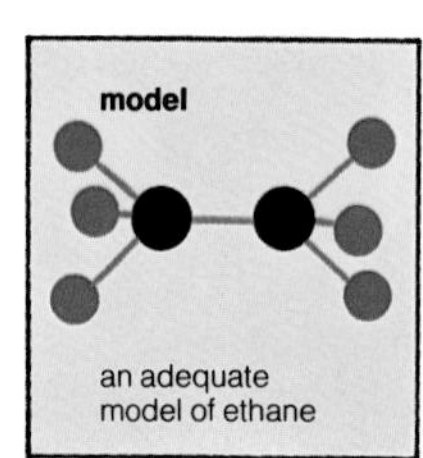

an adequate model of ethane

model (*n*) a device (p.23) which represents (↓) an object and allows a person to think about the object when the object itself cannot be observed (p.42). A **scale model** is an exact (p.79) representation with each part a known fraction, in size, of the object, e.g. a scale model of a house. An **adequate model** has sufficient (p.231) detail for a particular purpose, e.g. an adequate model of an atom describes the nucleus (p.110) and extra-nuclear electrons (p.110). An **analogue model** uses only similar characteristics, e.g. a permeable membrane is represented as a sieve with holes.

represent (*v*) to make a drawing which is like an object and allows a person to remember or to think about the object, e.g. the drawings of apparatus on page 23 represent each piece of apparatus; they are not the same as pictures of the apparatus, but the representation allows a person to imagine the actual piece of apparatus. **representation** (*n*).

apparent (*adj*) describes anything that appears to be correct when observed (p.42) by the senses, but can be shown by experiment, or is known, not to be correct, e.g. the apparent r.m.m. of ethanoic acid, when dissolved in some solvents, is twice the actual value, i.e. the measurement does not give a correct value.

recapitulate (*v*) to state (↑) facts a second time, usually in a different way, in order to clarify (↑) the information.

refer (*v*) to bring facts, statements, or information to a person's attention, e.g. when discussing electrovalency it is useful to refer to the atomic structure of elements, i.e. to direct a person's attention to the information on atomic structure.

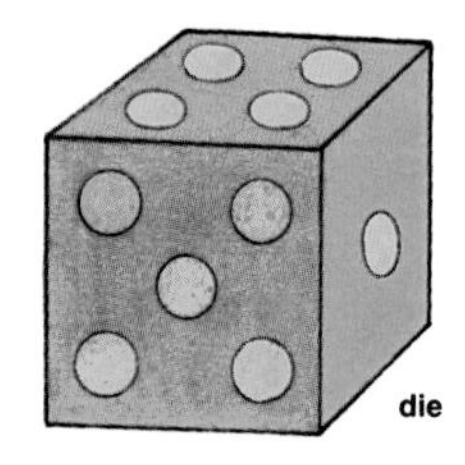
die

random (*n*) describes an event, a process or a state which exhibits (p.221) no order (p.221), e.g. it is not possible to predict (p.85) the path that the molecules of a gas will follow.

probability (*n*) the mathematical fraction which describes chance, e.g. when a die is thrown, there is an equal chance that any one of the six faces will appear; the probability of any one particular face appearing is 1/6.

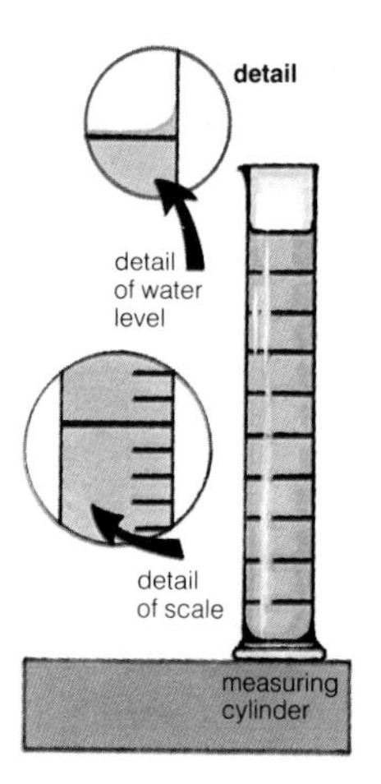

detail (*n*) a small part of a structure, object, or process; or a fact, which is not important, in a set of facts. For example, (a) in describing the structure of a molecule of ammonia, the angles between the bonds are a detail; (b) in describing the preparation of copper (II) sulphate from copper (II) oxide and sulphuric acid, an experimental detail is that the oxide should be added in small amounts (a part of a process, or a fact in a set of descriptive facts).

detailed (*adj*) describes a statement (p.222) which gives as many details as possible.

essential (*adj*) describes any part of a whole without which the whole loses its identity (p.225). An essential property or characteristic is one that must be possessed for the purpose of identification, e.g. (a) the amino group (p.186) is an essential part of the molecule of an amine (p.178); without it the substance is no longer an amine; (b) the ability to produce hydrogen ions in solution is an essential property of an acid; if a substance does not exhibit (p.221) that property it cannot be classified (p.120) as an acid. **essentially** (*adv*).

major (*adj*) describes any part, property, characteristic or fact which is important or is the most important, e.g. (a) the major use of chlorine is in the manufacture of plastics and synthetic rubber (fact); (b) the major source of bromine is sea water. **majority** (*n*).

minor (*adj*) describes any detail, property, characteristic or fact which is not important or is of lesser importance. For example, (a) a minor use of lead is in lead tetraethyl, added to petrol; (b) a minor detail in the process of distillation is controlling (p.221) the flow of water through the condenser to obtain the best conditions for condensation. **minority** (*n*).

definite (*adj*) describes a statement, a relation (p.232), a property or a characteristic about which there is no doubt, as previous experimental work has established (p.225) the facts, e.g. aluminium oxide has a definite giant covalent crystalline structure, i.e. the structure has been established (p.225).

degree (*n*) (1) a stage on a scale of quality (p.15), e.g. the degrees of a hot object are: lukewarm, warm, hot, boiling, red hot, white hot. A degree describes a simple method of measurement and can be used for quantities (p.81) as well as for qualities. (2) the extent reached by a process, e.g. the degree of ionization measures the extent to which a substance is ionized. This is an accurate (↓) measurement and is usually stated as a percentage or a decimal fraction.

appreciable (*adj*) describes a change in a quantity (p.81) or a quality (p.15) which is big enough or important enough to be considered, e.g. (a) during electrolysis there is an appreciable lowering of the concentration of ions around an electrode; because of this, the electrolyte is stirred; (b) carbon dioxide has an appreciable solubility in cold water, this is important in the carbon cycle (p.61).

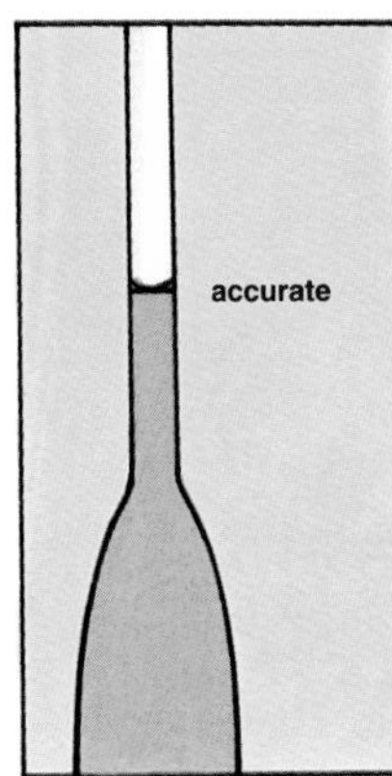

accurate (*adj*) describes a measurement made by the best instruments (p.23), or apparatus (p.23), available (p.85), e.g. a pipette (p.26) gives an accurate measurement of volume of a liquid at the temperature marked on the pipette. **accuracy** (*n*).

accuracy (*n*) the quality of being accurate (↑). No measurement can be exact (p.79) as all instruments and scales (p.26) cannot be read to an exact value. In giving a measurement, the degree (↑) of accuracy must be stated (p.222). This is usually done by giving the limits of accuracy, e.g. the accurate value established (p.225) for the Avogadro constant is $(6.022\,52 \pm 0.000\,28) \times 10^{23}\,mol^{-1}$; the figure $\pm\,0.000\,28$ gives the limits of accuracy, and thus the degree of accuracy of the measurement.

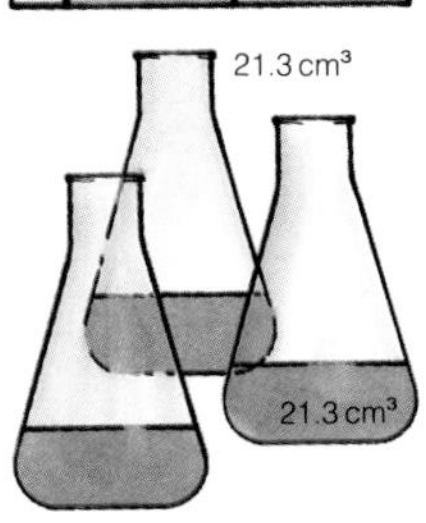

discrepancy (*n*) an appreciable (↑) difference between two measurements or two statements when the measurements should have been the same or the statements should have had the same meaning, e.g. if the results for three titrations of the same solutions are: $21.3\,cm^3$; $22.8\,cm^3$; $21.3\,cm^3$, then there is a discrepancy between the second result and the other two.

define (*v*) to state clearly, in known words, a description of a term which excludes (p.230) similar terms and allows the terms to be discussed (↓) without misunderstanding. Examples of terms are quantities, units of measurement, categories of objects and substances, properties and characteristics. **definition** (*n*).

definition (*n*) the result of defining a term. It can also include giving a name to a new idea, such as the naming of the neutron. Examples are: the definition (a) of an ion using the words: atom, electric charge, positive and negative, electron; (b) of electrochemical equivalent using the words: mass, deposit, coulomb.

discuss (*v*) to examine a statement and to give reasons, in writing or by speaking, for the statement, or against it, e.g. to discuss the statement that fractional distillation is not necessary to separate two liquids if their boiling points are 10°C apart (there are reasons for and against the statement). **discussion** (*n*).

comment (*v*) to explain, in writing or by speaking, why a statement is true or untrue, giving reasons for what is said, e.g. to comment on Boyle's law. **comment** (*n*).

exemplify (*v*) to give examples in order to clarify (p.222) a statement, e.g. the dehydrating properties of concentrated sulphuric acid can be exemplified by its action on sucrose, in which the sugar is converted to carbon. **exemplification** (*n*).

complex (*adj*) describes a structure, process or system formed from parts connected or combined, in an order that can be recognized, to form a whole, e.g. a polymer (p.207) molecule is complex; it consists of small molecules (the parts) combined into a whole, and the structure of the whole molecule can be recognized from the combination of the parts. **complexity** (*n*), **complex** (*n*), **complex** (*v*).

simple (*adj*) describes anything that is not made from parts, is not complex (↑), is easy to understand, e.g. compare a complex ion with a simple ion. **simplicity** (*n*).

complex

complex structure of a molecule

HO

CH_3

O

parts are 1

2

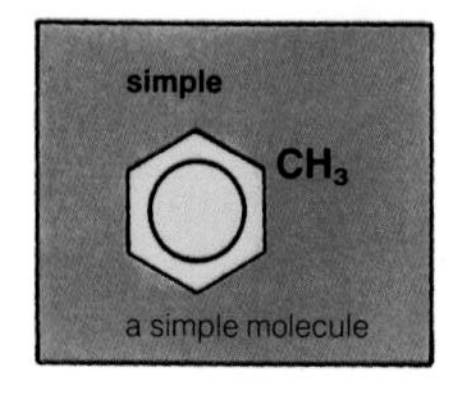

a simple molecule

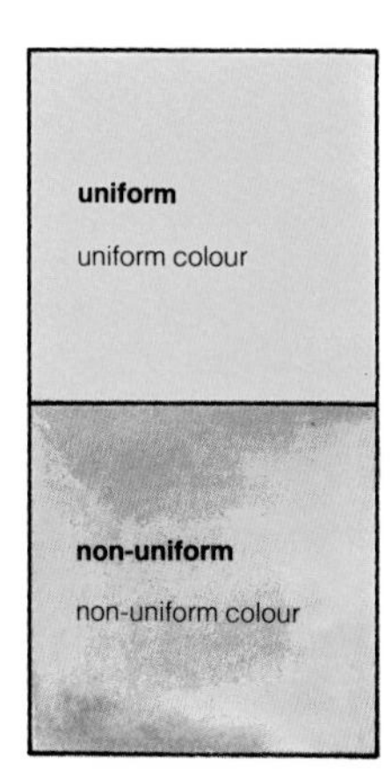

uniform (*adj*) describes a constant (p.106) value of a quantity, or quality, which is spread over space or time, e.g. (a) a uniform colour is a constant colour spread over a surface; (b) a uniform acceleration is a constant acceleration spread over a period of time. To contrast *uniform* and *constant:* there is a *constant* concentration of carbon dioxide in the air (it is 0.03%); there is a *uniform* concentration of carbon dioxide in the atmosphere (the same concentration spread over the surface of the Earth). **uniformity** (*n*).

steady (*adj*) describes a value of a quantity which is kept constant (p.106) although there is a tendency (p.216) for the value to vary, e.g. to maintain a steady vacuum in vacuum distillation (the pressure tends to rise).

standard¹ (*adj*) (1) describes the value of a quantity which is accepted by everyone as a constant (p.106) value, e.g. (a) standard atmospheric pressure is accepted as 760 mm of mercury; (b) standard electrode potentials (p.130) are those defined (↑) under known conditions of concentration and temperature. (2) describes an instrument (p.23) or a piece of apparatus (p.23) having a stated value of a quantity, or an accurate scale for a quantity, e.g. a standard graduated flask with an accurate volume at a given temperature. **standard** (*n*), **standardization** (*n*), **standardize** (*v*).

standard² (*n*) an accepted value of a quantity given in a definition, e.g. the standard for length is the metre, defined in S.I. units (p.212).

normal (*adj*) describes conditions, particularly room conditions, which are between the expected limits of variation, e.g. in northern countries, the normal room temperature is accepted to be 18°C; the temperature varies daily above or below this value, but on most days the temperature will not be too different from this value. **normality** (*n*).

abnormal (*adj*) describes conditions which are greatly different from normal (↑) conditions, e.g. in northern countries, a room temperature of 45°C would be considered abnormal. **abnormality** (*n*).

exception (*n*) anything left out of a statement or description, e.g. the chlorides of the alkaline earth metals (p.117) are not hydrolyzed (p.66) in solution with the exception of beryllium and magnesium chlorides (these two chlorides are hydrolyzed). **except** (*v*).

exclude (*v*) to keep out, or to put out, e.g. (a) in the preparation of iron (II) sulphate crystals, atmospheric oxygen is excluded to prevent oxidation of the salt (the air is kept out of the apparatus); (b) in the reduction of copper (II) oxide by hydrogen, air is excluded by the hydrogen before heating takes place (hydrogen pushes the air out). **exclusion** (*n*), **exclusive** (*adj*).

excess (*n*) an amount that is more than the amount needed, e.g. if 100 cm^3 of hydrochloric acid is the exact amount of acid needed to dissolve 5 g of zinc, then 150 cm^3 of the acid is added to the zinc to make sure all the metal is dissolved. Excess acid has been added to the zinc, and there is an *excess* of 50 cm^3 of hydrochloric acid. A volume of 150 cm^3 of hydrochloric acid is 50 cm^3 *in excess* of the 100 cm^3 of acid needed to dissolve 5 g of zinc. **excessive** (*adj*).

excess excess liquid in a flask

excess liquid
graduation mark
flask

intense (*adj*) describes a high degree (p.227) of a quantity, e.g. (a) an intense heat is at a very high temperature; (b) an intense radiation is a very powerful radiation. **intensity** (*n*).

converse (*n*) an equal and opposite action, or an equal and opposite statement, e.g. (a) condensation is the converse of evaporation, the two actions are equal and opposite; (b) the statement *'ideal gases obey Boyle's law'* has as its converse 'non-ideal gases do not obey Boyle's law'. **converse** (*adj*).

introduce (*v*) to put a solid, a liquid, or a gas into a vessel (p.25) when skill is needed in the technique (p.43), e.g. to introduce a small quantity of a volatile liquid into a mercury barometer tube when measuring vapour pressure. **introduction** (*n*).

insert (*v*) to put an object in a fixed position, e.g. a thermometer in a cork.

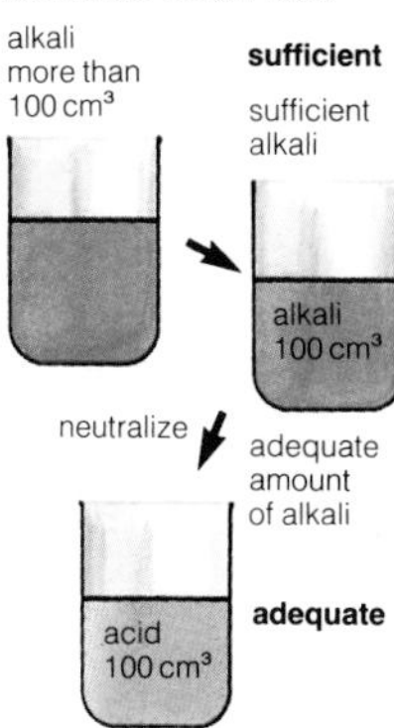

abundant (*adj*) describes something occurring in great quantity, especially spread over a large area, but which is not necessarily going to be used, e.g. coal is abundant in Europe.

plentiful (*adj*) describes something which is in great quantity and is available (p.85) for use, e.g. (a) petrol is plentiful in Arabia (there is a great quantity of it and it is put to use); (b) coal burned in a plentiful supply of air forms carbon dioxide and water. *Contrast limited* (↓).

adequate (*adj*) describes an amount that is equal to the amount needed for a particular purpose. The actual amount may not be stated, but is assumed (p.222) to be known, e.g. add an adequate amount of alkali to make the mixture alkaline (the amount is not known). **adequacy** (*n*).

sufficient (*adj*) describes an amount that is equal to the amount needed for a particular purpose, and is used with the name of the material or substance. To contrast *sufficient* and *adequate*: there is *sufficient* acid to neutralize the alkali; there is an *adequate* amount of acid to neutralize the alkali. **sufficiency** (*n*), **suffice** (*v*).

insufficient (*adj*) describes a quantity that is not sufficient (↑).

inadequate (*adj*) describes an amount that is not adequate (↑) or anything that is needed and is not adequate, e.g. (a) the amount of acid was inadequate to neutralize the alkali; (b) the voltage of the electrical supply was inadequate to carry out the electrolysis; (c) his knowledge of mathematics was inadequate for his work in chemistry. **inadequacy** (*n*).

limited² (*adj*) (1) describes something which is only available (p.85) for use in an inadequate (↑) amount. It is the opposite of plentiful (↑), e.g. coal burned in a limited supply of air forms carbon monoxide and water. (2) describes the application of theories and laws and the use of instruments, devices and apparatus when there are limits, e.g. (a) the law of constant composition is limited to stoichiometrical compounds (p.82); (b) ammeters have a limited range of measurement.

deficient (*adj*) describes a material, object, or an idea that lacks a part or is lacking in quantity in a part of it, e.g. (a) if rubber is deficient in its sulphur content it will be too soft for many purposes; (b) the laboratory is deficient in fume cupboards. **deficiency** (*n*), **deficit** (*n*).

supplementary (*adj*) (1) describes an amount which is in addition to a previous (p.220) amount and is needed to complete or to improve a reaction or process. (2) describes an angle, which with another angle adds up to 180°, e.g. an angle of 124° is the supplementary angle of 56°. **supplement** (*n*), **supplement** (*v*).

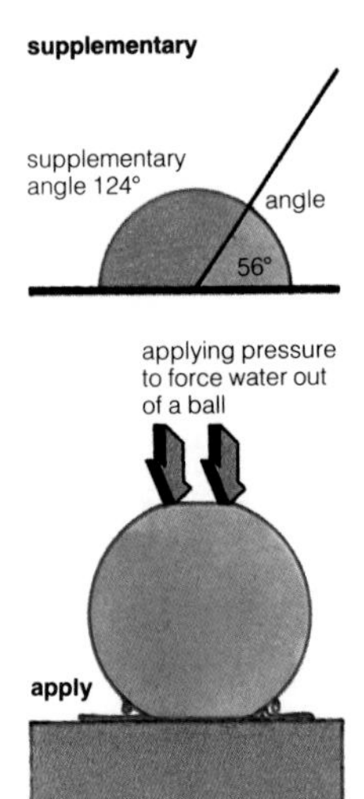

apply (*v*) (1) to cause a force or a potential to act at a point or place, e.g. (a) to apply pressure to a gas in a vessel using a column of mercury; (b) to apply a voltage of 12 volts to the electrodes in a voltameter. (2) to use a theory or a law, e.g. Boyle's law is applied to all gases, but is only valid (↓) when applied to ideal gases (p.107). **application** (*n*).

valid (*adj*) describes a statement (p.222) or experiment that is accurate and is in agreement with scientific experience, e.g. the gas laws are valid for predictions of volume changes at low pressures.

relation (*n*) a connection: between quantities that can vary; between cause and effect (p.214); or between objects, e.g. (a) there is a relation between the mass of a metal deposited during electrolysis and the quantity of electric charge passed through the electrolyte; (b) there is a spatial (p.211) relation between the atoms in a molecule. **relate** (*v*), **related** (*adj*).

relative (*adj*) describes the relation (↑) between one physical property of a substance and the same physical property of a standard (p.229) substance, e.g. (a) relative vapour density is the ratio of the density of a gas divided by the density of hydrogen; (b) relative molecular mass is the ratio of the mass of one molecule of a substance divided by 1/12 of the mass of one atom of carbon-12.

respective (*adj*) describes the order relating objects to their descriptions when there are more than two such objects, e.g. the respective states of matter of oxygen, bromine and copper are gas, liquid and solid. **respectively** (*adv*).

correspond (*v*) to be similar (↓) in some part, function, structure or situation, without being identical, e.g. (a) the hydroxyl group in an alcohol corresponds to the hydroxyl group in water; (b) the nitrate radical (p.52) and the carbonate radical (p.49) correspond in structure. **correspondence** (*n*).

similar (*adj*) two things are similar if they have many like properties, structures, qualities or characteristics, but have properties, structures, qualities or characteristics that distinguish (p.224) them. For example the characteristics of the transitional metals (p.121) are similar, yet each metal can be distinguished from the others. **similarity** (*n*).

identical (*adj*) describes two things which have the same number of properties and characteristics and these properties and characteristics are exactly the same. It is not possible to distinguish between identical objects except by the space they occupy (↓).

direct (*adj*) (1) describes a relation in which an increase in one quantity is related to the increase in a related quantity. (2) describes an inference (p.43) in which the properties of a subset are known and the inference is that the properties of the whole set will be the same.

inverse (*adj*) (1) describes a relation in which an increase in one quantity is related to a decrease in another quantity. (2) describes an inference (p.43) in which the properties of a whole set are known and the inference is that the properties of a subset will be the same.

occupy (*v*) to be in a particular place so that another object cannot be in that place, e.g. (a) the ions in a crystal (p.91) occupy positions in the lattice (p.92); (b) the alkali metals (p.117) occupy places in group I of the periodic system (p.119). **occupation** (*n*), **occupant** (*n*), **occupied** (*adj*).

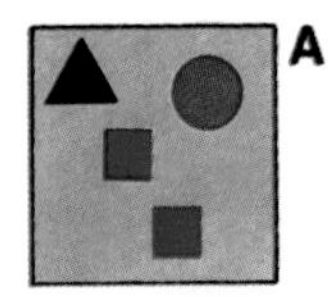

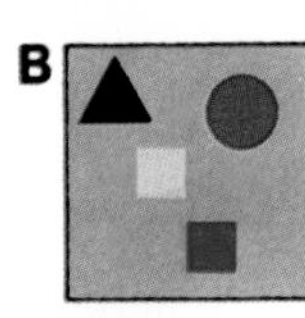

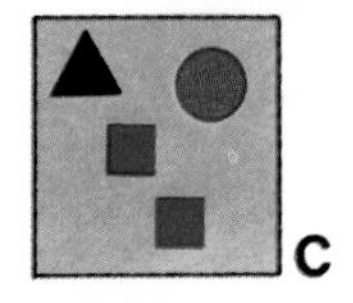

correspond

the black triangles — are in corresponding positions in A, B and C

similar

A, B are similar

identical

A, C are identical

occupy

the green colour occupies the same position in A, B, C

International System of Units (SI)

PREFIXES FOR SI UNITS

MULTIPLE	FIGURE	PREFIX	SYMBOL
10^{12}	1 000 000 000 000	tera	T
10^{9}	1 000 000 000	giga	G
10^{6}	1 000 000	mega	M
10^{3}	1 000	kilo	k
10^{-3}	0.001	milli	m
10^{-6}	0.000 001	micro	μ
10^{-9}	0.000 000 001	nano	n
10^{-12}	0.000 000 000 001	pico	p

BASIC UNITS

Système International d'Unités (SI) is based on seven basic units; the seventh, the candela, is not included in the list below, as it is not used in chemistry.

metre *(unit of length)* *symbol:* m
Defined from a wavelength in the spectrum of krypton.

kilogram *(unit of mass)* *symbol:* kg
The mass of the international prototype kept at Sèvres.

second *(unit of time)* *symbol:* s
Defined from a frequency in the spectrum of the caesium-133 atom.

kelvin *(unit of temperature)* *symbol:* K
The fraction 1/273.16 of the temperature of the triple point of water.

mole *(unit of amount)* *symbol:* mol
The amount of substance which contains as many elementary units as there are atoms in 0.012 kilogram of carbon-12. The elementary unit must be specified and may be an atom, a molecule, an ion, an electron, a radical, etc.

ampere *(unit of electric current)* *symbol:* A
Defined from a newton and a metre.

DERIVED UNITS

QUANTITY	SYMBOL FOR QUANTITY	UNIT	SYMBOL FOR UNIT	BRIEF DEFINITION
velocity	v	m/s	—	displacement/time
acceleration	a	m/s^2	—	velocity/time
force	F	newton	N	mass × acceleration
energy	E	joule	J	force × distance
pressure	p	pascal	Pa	force/unit area
density	ρ	kg/m^3	—	kilogram/cubic metre
frequency	f	hertz	Hz	number of oscillations/time
concentration	—	mol/dm^3	M	moles/cubic decimetre
electric charge	Q	coulomb	C	amperes × time
electric potential	V	volt	V	joule/coulomb
electric resistance	R	ohm	Ω	volt/amperes
volume	V	dm^3	—	cubic decimetre

Symbols used in chemistry

LETTER/SYMBOL	QUANTITY, OBJECT OR OPERATOR
A	mass number. A_r relative atomic mass
E	energy, electromotive force
e	electron. $^{0}_{-1}e$ charge and mass of electron
F	Faraday constant
f	frequency
H	heat of reaction
I	electric current
k	a constant
L	Avogadro constant
M	concentration in moles per cubic decimetre M_r relative molecular mass
m	mass
N	number of molecules
N	neutron number
n	any number; mole fraction; number of moles
n	a neutron. $^{1}_{0}n$ charge and mass of neutron
p	pressure
p	a proton. $^{1}_{1}p$ charge and mass of proton
Q	quantity of electric charge
R	molar gas constant, resistance
r	gas constant, radius
T	thermodynamic temperature (measured in kelvin)
t	time. $t_{\frac{1}{2}}$ half life
V	volume, electric potential difference V_m molar volume
Z	atomic number
$\triangle$	a change, e.g. $\triangle H$ change in heat
θ	temperature difference, temperature (Celsius scale)
ρ	density

Important constants

s.t.p. = standard temperature and pressure which is 1.00 atmospheres or 760 mm of mercury or 101 kPa and 273 K or 0°C.

Temperature of the triple point of water is 273.16 K.

Absolute zero of temperature is 0 K or − 273°C.

Volume of 1 mole of gas at s.t.p. is 22.4 dm^3 (molar volume).

Molar gas constant, 8.314 $J\,K^{-1}\,mol^{-1}$.

Avogadro constant, L = $6.02 \times 10^{23}\,mol^{-1}$.

Faraday constant, F, $9.65 \times 10^{4}\,C\,mol^{-1}$.

Mass of electron, 9.11×10^{-31} kg.

Ratio of masses, proton/electron, 1840.

Ratio of masses, neutron/electron, 1840.

Charge on an electron, 1.6×10^{-19} C.

Ionic product of water $K_w = 1.008 \times 10^{-14}\,mol^2\,dm^{-6}$ (298 K or 25°C).

1 calorie = 4.18 joules.

1 electron-volt (1 eV) = 1.6×10^{-19} J.

Common alloys

NAMES OF ALLOY	APPROXIMATE COMPOSITION	USES
brass	zinc 35%–10%, copper 65%–90%	decorative metal work
bronze		
– common	zinc 2%, tin 6%, copper 92%	machinery, decorative work
– aluminium	aluminium 10%, copper 90%	machinery castings
– coinage	zinc 1%, tin 4%, copper 95%	coins
dentist's amalgam	copper 30%, mercury 70%	dental fillings
duralumin	magnesium 0.5%, manganese 0.5%, copper 5%, aluminium 95%	framework of aeroplanes
gold		
– coinage	copper 10%, gold 90%	coins
– dental	silver 28%–14%, copper 14%–28%, gold 58%	dental fillings
lead, battery plate	antimony 6%, lead 94%	accumulators
manganin	nickel 1.5%, manganese 16%, copper 82.5%	resistance wire
nichrome	chromium 20%, nickel 80%	heating elements, resistance wire
pewter	lead 20%, tin 80%	utensils
silver		
– coinage	copper 10%, silver 90%	coins
solder	tin 50%, lead 50%	joining iron objects
steel		
– stainless	nickel 8%–20%, chromium 10%–20%, iron 80%–60%	utensils
– armour	nickel 1%–4%, chromium 0.5%–2%, iron 98%–95%	armour plating
– tool	chromium 2%–4%, molybdenum 6%–7%, iron 95%–90%	tools

Common abbreviations in chemistry

abs.	absolute
anhyd.	anhydrous
approx.	approximately
aq.	aqueous
b.p.	boiling point
conc.	concentrated
concn.	concentration
const.	constant
crit.	critical
cryst.	crystalline
d.	decomposed
decomp.	decomposition
dil.	dilute
dist.	distilled
e.g.	for example
e.m.f.	electromotive force
eqn.	equation
expt.	experiment
fig.	figure (diagram)
f.p.	freezing point
hyd.	hydrated
i.e.	that is
insol.	insoluble
liq.	liquid
max.	maximum
min.	minimum
m.p.	melting point
p.d.	potential difference
ppt.	precipitate
r.a.m.	relative atomic mass
r.m.m.	relative molecular mass
sol.	soluble
soln.	solution
sp.	specific
s.t.p.	standard temperature and pressure
temp.	temperature
vac.	vacuum
v.d.	relative vapour density
wt.	weight
°	degree (Celsius)

Understanding scientific words

New words can be made by adding **prefixes** or **suffixes** to a shorter word. Prefixes are put at the front of the shorter word and suffixes are put at the back of the word. Words can also be broken into parts, each of which can have a meaning, but cannot be used alone.

(i) correct→ *in*correct (adding a prefix)
correct→ correct*ness* (adding a suffix)
correct→ *in*correct*ness* (adding a prefix and a suffix)

(ii) **isomorphism** is broken into iso-morph-ism
iso- is a prefix which means 'identical in structure'
morph is a word part which means 'form or shape'
-ism is a suffix which means 'a condition'

Hence *isomorphism* means the condition of having identical forms or shapes; it describes the condition of two crystalline substances.

Prefixes describing numbers or quantities are taken from Greek or Latin words. The following table shows the common prefixes from these two languages. Prefixes, suffixes, and word parts are listed alphabetically in separate sections after the table.

	GREEK PREFIX	LATIN PREFIX	PREFIX	MEANING	
1	mono-	uni-	hemi-	half	Gr
2	di-	bi-	semi-	half	L
3	tri-	ter-	poly-	many	Gr
4	tetra-	quad-	multi-	many	L
5	penta-	quinq-	omni-	all	L
6	hexa-	sex-	dupli-	twice	L
7	hepta-	sept-	tripli-	three times	Gr
8	octo-	oct-	hypo-	less, under	Gr
9	nona-	novem-	hyper-	more, over	Gr
10	deca-	deci-	sub-	under	L
100	hecta-	centi-	super-	over	L
1000	kilo-	milli-	iso-	same, equal, identical	Gr

PREFIXES

a- without, lacking, lacking in, e.g. *a*morphous, being without shape; *a*symmetrical, without symmetry, or lacking in symmetry.

allo- different, or different kinds, e.g. *allo*tropy, the existence of an element in two or more different forms.

amphi- on both sides, e.g. *ampho*teric, having the nature of both an acid and a base.

an- the same prefix as **a-**, used in front of words beginning with a vowel, or the letter *h*, e.g. *an*isotropic, not having the same properties in all directions; *an*hydrous, being without, or lacking, water, in a crystal.

anti- opposite in direction, or in position, e.g. *anti*catalyst, a catalyst which slows down a chemical reaction, i.e. works in the opposite direction to a catalyst.

auto- caused by itself, e.g. *auto*xidation, reaction of a substance with atmospheric oxygen at room temperature, the substance oxidizes itself; *auto*catalysis, a chemical reaction in which the products act as catalysts for the reaction.

***cis*-** on the same side, e.g. *cis*-compound, an isomer in which two like groups are on the same side of the double bond in the compound. See *trans*-.

co- acting together, with, e.g. *co*hesion, the force holding two or more objects together.

counter- acting against, acting in the opposite direction, e.g. *counter*act, to act against, such as a mild alkali counteracts the effect of acid on skin; *counter*clockwise, turning in the opposite direction to the hands of a clock.

de- opposite action, e.g. *de*compression, the lessening of a pressure, it is the opposite action to compression; *de*activate, to make less active, it is the opposite of activate.

dia- through, across, e.g. *dia*meter, the line going across a circle, through the centre.

dis- opposite action, e.g. *dis*charge, to take an electric charge away from a charged body, the opposite of charge; *dis*connect, to break, or open, a connection, the opposite of connect.

equi- having the same number, equal, e.g. *equi*molecular, having the same number of molecules; *equi*librium, the condition of two rates of reaction being equal and opposite, so that there is no further change in a reversible reaction.

im- the opposite, not. (Used with words beginning with b, m, p.) For example, *im*perfect, not perfect, the opposite of perfect; *im*permeable, not permeable.

in- the opposite, not. (Used with all words other than those beginning with b, m, p.) For example, *in*active, the opposite of active; *in*adequate, not adequate.

infra- below, e.g. *infra*molecular, having a size smaller than a molecule, so the size is below molecular size.

inter- between, among, e.g. *inter*face, a common surface between two liquids or two solids; *inter*stice, a narrow space between two solid objects.

macro- great, large, e.g. *macro*molecule, a large molecule composed of many smaller molecules, as in a polymer.

micro- small, especially if too small to be seen by the human eye alone, e.g. *micro*balance, a balance used for measuring masses of less than 1 mg; *micro*analysis, analysis using very small amounts of substances.

non- not, e.g. *non*-electrolyte, a substance which is not an electrolyte; *non*-ferrous, any metal other than iron.

ortho- straight, right-angled, upright, e.g. *ortho*gonal, with parts at right-angles; *ortho*rhombic, a crystal system with three unequal axes at right-angles.

pan- all, complete, every, e.g. *pan*chromatic, covering all wavelengths of light in the spectrum.

para- at the side of, by, e.g. *para*casein, an insoluble form of casein, formed when soluble casein coagulates.

pseudo- has the same appearance, but is false, e.g. *pseudo*alum, a substance which has the appearance of an alum, but is not an alum.

re- again, e.g. *re*activate, to make something activated again; *re*crystallize, to crystallize again.

syn, sym- joined together, united, e.g. *syn*thesis, combining elements or compounds to make new compounds.

trans- across, on the opposite side of, e.g. *trans*-compound, an isomer in which two like groups are on opposite sides of the double bond in the compound. See *cis-*.

ultra- beyond, e.g. *ultra*filter, a filter which has holes so small it filters out colloids; it thus has uses beyond those of the ordinary filter.

un- not, the opposite, e.g. *un*saturated, means not saturated; *un*stable, means not stable; *un*paired, means not in a pair, and so by itself.

SUFFIXES

-able forms an adjective which shows an action can possibly take place, e.g. change*able*, something which can change; transform*able*, something which it is possible to transform.

-al of, or to do with; forms a general adjective, e.g. experiment*al*, of, or to do with, experiment; fraction*al*, of, or to do with, fractions; therm*al*, of, or to do with, heat.

-ed forms the past participle of a verb, can be used as an adjective; it shows an action under the control of an experimenter, e.g. vari*ed*, describes a quantity changed by an experimenter; dehydrat*ed*, describes a substance from which water has been removed under the control of an observer.

-er (-or) forms a noun from a verb and describes an agent, e.g. mix*er*, a device which mixes; desiccat*or*, a device that desiccates; generat*or*, a device that generates a gas.

-gram forms a noun describing a record which is written or drawn, e.g. chromato*gram*, the recorded result from an experiment on chromatography; tele*gram*, the written message recorded by telegraph.

-graph forms a noun describing an instrument or device that records variation in a quantity, or other information, e.g. thermo*graph*, a kind of thermometer which records changes of temperature over a period of time; tele*graph*, a device which records information in words.

-ic of, or to do with; forms a general adjective, e.g. bas*ic*, of, or to do with, a base; cycl*ic*, of, or to do with, a cycle; ion*ic*, of, or to do with, ions.

-ify forms a verb which is causative in action, e.g. pur*ify*, to cause to become pure; solid*ify*, to cause to become solid.

-ing forms the present participle of a verb, can be used as an adjective; it shows an action not under the control of an experimenter, e.g. fluctuat*ing*, describes a quantity varying above and below an average value, which cannot be controlled by an observer; disintegrat*ing*, describes a radioactive substance undergoing disintegration, as the process cannot be controlled by an observer.

-ity forms a noun of a state or quality, e.g. pur*ity*, the quality or state of being pure; acid*ity*, the quality of being acid.

-ive forms an adjective by replacing *-ion* in nouns; the adjective describes an agent producing the effect described by the noun, e.g. inhibit*ion* → inhibit*ive*, describes an agent causing inhibition; oxidat*ion* → oxidat*ive*, describes a process causing oxidation; explos*ion* → explos*ive*, describes an agent causing an explosion.

-ize forms a verb which is causative in the formation of something, e.g. ion*ize*, to cause ions to be formed; polymer*ize*, to cause polymers to be formed.

-lysis forms a noun describing the action of breaking down into simpler parts, e.g. hydro*lysis*, the decomposition of a compound by the action of water; electro*lysis*, the decomposition of a substance by an electric current.

-meter forms a noun describing an instrument which measures quantitatively, e.g. thermo*meter*, an instrument which measures temperature accurately; volt*meter*, an instrument which measured electric potential in volts.

-metry forms a noun describing a particular science of accurate measurement, e.g. thermo*metry*, the science of measuring temperature; hydro*metry*, the science of measuring the density of liquids.

-ness forms an abstract noun of state or quality, e.g. sweet*ness*, the quality of being sweet; soft*ness*, the quality of being soft.

-ous forms an adjective showing possession, or describing a state, e.g. anhydr*ous*, being in the state of not possessing water, homolog*ous*, in the state of being a homologue; homogen*ous*, in the state of having the same properties throughout a substance.

-philic forms an adjective describing a liking for something, e.g. proto*philic*, describes a substance which accepts protons.

-phobic forms an adjective describing a dislike for something, e.g. lyo*phobic*, describes a colloid which does not go readily into solution.

-scope forms a noun describing an instrument which measures qualitatively, e.g. spectro*scope*, an instrument by which spectra can be observed qualitatively; hygro*scope*, an instrument which measures qualitatively the humidity of the atmosphere.

-scopy forms a noun describing the use of instruments for observation in science, e.g. micro*scopy*, the use of microscopes for scientific observation.

-stat forms a noun describing a device which keeps a quantity constant, e.g. hydro*stat*, a device which keeps water in a boiler at a constant level; thermo*stat*, a device which keeps a liquid, or an object, at a constant temperature.

-tion forms an abstract noun. With *-ation*, it forms a noun of action, e.g. pollu*tion*, the result of polluting; concentra*tion*, the degree to which a solution is concentrated; distill*ation*, the noun of action from distil; precipit*ation*, the noun of action from precipitate.

WORD PARTS

aqua water, to do with water, e.g. *aque*ous, a solution containing water; *aqua*ion, an ion with molecules of water associated with it.

chrom colour, to do with colour, e.g. pan*chrom*atic, all the colours, and hence all the wavelengths of the visible spectrum; *chrom*atography, the analysis of complex substances in which a coloured record of the analysis is produced.

gen to produce, e.g. homo*gen*ize, to make a mixture of solid and liquid substances into a viscous liquid of the same texture throughout; *gen*erate, to produce energy or a flow of gas.

hydr water or liquids, e.g. de*hydr*ate, to remove water; an*hydr*ous, describes a substance without water.

hygro damp or humid, e.g. *hygro*scopic, attracting water from the atmosphere to become damp; *hygro*meter, an instrument that measures the relative humidity of the atmosphere.

morph shape or form, e.g. a*morph*ous, describes a substance which is without a crystalline form; poly*morph*ism, existing in different forms.

photo light, e.g. *photo*lysis, decomposition caused by light; *photo*halide, any halide which is decomposed by light.

pneumo air or gas, e.g. *pneum*atic trough, a trough for the collection of gases.

pyro great heat, e.g. *pyro*lysis, decomposition caused by heating; *pyro*meter, a special kind of thermometer for measuring very high temperatures.

therm heat, e.g. *thermo*stable, stable when heated; *therm*al, of, or to do with heat; *thermo*meter, an instrument for the quantitative measure of temperature.

Index

abnormal 229
abrasive 14
absolute alcohol 203
absolute scale 102
absorption 35
abundant 231
accelerate 219
acceptor 136
accompany 213
accumulate 219
accuracy 227
accurate 227
acetylene 174
acetylenes 174
acetylide 49
acid 45
acid anhydride 176
acidic 45
acidify 38
acid radical 45
acid salt 47
acrid 22
action 19
activated 21
activated state 152
activation energy 152
active 19
active electrode 129
act on 19
acyl 180
acyl chloride 176
addition 188
addition polymerization 208
adduct 90
adequate 231
aerosol 100
affinity 69
agent 63
air 56
alcohol[1] 175
alcohol[2] 175
aldehyde 175
aldol addition 200
aldose 205
alicyclic 179
aliphatic 179
alkali 45
alkali metal 117
alkaline 45
alkaline earth metal 117
alkane 172
alkene 173
alkyl 180
alkylation 189
alkyl halide 177
alkyne 174
allotrope 118
allotropy 118
alloy 55
alpha emission 140
alpha particle 139
alpha ray 139
alter 219
amalgam 55
ambient 103
amide 177
amido group 186
amine 178
amino acid 178
amino group 186
ammeter 123
amorphous 15
amount 81
amphoteric 46
analysis 82
anhydride 48
anhydrous 48
anion 125
anode 123
anodic 128
anodic oxidation 128
anodize 128
apparatus 23
apparent 223
apply 232
appreciable 227
approximate 79
aqua-ion 132
aqueous 88
arbitrary 79
aromatic 179
artificial radioactivity 138
aryl 180
ash 164
asphalt 161
aspirator 24
assay 155
assume 222
asymmetric carbon atom 183
atmospheric pressure 102
atom 110
atomicity 104
atomic mass unit 114
atomic number 113
atomic structure 113
atomic theory 76
atomic weight 114
atomize 101
attract 124
autocatalysis 72
autoxidation 71
auxochrome 187
available 85
average 79
Avogadro constant 80
Avogadro's hypothesis 108
Avogadro's principle 108
axis 92
azo group 186

balance[1] (*n*) 27
balance[2] (*v*) 78

base 46
base exchange 68
base metal 117
basic 46
basicity 46
basic salt 47
beaker 25
Becquerel rays 139
beehive 24
Benedict's test 196
benzene 179
Bessemer process 165
beta emission 140
beta particle 139
beta ray 139
bicarbonate 49
binary compound 44
bisulphate 51
bisulphite 51
bitumen 156
Biuret test 196
bland 21
blast furnace 163
bleach 73
blende 155
blowpipe 29
body-centred lattice 97
boil 10
boiled 10
boiling point 12
bombard 143
bomb calorimeter 148
bond 133
bond energy 136
bore 29
Bosch process 168
Boyle's law 105
branched chain 182
brass 55
brine 169
brittle 14
bromide 50
bromination 192
bromo group 187
Brownian motion 109
bubble[1] (*n*) 40
bubble[2] (*v*) 40
bumping 33
bung 24
burette 26
burn 59
burst into 60
butyl 184
by-product 157

calcine 32
calculate 79
calibrate 26
calorie 153
Cannizzaro reaction 199
capture 217
carbide 49
carbohydrate 205
carbon 118
carbonaceous 156
carbonate 49
carbon cycle 61
carboniferous 156
carbonyl group 185
carboxyl group 186
carboxylic acid 176
carry out 157
cast iron 163
Castner-Kellner process 169
catalysis 72
catalyst 72
cataphoresis 101
cathode 123
cathodic 128
cathodic reduction 128
cation 125
cause 214
caustic 21
cellulose 207
Celsius scale 102
ceramics 171
chain 182
chain reaction 64
chalk 155
chamber 143
change of state 9
char 59
characteristic 9
charcoal 156
charge 138
Charles' law 105
chemical[1] (*adj*) 20
chemical[2] (*n*) 20
chemical change 19
chemical property 19
chemical reaction 62
chip 13
chloride 50
chlorination 192
chloro group 187
choking 22
chromate 52
chromate (VI) 52
chromatogram 36
chromatography 36
chromophore 187
circumstances 212
cis-configuration 184
clarify 222
classify 120
clear 17
cleavage plane 94
clinker 163
close packing 95
cloud chamber 143
coagulate 99
coal 156
coarse 13
coat 127
coinage metal 117
coke 156
collect 41
colloid 98
coloured 15
colourless 15
column chromatography 37
combination 64
combining weight 77
combustion 58
comment 228
common salt 155
compare 224
complex 228
complex ion 132
complex salt 47
composition 82
compound 8
concentrate 32
concentrated 88
concentration 81
conclusion 43
condensation[1] 11
condensation[2] 104
condensation[3] 191
condensation polymerization 208
condense 11
condenser 28

conditions 103
conduct 122
conductivity 122
confirmatory 42
conform 107
conical flask 25
connect 24
consist of 55
constant[1] (*n*) 106
constant[2] (*adj*) 106
constituent 54
constitution 82
construct 211
contact 217
contact process 166
contain 55
contaminate 20
contamination 57
content 85
contrast[1] (*v*) 224
contrast[2] (*n*) 224
control 221
converse 230
convert 73
converter 164
coordinate bond 136
copolymerization 208
cork 24
correspond 233
corrode 61
corrosive 21
counteract 216
covalency 136
covalent bond 136
cracking 194
creamy 17
critical pressure 104
critical
temperature 104
crucible 27
crystal 91
crystal face 93
crystal lattice 92
crystalline 15
crystallization 91
crystalloid 91
crystal symmetry 93
crystal systems 96
cubic close
packing 95
cubic system 96
curie 142
current 122
curve 106
cyano group 186
cycle 64
cyclic chain 182

Dalton's atomic
theory 76
Dalton's law 108
dative bond 136
dative covalent
bond 136
d-block elements 121
decant 31
decolorize 73
decomposition 65
decomposition
voltage 126
decrease 219
decrepitate 74
deduce 222
deficient 232
define 228
definite 226
definition 228
deflagration 32
degree 227
dehydrate 66
dehydration 190
dehydrogenation 190
d-electron 112
deliquesce 67
delivery tube 24
demonstrate 42
dense 22
density 12
deposit 154
derivative 200
derive 106
description 9
desiccant 66
desiccate 66
destructive
distillation 203
detail 226
detailed 226
detect 225
detergent 171
determine 222
detonate 74
development 36
device 23
diagram 29
dialysis 34
diamond 118
diatomic 104
diazonium salt 180
dibasic 46
dicarboxylic acid 176
dichromate 52
dichromate (VI) 52
differentiate 225
diffusion 35
digest 32
dihydric 185
diluent 56
dilute 88
dilution 81
dimer 207
direct 233
disaccharide 206
discharge 124
disconnect 24
discrepancy 227
discuss 228
disintegrate 65
disintegration
series 142
disperse 99
disperse phase 99
dispersion medium 99
displacement 68
disproportionation 71
dissociation 65
dissolve 30
distil 33
distillate 201
distillation 33
distillation flask 28
distilled 33
distinct 224
distinction 224
distinctive 224
distinguish 224
divalent 137
doctor solution 161
dolomite 155
donor 136
d-orbital 112
double bond 181
double salt 47
dross 159
dry distillation 203
ductile 14

dull 16
duplicate 214
duration 220
dye 162
dynamic allotropy 118
dynamic equilibrium 150

effect 214
effective 214
effervesce 40
efficient 214
effloresce 67
effort 213
effusion 35
elastic 14
electrochemical 128
electrochemical equivalent 128
electrochemical series 130
electrode 122
electrodeposit 127
electrode potential 128
electrolysis 122
electrolyte 122
electrolytic 122
electrolytic cell 122
electrolyze 123
electromotive force 129
electron 110
electron pair 133
electron-volt 153
electrophoresis 101
electroplating 127
electrovalency 134
electrovalent bond 134
element[1] 8
element[2] 116
elimination 189
eluent 37
emit 138
empirical formula 181
emulsifying agent 101
emulsion 100
enantiomorph 183
enantiotropy 92
endothermic 148
end point 39
end product 158
energy 135
energy barrier 152
energy level 152
enzymatic 72
equation 78
equilibrium 150
equilibrium constant 150
equilibrium mixture 150
equivalent weight 89
Erlenmeyer flask 25
error 79
essential 226
essential oil 204
establish 225
ester 177
esterification 191
ethane 173
ethene 174
ether 177
ethyl 184
ethylene 174
ethyne 174
eudiometer 24
evaporate[1] 11
evaporate[2] 32
evaporating basin 32
event 220
evolve 40
exact 79
exception 230
excess 230
exchange 215
excitation 152
exclude 230
exemplify 228
exert 106
exhibit 221
exist 213
exit 215
exothermic 148
experiment 42
explosion 58
explosive 204
extend 213
extensive property 9
extent 213
extinguish 60
extract 164
extraction 34
extranuclear 113
face-centred lattice 97
facilitate 214
factor 103
Fajans and Soddy law 142
faraday 129
fat 177
favour 214
feature 9
f-block elements 121
Fehling's test 196
fermentation 194
ferricyanide 53
ferrocyanide 53
filings 13
film 18
filter 30
filtrate 30
final 85
fine 13
fine chemical 20
finely divided 13
fissile 141
Fittig reaction 197
fixed 79
fixed oil 204
flake 13
flame 58
flammable 21
flash point 204
flask 25
flocculent 17
flow 217
flue 164
fluid 11
fluorescent 140
foam 100
form 41
formula 78
formula weight 78
fraction 202
fractional crystallization 91
fractional distillation 201
fractionating column 201
fragrant 22
freeze 10
freezing point 12
Friedel-Crafts reaction 199

froth 100
froth flotation 158
fructose 206
fuel 160
fumes 33
functional group 185
funnel 27
furnace 164
fuse 32

galvanize 166
gamma radiation 139
gamma ray 139
gas 11
gaseous 11
gas equation 106
gas-jar 24
gas laws 109
gas oil 161
gasoline 160
gasometer 162
Gattermann reaction[1] 198
Gattermann reaction[2] 199
Gay-Lussac's law 109
Geiger counter 141
gel 100
general 212
general formula 181
generate 41
generator 27
geometrical isomerism 184
giant molecular crystal 95
giant structure 94
give off 41
glasses 209
glow 59
glucose 206
graduated flask 26
graduation 26
Graham's law 107
grain 13
gram-molecular volume 104
gram molecule 89
granular 16
granule 13
graph 39
graphic formula 181
graphite 118
gravimetric analysis 82
Grignard reagent 198
grind 38
ground glass 29
ground state 152
group 119

Haber process 170
half-life 141
halide 50
halogen 117
halogenation 192
hard water 57
hearth 164
heat 59
heat of combustion 146
heat of dilution 147
heat of formation 147
heat of ionization 147
heat of neutralization 146
heat of reaction 146
heat of solution 147
heavy 17
heavy chemical 171
Hess's law 148
heterocyclic 179
heterogenous 54
hexacyanoferrate (II) 53
hexacyanoferrate (III) 53
hexagonal close packing 95
hexagonal system 96
hexose 205
higher oxide 48
homogenous 54
homologous 172
hydrate 90
hydrated 48
hydration 90
hydrocarbon 172
hydrocarbon oil 204
hydrogenation 188
hydrogen carbonate 49
hydrogen electrode 130
hydrogen sulphate 51
hydrogen sulphite 51
hydrolysis[1] 66
hydrolysis[2] 190
hydrophilic 101
hydrophobic 101
hydrosol 100
hydroxide 48
hydroxyl group 185
hydroxyl ion 132
hygroscopic 67
hypothesis 108

ideal 107
identical 233
identification 225
identify 225
identity 225
ignite 32
ignition point 204
immediate 221
immiscible 18
impart 15
impure 20
impurity 20
inactive 19
inactive electrode 129
inadequate 231
incandescent 60
incombustible 58
indicate 38
indicator 38
industrial 157
inert 19
inference 43
inflammable 21
information 222
ingredient 54
inhibitor 72
initial 85
initiate 74
inorganic 55
insert 230
insoluble 17
instantaneous 75
instrument 23
insufficient 231
intense 230
intensive property 9
interchange 215
interface 18
interfere 216
intermediate 85

interstice 93
interval 220
introduce 230
inverse 233
investigate 42
iodide 50
iodination 193
iodine test 196
iodo group 187
ion 123
ionic bond 134
ionic theory 124
ionization 123
irregular 93
irreversible reaction 64
irritate 22
isocyanide 178
isolate 43
isomer 182
isomerism 182
isomorphism 92
isotope 114
isotopic ratio 114
isotopic weight 114

jet 29
joule 153

Kellner-Solvay process 169
kelvin 102
kerosene 160
ketone 176
ketose 206
kiln 170
kinetic theory 108
Kipp's apparatus 27
Kolbe electrolytic reaction 200
K-shell 113

labile 75
laboratory 23
lag 28
Lassaigne test 196
lather 57
lattice 92
law 109
law of constant proportions 76
law of mass action 149
layer 18
leaching 158
lead-chamber process 167
leak 25
Le Chatelier's principle 151
leuco base 187
leuco compound 187
liberate 69
light[1] (*n*) 60
light[2] (*v*) 60
lime 169
limestone 155
limit[1] (*n*) 211
limit[2] (*v*) 213
limited[1] 109
limited[2] 231
linear 83
Linz-Donawitz process 165
liquation 159
liquefaction 104
liquefy 11
liquid 10
liquid air 156
liquid extraction 202
lixiviation 158
local 212
location 37
lode 154
lone pair 133
L-shell 113
lubricants 161
lump 13
lustre 16
lyophilic 101
lyophobic 101

main product 157
major 226
malleable 14
malodorous 22
maltose 206
manganate 53
manganate (VI) 53
manganate (VII) 53
manufacture 157
marble 155
mass 12
massive 14
mass number 113
mass spectrograph 145
mass spectrometer 145
material 8
maximum 218
mean 79
measuring cylinder 26
mechanism 195
melt 10
melting point 12
membrane 99
mercury cathode cell 168
meson 110
meta-directing 194
metal 116
metal crystal 95
metallic bond 137
metalloid 117
metallurgy 164
methane 172
method 221
methyl 184
methylated spirit 203
methylation 194
mild 21
milky 17
mill 158
mine 154
mineral 154
mineral acid 55
mineral oil 156
minimum 218
minor 226
miscible 18
mixture 54
mobile 18
model 223
moderate 69
modify 219
molality 88
molar concentration 88
molarity 88
molar volume 80
mole 80
molecular crystal 94
molecular formula 181
molecular structure 83
molecule 77
mole fraction 80
molten 10

monatomic 104
monobasic 46
monoclinic system 96
monohydric 185
monomer 207
monosaccharide 205
monotropy 92
monovalent 137
mordant 162
mortar 38
mother liquor 90
mould 210
M-shell 113
multiple 79
M-value 88

naphtha 161
naphthalene 179
nascent 75
native 155
natural radioactivity 138
nature 19
negative catalyst 72
neutral 45
neutralization 67
neutron 110
nitrate 52
nitration 193
nitride 52
nitrile 178
nitrite 52
nitro compound 180
nitro group 186
noble 56
noble gas 104
nomenclature 44
non-aqueous 88
non-electrolyte 122
non-ideal 107
non-inflammable 21
non-linear 83
non-metal 116
non-polar 137
non-reducing sugar 206
normal 229
normality 89
normal salt 46
normal solution 89
nuclear 110
nuclear fission 141
nuclear fusion 142
nuclear reaction 144
nucleus 110
nuclide 142

obey 107
observation 42
obsolete 171
obstruct 217
occupy 233
occur[1] 63
occur[2] 154
octahedral 83
octet 133
odoriferous 22
odour 15
odourless 15
oil 204
olefine 173
oleum 167
opaque 16
open-hearth furnace 164
open-hearth process 165
operate 157
optical isomerism 183
optimum 159
orbit 110
orbital 111
order 221
ore 154
organic 55
orientation 93
original 220
ortho-para- directing 194
orthorhombic system 96
overcome 213
overlap 218
overvoltage 127
oxidant 71
oxidation 70
oxidation number 78
oxidation state 135
oxide 48
oxidizing agent 71
ozonolysis 195

paper chromatography 36
paraffin[1] 161
paraffin[2] 173
parallel 129
partial pressure 108
particle[1] 13
particle[2] 110
passive 21
pass over 41
pattern 93
p-block elements 121
p-electron 112
penetrate 144
pentose 205
pentyl 184
peptide 209
peptide bond 209
perceptible 42
perfect crystal 93
period 120
periodicity 120
periodic system 119
periodic table 119
permanent gas 104
permanent hardness 57
permanganate 53
permeable 99
peroxide 48
pestle 38
petrol 160
petroleum 160
pH 38
pharmaceutical chemical 20
phenol 180
photochemical 65
photosynthesis 61
pH value 38
physical change 13
physical property 9
pickling 165
pig iron 163
pigment 162
pinchcock 23
pipette 26
pitch 161
plane 92
plane of symmetry 93
plant 157
plastic[1] (*adj*) 14
plastic[2] (*n*) 210

plasticizer 210
plate 127
platinum electrode 126
plentiful 231
plot 39
pneumatic trough 24
poison 72
polar 137
polarization 127
pollution 56
polyatomic 104
polymer 207
polymerization 207
polymorphism 92
polysaccharide 207
polythene 208
polyunsaturated 185
p-orbital 112
porous 15
porous pot 27
position 211
positron 110
powder 13
practical 23
precede 220
precipitate 30
predict 85
preliminary 220
preparation 43
present[1] (*v*) 217
present[2] (*adj*) 217
pressure 102
pressure law 105
prevent 216
previous 220
primary cell 129
primitive structure 97
probability 223
process 157
producer gas 168
product 62
progress 221
promoter 72
property 9
proportion 76
propyl 184
protein 209
proton 110
puddling process 163
pulverize 158
pungent 22
pure 20
purification 43
pyramidal 84
pyrites 155
pyrolysis 33

qualitative 85
quality 15
quantitative 85
quantity 81
quench 165
quick-fit 29

racemate 183
radiation 138
radical 45
radioactive 138
radioactive decay 141
radioactive
disintegration 141
radioactive series 142
radioactivity 138
radiology 144
radio opaque 144
radius 111
random 223
range 140
rapid combustion 58
rate constant 149
rate of reaction 149
ratio 79
raw materials 154
ray 138
react 62
reactant 62
reaction 62
reaction profile 152
reading 39
reagent 63
real 107
recapitulate 223
receiver 28
receptacle 25
record 39
recrystallization 91
recur 217
red-hot 60
redox potential 131
redox process 70
redox series 131
reduce 219
reducing agent 71
reducing sugar 206
reduction[1] 70
reduction[2] 193
refer 223
refine 159
reflux 201
reflux condenser 201
reforming 201
refractory 14
Reimer-Tiemann
reaction 199
relation 232
relative 232
relative atomic
mass 113
relative density 12
relative formula
mass 78
relative isotopic
mass 114
relative molecular
mass 114
relative vapour
density 12
release 69
remove 215
repel 124
replace 68
replaceable 68
replenish 159
represent 223
repulsion 124
reserve 154
residue 31
resist 213
respective 233
respiration 61
result 39
retain 215
retarder 72
retort 28
return 215
reverberatory
furnace 164
reverse 216
reversible
reaction 64
revert 215
rim 25
ring chain 182
roast 163
rock salt 155
rotate 218

rubber 209
rust 61

salt 46
salting out 162
Sandmeyer
reaction 198
saponification 192
saturated[1] 87
saturated[2] 185
saturated water
vapour pressure 56
s-block elements 121
scale 26
Schiff's reagent 196
scintillation 140
scrap 171
scrubber 162
scum 158
seam 154
sediment 31
s-electron 112
semipolar bond 136
semi-water gas 168
separating funnel 27
separation 34
series[1] 129
series[2] 172
set 10
settle 31
shared electron 133
shatter 94
shell 111
sherardize 166
Siemens-Martin
process 165
silica 156
similar 233
simple 228
simultaneous 221
single bond 181
slag 163
slaking 158
slightly 17
slip plane 94
slit 211
slow combustion 58
sludge 158
smelt 164
smoke 100
smoulder 59
soap 162
soda 169
soft water 57
sol 100
solid 10
solidify 10
solubility 87
soluble 17
solute 86
solution 86
solvation 90
Solvay process 169
solvent 86
solvent front 36
s-orbital 112
source 138
sparingly 17
spatial 211
spatula 29
special 212
specimen 43
spelter 166
spent 170
spent oxide 167
sphere 111
spinthariscope 140
spirit 203
spontaneous 75
spontaneous
combustion 58
spout 25
spraying 158
square planar 84
square pyramidal 84
stabile 74
stabilize 101
stable 74
stage 159
standard[1] (*adj*) 229
standard[2] (*n*) 229
standard atmospheric
pressure 102
standard electrode
potential 130
standard solution 89
starch 207
state of division 13
state of matter 9
statement 222
steady 229
steam distillation 202
steel 165
stereoisomerism 183
still 201
stir 32
stoichiometric 82
stopcock 23
stoppage 217
stopper 24
stout-walled 29
s.t.p. 102
straight chain 182
strength 124
strong electrolyte 125
structural 211
structural formula 181
structural
isomerism 182
structure[1] 82
structure[2] 211
subatomic 110
sublimate 33
sublimation 33
sublime 33
subsequent 220
substance 8
substitute 215
substitution 188
sucrose 206
sufficient 231
sugar 205
sulphate 51
sulphide 51
sulphite 51
sulphonate group 186
sulphonation 193
sulphur 155
supernatant 90
supersaturated 87
supplementary 232
supply 154
support 28
surface 16
surroundings 103
suspended 31
suspension 86
symbol 77
symmetry 93
synthesis 200
synthetic 200
system 212
systematic name 44

tabulate 39
take place 63

tamp 171
tank 23
tap 23
tar 162
tarnish 61
tautomer 184
tautomerism 184
technical chemical 20
technique 43
temper 165
temperature 102
temporary hardness 57
tend 216
terminate 214
tervalent 137
test 42
test paper 42
tetragonal system 96
tetrahedral 83
texture 14
theoretical 23
theory 76
thermal 65
thermal decomposition 65
thermal dissociation 65
Thermit process 171
thermochemical equation 147
thermochemistry 148
thermolabile 75
thermometer 28
thermoplastic 210
thermosetting plastic 210
thermostable 75
thermostat 27
thin-walled 29
thiosulphate 51
thixotropy 101
tin 166
titration 39
titre 39
tongs 29
tool 23
trace 20
track 143
traditional name 44
trans-configuration 184
transfer 132
transform 144
transition element 121
transition point 92
translucent 16
transmute 73
transparent 16
trap 29
treat 38
trend 216
tribasic 46
triclinic system 96
trigonal bipyramidal 84
trigonal planar 83
trigonal pyramidal 84
trihydric 185
triple bond 181
triturate 38
trivial name 44
trough 24
tube 29
tubing 29
turnings 13
tuyere 163

Ullmann reaction 197
undergo 213
uniform 229
unit 212
universal 212
unsaturated[1] 87
unsaturated[2] 185
unstable 75
U-tube 25

vacuum distillation 202
valency 133
valency electron 133
valency shell 133
valid 232
van der Waals' bond 137
vaporize 11
vapour 11
vapour density 12
vapour pressure 103
variable[1] (*n*) 218
variable[2] (*adj*) 218
variation 218
vary 218
vat 162
velocity constant 149
verify 222
vessel 25
violent 69
viscosity 18
viscous 18
visible 42
volatile 18
voltage 126
voltameter 129
voltmeter 126
volumetric analysis 82
vulcanization 171

warm 59
waste 170
waste product 157
water 56
water-bath 59
water cycle 56
water gas 168
water of crystallization 90
water softening 57
water vapour pressure 56
weak electrolyte 125
weight 12
weld 164
Williamson's synthesis 198
Woulfe's bottle 25
wrought iron 163
Wurtz reaction 197
Wurtz-Fittig reaction 197

yield 159

zeolite process 57

α-particle 139
β-particle 139
γ-radiation 139

key

metals
non-metals
noble gases
a b c d

a – atomic number
b – name
c – symbol
d – relative atomic mass

• Lanthanide elements

•• Actinide elements

period	group 1	group 2		group 3
1	1 Hydrogen H 1.01			
2	3 Lithium Li 6.94	4 Beryllium Be 9.01		5 Boron B 10.81
3	11 Sodium Na 22.99	12 Magnesium Mg 24.31		13 Aluminiu Al 26.98
4	19 Potassium K 39.10	20 Calcium Ca 40.08	transition elements	31 Gallium Ga 69.72
5	37 Rubidium Rb 85.47	38 Strontium Sr 87.62		49 Indium In 114.82
6	55 Cesium Cs 132.91	56 Barium Ba 137.33		81 Thallium Tl 204.37
7	87 Francium Fr (223)	88 Radium Ra 226.03		

transition elements

4	21 Scandium Sc 44.96	22 Titanium Ti 47.90	23 Vanadium V 50.94	24 Chromium Cr 52.00	25 Mangane Mn 54.94
5	39 Yttrium Y 88.91	40 Zirconium Zr 91.22	41 Niobium Nb 92.91	42 Molybdenum Mo 95.94	43 Technet Tc 98.91
6	57 • Lanthanum La 138.91	72 Hafnium Hf 178.49	73 Tantalum Ta 180.95	74 Tungsten W 183.85	75 Rheniu Re 186.2
7	89 •• Actinium Ac 227.03	104 Rutherfordium Rf (261)	105 Hahnium Hn (260)	106 (263)	